Gallium Nitride
and
Silicon Carbide
Power Devices

Gallium Nitride
and
Silicon Carbide
Power Devices

B Jayant Baliga
North Carolina State University, USA

World Scientific

NEW JERSEY · LONDON · SINGAPORE · BEIJING · SHANGHAI · HONG KONG · TAIPEI · CHENNAI · TOKYO

Published by

World Scientific Publishing Co. Pte. Ltd.

5 Toh Tuck Link, Singapore 596224

USA office: 27 Warren Street, Suite 401-402, Hackensack, NJ 07601

UK office: 57 Shelton Street, Covent Garden, London WC2H 9HE

Library of Congress Cataloging-in-Publication Data
Names: Baliga, B. Jayant, 1948–   author.
Title: Gallium nitride and silicon carbide power devices / B. Jayant Baliga,
    North Carolina State University, USA.
Description: New Jersey : World Scientific, [2016] |
    Includes bibliographical references and index.
Identifiers: LCCN 2016032951 | ISBN 9789813109407 (hc : alk. paper)
Subjects: LCSH: Power electronics. | Gallium nitride--Electric properties. |
    Silicon carbide--Electric properties.
Classification: LCC TK7881.15 .B349 1992 | DDC 621.31/7--dc23
LC record available at https://lccn.loc.gov/2016032951

British Library Cataloguing-in-Publication Data
A catalogue record for this book is available from the British Library.

Desk Editor: Anthony Alexander

Typeset by Stallion Press
Email: enquiries@stallionpress.com

Printed in Singapore

Dedication

The author proposed the development of wide band gap semiconductor based power devices in 1979. He would like to dedicate this book to the management at the General Electric Company for supporting his vision by providing resources to make the first wide band gap power devices out of Gallium Arsenide in the 1980s. He would also like to dedicate the book to the companies that sponsored his Power Semiconductor Research Center in the 1990s allowing demonstration of the first high performance silicon carbide diodes and MOSFETs.

About the Author

Professor Baliga is internationally recognized for his leadership in the area of power semiconductor devices. In addition to over 550 publications in international journals and conference digests, he has authored and edited 19 books ("*Power Transistors*", IEEE Press 1984; "*Epitaxial Silicon Technology*", Academic Press 1986; "*Modern Power Devices*", John Wiley 1987; "*High Voltage Integrated Circuits*", IEEE Press 1988; "*Solution Manual: Modern Power Devices*", John Wiley 1988; "*Proceedings of the 3rd Int. Symposium on Power Devices and ICs*", IEEE Press 1991; "*Modern Power Devices*", Krieger Publishing Co. 1992; "*Proceedings of the 5th Int. Symposium on Power Devices and ICs*", IEEE Press 1993; "*Power Semiconductor Devices*"; PWS Publishing Company 1995; "*Solution Manual: Power Semiconductor Devices*"; PWS Publishing Company 1996; "*Cryogenic Operation of Power Devices*", Kluwer Press 1998; "*Silicon RF Power MOSFETs*", World Scientific Publishing Company 2005; "*Silicon Carbide Power*

Devices", World Scientific Publishing Company 2006; *"Fundamentals of Power Semiconductor Devices"*, Springer Science, 2008; *"Solution Manual: Fundamentals of Power Semiconductor Devices"*, Springer Science, 2008; *"Advanced Power Rectifier Concepts"*, Springer Science, 2009; *"Advanced Power MOSFET Concepts"*, Springer Science, 2010; *"Advanced High Voltage Power Device Concepts"*, Springer Science, 2011. In addition, he has contributed chapters to another twenty books. He holds 120 U.S. Patents in the solid-state area. In 1995, one of his inventions was selected for the *B.F. Goodrich Collegiate Inventors Award* presented at the *Inventors Hall of Fame*.

Professor Baliga obtained his Bachelor of Technology degree in 1969 from the Indian Institute of Technology, Madras, India. He was the recipient of the *Philips India Medal* and the *Special Merit Medal (as Valedictorian)* at I.I.T, Madras. He obtained his Masters and Ph.D. degrees from Rensselaer Polytechnic Institute, Troy NY, in 1971 and 1974, respectively. His thesis work involved Gallium Arsenide diffusion mechanisms and pioneering work on the growth of InAs and GaInAs layers using Organometallic CVD techniques. At R.P.I., he was the recipient of the *IBM Fellowship* in 1972 and the *Allen B. Dumont Prize* in 1974.

From 1974 to 1988, Dr. Baliga performed research and directed a group of 40 scientists at the General Electric Research and Development Center in Schenectady, NY, in the area of Power Semiconductor Devices and High Voltage Integrated Circuits. During this time, he pioneered the concept of combining MOS and Bipolar physics to create a new family of discrete devices. He is the *inventor of the IGBT* which is now in production by many International Semiconductor companies. This invention is widely used around the globe for air-conditioning, home appliance (washing machines, refrigerators, mixers, etc) control, factory automation (robotics), medical systems (CAT scanners, uninterruptible power supplies), and electric street-cars/bullet-trains, as well as for the drive-train in electric and hybrid-electric cars. IGBT-based motor control improves efficiency by over 40 percent. The IGBT is essential for deployment of Compact Fluorescent Lamps (CFLs) to replace of incandescent lamps producing efficiency improvement by 75 percent. Since two-thirds of the electricity in the world is used to run motors and twenty percent of the electricity in the world is used for lighting, the availability of IGBTs has produces a cumulative electrical energy savings of more than 73,000 Terra-Watt-Hours. In addition, the IGBT enabled the introduction of the electronic ignition system for running spark plugs in the internal combustion engine of gasoline powered cars and trucks. The resulting 10 percent improvement in fuel efficiency has saved consumers more than 1.5 trillion gallons of gasoline during the last 20 years. The cumulative impact of these electrical energy and gasoline savings is a cost savings of over $ 23.7 Trillion for consumers, and a reduction in Carbon Dioxide emissions from Coal-Fired power plants by over 109 Trillion pounds. For this achievement, he has been labeled the "man with the largest negative carbon footprint in the world". Most recently, the IGBT has enabled creation of very compact, light-weight, and inexpensive defibrillators used to resuscitate cardiac arrest victims. Deployment of these portable defibrillators in fire-trucks, paramedic vans, in buildings, and on-board airlines, is projected by the American Medical Association (AMA) to save 100,000 lives per year in the US. For this work, *Scientific American Magazine* named him one of the *'Eight heroes of the semiconductor revolution'* in their 1997 special issue commemorating the Solid-State Century.

Dr. Baliga is also the originator of the concept of merging Schottky and p-n junction physics to create a new family of JBS power rectifiers that are commercially available from various companies. This concept originally implemented for silicon devices has become an essential concept for the commercialization of silicon carbide high voltage Schottky rectifiers.

In 1979, Dr. Baliga developed a theoretical analysis resulting in the Baliga's Figure of Merit (BFOM) which relates the resistance within power rectifiers and FETs to the basic semiconductor properties. He predicted that the performance of Schottky power rectifiers and power MOSFETs could be enhanced by several orders of magnitude by replacing silicon with other materials such as gallium arsenide and silicon carbide. This is forming the basis of a new generation of power devices in the 21st Century.

In August 1988, Dr. Baliga joined the faculty of the Department of Electrical and Computer Engineering at North Carolina State University, Raleigh, North Carolina, as a Full Professor. At NCSU, in 1991 he established an international center called the *Power Semiconductor Research Center* (PSRC) for research in the area of power semiconductor devices and high voltage integrated circuits, and has served as its Founding Director. His research interests include the modeling of novel device concepts, device fabrication technology, and the investigation of the impact of new materials, such as GaAs and Silicon Carbide, on power devices. The first high performance SiC Schottky rectifiers and power MOSFETs were demonstrated at PSRC in the 1990s resulting in the release of products by many companies during the last 10 years.

In 1997, in recognition of his contributions to NCSU, he was given the highest university faculty rank of *Distinguished University Professor of Electrical Engineering*. In 2008, Professor Baliga was a key member of an NCSU team — partnered with four other universities — that was successful in being granted an Engineering Research Center from the National Science Foundation for the development of micro-grids that allow integration of renewable energy sources. Within this program, he is responsible for the fundamental sciences platform and the development of power devices from wide-band-gap semiconductors for utility applications.

Professor Baliga has received numerous awards in recognition for his contributions to semiconductor devices. These include two *IR 100 awards* (1983, 1984), the *Dushman* and *Coolidge Awards* at GE (1983), and being selected among the *100 Brightest Young Scientists in America* by Science Digest Magazine (1984). He was elected *Fellow of the IEEE* in 1983 at the age of 35 for his contributions to power semiconductor devices. In 1984, he was given the *Applied Sciences Award* by the world famous sitar maestro Ravi Shankar at the Third Convention of Asians in North America. He received the 1991 IEEE William E. Newell Award, the highest honor given by the Power Electronics Society, followed by the 1993 IEEE Morris E. Liebman Award for his contributions to the emerging *Smart Power Technology*. In 1992, he was the first recipient of the BSS Society's *Pride of India Award*. At the age of 45, he was elected as Foreign Affiliate to the prestigious National Academy of Engineering, and was one of only 4 citizens of India to have the honor at that time (converted to regular Member in 2000 after taking U.S. Citizenship). In 1998, the University of North Carolina system selected him for the *O. Max Gardner Award*, which recognizes the faculty member among the 16 constituent universities who has made the greatest contribution to the welfare of the human race. In December 1998, he received the IEEE J.J. Ebers Award, the highest recognition given by the IEEE Electron Devices Society for his technical contributions to the Solid-State area. In June 1999, he was honored at the Whitehall Palace in London with the IEEE Lamme Medal, one of the highest forms of recognition given by the IEEE Board of Governors, for his contributions to development of an apparatus/technology of benefit to society. In April 2000, he was honored by his Alma Mater as a *Distinguished Alumnus*. In November 2000, he received the *R.J. Reynolds Tobacco Company Award for Excellence in Teaching, Research, and Extension* for his contributions to the College of Engineering at North Carolina State University. In 2011, Dr. Baliga was selected to receive the *Alexander Quarles Holladay Medal for Excellence*, which recognizes members of the NCSU faculty who over their careers have made outstanding contributions to the University through their research, teaching, and extension services.

In 1999, Prof. Baliga founded a company, *Giant Semiconductor Corporation*, with seed investment from Centennial Venture Partners, to acquire an exclusive license for his patented technology from North Carolina State University with the goal of bringing his NCSU inventions to the marketplace. A company, *Micro-Ohm Corporation*, subsequently formed by him in 1999, has been successful in licensing the GD-TMBS power rectifier technology to several major semiconductor companies for world-wide distribution. These devices have application in power supplies, battery chargers, and automotive electronics. In June 2000, Prof. Baliga founded another company, *Silicon Wireless Corporation*, to commercialize a novel super-linear silicon RF transistor that he invented for application in cellular base-stations and grew it to 41 employees. This company (renamed *Silicon Semiconductor Corporation*) is located at Research Triangle Park, N.C. It received an investment of $ 10 Million from *Fairchild Semiconductor Corporation* in December 2000 to co-develop and market this technology. Based upon his additional inventions, this company has also produced a new generation of Power MOSFETs for delivering power to microprocessors in notebooks and servers. This technology was licensed by his company to Linear Technologies Corporation with transfer of the know-how and manufacturing process. Voltage Regulator Modules (VRMs) using his transistors are currently available in the market for powering microprocessor and graphics chips in laptops and servers.

In 2010, Dr. Baliga was inducted into the Engineering Design Magazine's "Engineering Hall of Fame" for his invention, development, and commercialization of the Insulated Gate Bipolar Transistor (IGBT), joining well known luminaries (e.g. Edison, Tesla, and Marconi) in the electrical engineering field. The award announcement states: *"While working at General Electric in the late 1970s, Baliga conceived the idea of a functional integration of MOS technology and bipolar physics that directly led to the IGBT's development... it remains undeniable that Baliga's vision and leadership played a critical role in moving the IGBT from a paper-based concept to a viable product with many practical applications."*

President Obama personally presented Dr. B. Jayant Baliga with the National Medal of Technology and Innovation, the highest form of recognition given by the United States Government to an Engineer, in a ceremony at the White House on October 21, 2011. Dr. Baliga's award citation reads: *For development and commercialization of the Insulated Gate Bipolar Transistor and other power semiconductor devices that are extensively used in transportation, lighting, medicine, defense, and renewable energy generation systems.* His IGBT innovation has saved world-wide consumers $ 23.7 Trillion while reducing carbon dioxide emissions by 109 Trillion pounds over the last 25 years.

In October 2012, Governor Beverly Purdue presented Dr. Baliga the North Carolina Award for Science. This is the highest award given by the State of North Carolina and the Governor to a civilian. On October 4, 2013, he was inducted into the Rensselaer Alumni Hall of Fame by Rensselaer Polytechnic Institute President Shirley Jackson. The ceremony included unveiling his portrait etched on a window in Thomsen Hall in the Darrin Communications Center.

On August 23, 2014, Dr. Baliga received the IEEE Medal of Honor *'For the invention, implementation, and commercialization of power semiconductor devices with widespread benefits to society'* in a ceremony held in Amsterdam, The Netherlands. This award has been given since 1917 to recognize great achievements in the field of electrical engineering.

On June 19, 2015, Dr. Baliga was honored by the Russian Federation's highest international award — *the Global Energy Prize* — for his *"invention, implementation, and commercialization of the Insulated Gate Bipolar Transistor — one of the most important innovations for the control and distribution of energy"*. This award was established by President Putin in 2008 to recognize important innovations in the generation, management and conservation of energy.

On May 6, 2016, Dr. Baliga was inducted into the *National Inventors hall of Fame* as the sole inventor of the Insulated Gate Bipolar Transistors. His invention is among only about 500 U.S. patents out of over 9 Million issued to date that have been given this recognition. His work was commemorated with a display at the *National Inventors Hall of Fame Museum* in Alexandria, VA near Washington DC.

Preface

Power semiconductor devices are a key component of all power electronic systems. It is estimated that at least 50 percent of the electricity used in the world is controlled by power devices. With the wide spread use of electronics in the consumer, industrial, medical, and transportation sectors, power devices have a major impact on the economy because they determine the cost and efficiency of systems. After the initial replacement of vacuum tubes by solid state devices in the 1950s, semiconductor power devices have taken a dominant role with silicon serving as the base material.

Bipolar power devices, such as bipolar transistors and thyristors, were first developed in the 1950s. Their power ratings and switching frequency increased with advancements in the understanding of the operating physics and availability of more advanced lithography capability. The physics underlying the current conduction and switching speed of these devices has been described in my textbook[1]. Since the thyristors were developed for high voltage DC transmission and electric locomotive drives, the emphasis was on increasing the voltage rating and current handling capability. The ability to use neutron transmutation doping to produce high resistivity n-type silicon with improved uniformity across large diameter wafers enabled increasing the blocking voltage of thyristors to over 5000 volts while being able to handle over 2000 amperes of current in a single device. Meanwhile, bipolar power transistors were developed with the goal of increasing the switching frequency in medium power systems. Unfortunately, the current gain of bipolar transistors becomes low when it is designed for high voltage operation at high current density. The popular solution to this problem, using the Darlington configuration, had the disadvantage of increasing the on-state voltage drop resulting in an increase in the power dissipation. In addition to the large control currents required for bipolar transistors, they suffered from second breakdown failure modes. These issues produced a cumbersome design with snubber networks, which raised the cost and degraded the efficiency of the power control system.

In the 1970s, the power MOSFET product was first introduced by International Rectifier Corporation. Although initially hailed as a replacement for all bipolar power devices due to its high input impedance and fast switching speed, the power MOSFET has successfully cornered the market for low voltage (< 100 V) and high switching speed (> 100 kHz) applications but failed to make serious inroads in the high voltage arena. This is because the on-state resistance of power MOSFETs increases very rapidly with increase in the breakdown voltage. The resulting high conduction losses, even when using larger more expensive die, degrade the overall system efficiency.

In recognition of these issues, I proposed two new thrusts in 1979 for the power device field. The first was based upon the merging of MOS and bipolar device physics to create a new category of power devices[2]. My most successful innovation among MOS-Bipolar devices has been the *Insulated Gate Bipolar Transistor (IGBT)*. Soon after commercial introduction in the early 1980s, the IGBT was adopted for all medium power electronics applications[3]. Today, it is manufactured by more than a dozen companies around the world for consumer, industrial, medical, and other applications that benefit society. The triumph of the IGBT is associated with its huge power gain, high input impedance, wide safe operating area, and a switching speed that can be tailored for applications depending upon their operating frequency.

The second approach that I suggested in 1979 for enhancing the performance of power devices was to replace silicon with wide band gap semiconductors. The basis for this approach was an equation that I derived relating the on-resistance of the drift region in unipolar power devices to the basic properties of the semiconductor material. This equation has since been referred to as *Baliga's Figure of Merit (BFOM)*. In addition to the expected reduction in the on-state resistance with higher carrier mobility, the equation predicts a reduction in on-resistance as the inverse of the cube of the breakdown electric field strength of the semiconductor material.

In the 1970s, there was a dearth of knowledge of the impact ionization coefficients of semiconductors. Consequently, an association of the breakdown electric field strength was made with the energy band gap of the semiconductor[4]. This led to the conclusion that wide band gap semiconductors offer the opportunity to greatly reduce the on-state resistance of the drift region in power devices. With a sufficiently low on-state resistance, it became possible to postulate that unipolar power devices could be constructed from wide band gap semiconductors with lower on-state voltage drop than bipolar devices made out of silicon. Since unipolar devices exhibit much faster switching speed than bipolar devices because of the absence of minority carrier stored charge, wide band gap based power devices offered a much superior alternative to silicon bipolar devices for medium and high power applications. Device structures that were particularly suitable for development were identified as Schottky rectifiers to replace silicon P-i-N rectifiers, and power Field Effect Transistors to replace the bipolar transistors and thyristors prevalent in the 1970s.

The first attempt to develop wide band gap based power devices was undertaken at the General Electric Corporate Research and Development Center, Schenectady, NY, under my direction. The goal was to leverage a 13-fold reduction in specific on-resistance for the drift region predicted by the BFOM for Gallium Arsenide. A team of 10 scientists was assembled to tackle the difficult problems of the growth of high resistivity epitaxial layers, the fabrication of low resistivity ohmic contacts, low leakage Schottky contacts, and the passivation of the GaAs surface. This led to an enhanced understanding of the breakdown strength[5] for GaAs and the successful fabrication of high performance Schottky rectifiers[6] and MESFETs[7]. Experimental verification of the basic thesis of the analysis represented by BFOM was therefore demonstrated during this period. Commercial GaAs based Schottky rectifier products were subsequently introduced in the market by several companies.

In the later half of the 1980s, the technology for the growth of silicon carbide was developed with the culmination of commercial availability of wafers from CREE Research Corporation. Although data on the impact ionization coefficients of SiC was not available, early reports on the breakdown voltage of diodes enabled estimation of the breakdown electric field strength. Using these numbers in the BFOM predicted an impressive 100-200 fold reduction in the specific on-resistance of the drift region for SiC based unipolar devices. In 1988, I joined North Carolina State University and subsequently founded the *Power Semiconductor Research Center (PSRC)* — an industrial consortium — with the objective of exploring ideas to enhance power device performance. Within the first year of the inception of the program, SiC Schottky barrier rectifiers with breakdown voltage of 400 volts were successfully fabricated with on-state voltage drop of about 1 volt and no reverse recovery transients[8]. By improving the edge termination of these diodes, the breakdown voltage was found to increase to 1000 volts. With the availability of epitaxial SiC material with lower doping concentrations, SiC Schottky rectifiers with breakdown voltages over 2.5 kV have been fabricated at PSRC[9]. These results have motivated many other groups around the world to develop SiC based power rectifiers. In this regard, it has been my privilege to assist in the establishment of national programs to fund research on silicon carbide technology in the United States, Japan, and Switzerland-Sweden. Meanwhile, accurate measurements of the impact ionization coefficients for 6H-SiC and 4H-SiC in defect free regions were performed at PSRC using an electron beam excitation method[10]. Using these coefficients, a BFOM of over 1000 is predicted for SiC providing even greater motivation to develop power devices from this material.

Although the fabrication of high performance, high voltage Schottky rectifiers has been relatively straight-forward, the development of a suitable silicon carbide MOSFET structure has been more problematic. The existing silicon power D-MOSFET and U-MOSFET structures do not directly translate to suitable structures in silicon carbide. The interface between SiC and silicon dioxide, as a gate dielectric, needed extensive investigation due to the large density of traps that prevent the formation of high conductivity inversion layers. Even after overcoming this hurdle, the much higher electric field in the silicon dioxide when compared with silicon devices, resulting from the much larger electric field in the underlying SiC, leads to reliability problems. Fortunately, a structural innovation, called the ACCUFET, to overcome both problems was proposed and demonstrated at PSRC[11]. In this structure, a buried P^+ region is used to shield the gate region from the high electric field within the SiC drift region. This concept is applicable to devices that utilize either accumulation channels or inversion channels. Devices with low specific on-resistance have been demonstrated at PSRC using both 6H-SiC and 4H-SiC with epitaxial material capable of supporting over 5000 volts[12]. This device structure has been subsequently emulated by several groups around the world.

Although many papers have been published on silicon carbide device structures and process technology, no comprehensive book written by a single author was available that provides a unified treatment of silicon carbide power device structures until I wrote and published a book in 2006[13]. This new book has been prepared to not only update the information on silicon carbide power devices but also include progress made with development of gallium nitride power devices.

The emphasis in the book is on the physics of operation of the devices. The analyses provide general guidelines for understanding the design and operation of the various device structures. For designs that may be pertinent to specific applications, the reader should refer to the papers published in the literature, the theses of my M.S. and Ph.D. students, as well as the work reported by other research groups. Comparison with silicon devices is provided to enable the reader to understand the benefits of gallium nitride and silicon carbide devices.

In the introduction chapter, the desired characteristics of power devices are described with a broad introduction to potential applications. The second chapter provides the properties of gallium nitride and silicon carbide that have relevance to the analysis and performance of power device structures. Issues pertinent to the fabrication of gallium nitride and silicon carbide devices are reviewed here because the structures analyzed in the book have been constructed with these process limitations in mind. The third chapter discusses breakdown voltage, which is the most unique distinguishing characteristic for power devices, together with edge termination structures. This analysis is pertinent to all the device structures discussed in subsequent chapters of the book.

The fourth chapter describes the basic thesis for development of wide bandgap power devices. The reduction in the specific on-resistance of the drift region is analyzed with design rules for choosing its doping and thickness. Information about superjunction silicon power MOSFFETs has been included here for comparison with the GaN and SiC devices.

Chapter 5 and 6 describe the physics of operation of Schottky rectifiers made from GaN and SiC. The advantages of the low drift region resistance are made evident here. The problem of high reverse leakage current due to Schottky barrier lowering and tunneling are described in Chapter 5. Chapter 6 describes solutions for this problem by using P-N junctions to shield the Schottky contact in the JBS rectifier structure.

A brief analysis of P-i-N rectifiers is given in the seventh chapter. It is shown here that these structure are valid only when the blocking voltage exceeds 10-kV. The unique problem of bipolar degradation in SiC P-i-N rectifiers is described here. Chapter 8 describes the MPS rectifier structure which allows significant improvement in the trade-off curve between on-state voltage drop and reverse recovery switching losses.

Chapter 9 discusses power JFET and MESFET structures with planar and trench gate regions. These normally-on structures can be used to construct the *Baliga-Pair* circuit[14], described in chapter 10, which has been shown to be ideally suitable for motor control applications. This represents a solution to taking advantage of the low specific on-resistance of the silicon carbide drift region while awaiting resolution of reliability issues with the gate oxide in silicon carbide power MOSFET structures.

Chapter 11 provides a description of silicon carbide planar power MOSFET structures with emphasis on problems associated with simply replicating structures originally developed in silicon technology. Innovative approaches to prevent high electric field in the gate oxide of silicon carbide power MOSFETs are then discussed here. Accumulation-mode structures are shown to provide the advantages of larger channel mobility and lower threshold voltage leading to a reduction in the specific on-resistance. In Chapter 12, issues with adopting the silicon trench-gate MOSFET structure to silicon carbide are enunciated followed by analysis of solutions to these problems. Once again, methods for shielding the gate oxide are shown to enable reduction of the electric field in the gate oxide to acceptable levels. The shielding is also demonstrates to ameliorate the base reach-through problem allowing a reduction of the base width and hence the channel resistance contribution.

Chapter 13 describes gallium nitride vertical device technology. This effort has been hampered until recently by lack of availability of GaN substrates. Unique power MOSFET structures that take advantage of the 2D-electron gas between GaN and AlGaN are discussed here.

Lateral gallium nitride HFET structures are described and analyzed in Chapter 14. These devices became possible with the demonstration of sufficiently high quality GaN layer growth on low cost silicon substrates. This approach has now become the dominant technology for the development and commercialization of GaN power transistors with blocking voltages ranging from 100 to 600-V. The best performance has been observed for normally-on structures which must be used in the Baliga-Pair circuit configuration. Structural modifications to create normally-off structures are described here as well.

Although the emphasis in this book is on unipolar device structures, a description of silicon carbide bipolar junction transistors and gate-turn-off thyristors is provided in chapters 15 and 16. These devices have been superseded by silicon carbide power MOSFETs. However, interest in silicon carbide IGBTs has been growing for utility scale applications where the blocking voltage exceeds 15-kV.

These devices are extensively analyzed in chapter 17. An optimized SiC IGBT structure is described here that prevent high [dV/dt] during the turn-off transient.

Chapter 18 gives a short description of promising applications for the silicon carbide and gallium nitride devices. It is shown that these devices enable increasing the operating frequency in power circuits leading to significant reduction in the size and weight of systems. A modest gain in efficiency of 1-5 percent over silicon devices has been reported by using these wide band gap power devices.

Experimental results are included throughout the book whenever pertinent to each chapter. This provides a historical context and a brief summary of the state of the art for gallium nitride and silicon carbide devices. However, the analytical models provided in the book are fundamental in nature and will not become obsolete.

I am hopeful that this book will be used for the teaching of courses on solid state devices and that it will make an essential reference for the power device industry. To facilitate this, analytical solutions are provided throughout the book that can be utilized to understand the underlying physics and model the structures. Homework problems have been included in the Appendix as an aid to instructors.

Ever since my identification of the benefits of utilizing wide band gap semiconductors for the development of superior power devices thirty-five years ago, it has been my mission to resolve issues that would impede their commercialization. I wish to thank the sponsors of the Power Semiconductor Research Center and the Office of Naval Research for supporting this mission during the 1990s. This has been essential to providing the resources required to create many breakthroughs in the silicon carbide technology which enabled commercialization of the technology. The establishment of the PowerAmerica Institute in 2015 is a major step forward in bringing down the cost of the wide band gap power device technology. This should accelerate its market penetration in years to come. I look forward to observing the benefits accrued to society by adopting the technology for conservation of fossil fuel usage resulting in reduced environmental pollution.

Prof. B. Jayant Baliga
June 2016

xx GaN and SiC Power Devices

References

[1] B.J. Baliga, "Power Semiconductor Devices", PWS Publishing Company, 1996.

[2] B.J. Baliga, "Evolution of MOS-Bipolar Power Semiconductor Technology", Proceedings IEEE, pp. 409-418, April 1988.

[3] B.J. Baliga, "The IGBT Device: Physics, Design, and Applications", Elsevier Press, 2015.

[4] B.J. Baliga, "Semiconductors for High Voltage Vertical Channel Field Effect Transistors", J. Applied Physics, Vol. 53, pp. 1759-1764, March 1982.

[5] B.J. Baliga, R. Ehle, J.R. Shealy and W. Garwacki, "Breakdown Characteristics of Gallium Arsenide", IEEE Electron Device Letters, Vol. EDL-2, pp. 302-304, November 1981.

[6] B. J. Baliga, A.R. Sears, M.M. Barnicle, P.M. Campbell, W. Garwacki and J.P. Walden, "Gallium Arsenide Schottky Power Rectifiers", IEEE Transactions on Electron Devices, Vol. ED-32, pp. 1130-1134, June 1985.

[7] P.M. Campbell, W. Garwacki, A.R. Sears, P. Menditto and B.J. Baliga, "Trapezoidal-Groove Schottky-Gate Vertical-Channel GaAs FET", IEEE Electron Device Letters, Vol. EDL-6, pp. 304-306, June 1985.

[8] M. Bhatnagar, P.K. McLarty and B.J. Baliga, "Silicon-Carbide High-Voltage (400 V) Schottky Barrier Diodes", IEEE Electron Device Letters, Vol. EDL-13, pp.501-503, October 1992.

[9] R.K. Chilukuri and B.J. Baliga, "High Voltage Ni/4H-SiC Schottky Rectifiers", International Symposium on Power Semiconductor Devices and ICs, pp. 161-164, May 1999.

[10] R. Raghunathan and B.J. Baliga, "Temperature dependence of Hole Impact Ionization Coefficients in 4H and 6H-SiC", Solid State Electronics, Vol. 43, pp. 199-211, February 1999.

[11] P.M. Shenoy and B.J. Baliga, "High Voltage Planar 6H-SiC ACCUFET", International Conference on Silicon Carbide, III-Nitrides, and Related Materials, Abstract Tu3b-3, pp. 158-159, August 1997.

[12] R.K. Chilukuri and B.J. Baliga, PSRC Technical Report TR-00-007, May 2000.

[13] B.J. Baliga, "Silicon Carbide Power Devices", World Scientific Press, 2006.

[14] B.J. Baliga, See Chapter 7 in "Power Semiconductor Devices", pp. 417-420, PWS Publishing Company, 1996.

Contents

Chapter 1

Introduction

Modern society is reliant upon electrical appliances for comfort, transportation, and healthcare. This has motivated great advances in power generation, power distribution and power management technologies based on enhancements in the performance of power devices that regulate the flow of electricity. After the displacement of vacuum tubes by solid state devices in the 1950s, the industry relied upon silicon bipolar devices, such as bipolar power transistors and thyristors. Although the ratings of these devices grew rapidly to serve an ever broader system need, their fundamental limitations in terms of the cumbersome control and protection circuitry led to bulky and costly solutions. The advent of MOS technology for digital electronics enabled the creation of a new class of devices in the 1970s for power switching applications as well. These silicon power MOSFETs have found extensive use in high frequency applications with relatively low operating voltages (under 100 volts). The merger of MOS and bipolar physics enabled creation of yet another class of devices in the 1980s. The most successful innovation in this class of devices has been the Insulated Gate Bipolar Transistor (IGBT)[1]. The high power density, simple interface, and ruggedness of the IGBT have made it the technology of choice for all medium and high power applications.

Power devices are required for systems that operate over a broad spectrum of power levels and frequencies. In Fig. 1.1, the applications for power devices are shown as a function of operating frequency. High power systems, such as HVDC power distribution and locomotive drives, requiring the control of megawatts of power operate at relatively low frequencies. As the operating frequency increases, the power ratings decrease for the devices with typical microwave devices handling about 100 watts. All of these applications are served by silicon devices today. Until recently, thyristors were the only devices available with sufficient voltage and current ratings

1

favored for the HVDC power distribution applications. The ratings of IGBTs have now grown to levels where they are now preferred to thyristors for voltage source converters and FACTs designs.

The medium frequency and power applications such as electric trains, hybrid-electric cars, home appliances, compact fluorescent lamps, medical equipment, and industrial motor drives also utilize the IGBT. Power MOSFETs are preferred for the high frequency applications operating from low power source voltages. These applications include power supplies for computers and laptops, power management in smart phones, and automotive electronics.

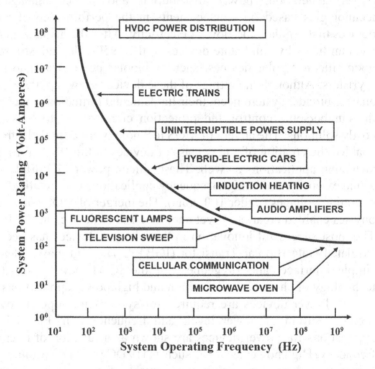

Fig. 1.1 Applications for power devices.

Another approach for classification of applications for power devices is based on their current and voltage handling requirements as shown in Fig. 1.2. On the high power end of the chart, thyristors are available that can individually handle over 6000 volts and 2000 amperes enabling the control of over 10 megawatts of power by a single monolithic device. These devices are suitable for the HVDC power transmission and locomotive drive (traction) applications. During last 10 years, IGBT modules have been developed with blocking voltages of up to 6500 volts and current handling capability above 1000 amperes. This has allowed the IGBT to replace thyristors in HVDC and traction applications.

For the broad range of systems that require operating voltages between 300 volts and 3000 volts with significant current handling capability, the IGBT was found to be the optimum solution since the 1990s. These applications span all sectors of the economy including consumer, industrial, transportation, lighting, medical, defense, and renewable energy generation[1].

When the current requirements fall below 1 ampere, it is feasible to integrate multiple devices on a single monolithic chip to provide greater functionality for systems such as telecommunications and display drives. However, when the current exceeds a few amperes, it is more cost effective to use discrete power MOSFETs with appropriate control ICs to serve applications such as automotive electronics and switch mode power supplies.

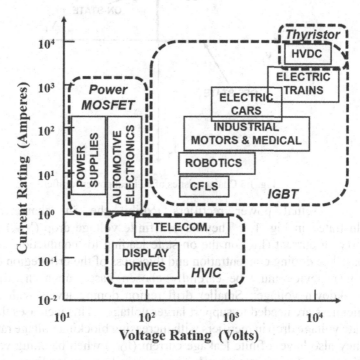

Fig. 1.2 System ratings for power devices.

1.1 Ideal and Typical Power Device Characteristics

An ideal power rectifier should exhibit the *i-v* characteristics shown in Fig. 1.3. In the forward conduction mode, the first quadrant of operation the figure, it should be able to carry any amount of current with zero on-state voltage drop. In the reverse blocking mode, the third quadrant of operation in the figure, it should be able to hold off any value of voltage with zero leakage current. Further, the ideal rectifier should be able to instantaneously switch between the on-state and the off-state with zero switching time. The ideal power rectifier would then dissipate no power while allowing control of the direction of current flow in circuits.

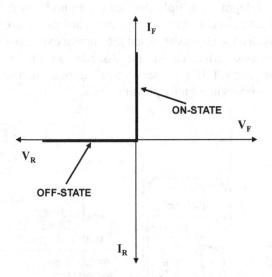

Fig. 1.3 Characteristics of an ideal power rectifier.

Actual power rectifiers exhibit the *i-v* characteristics illustrated in Fig. 1.4. They have a finite voltage drop (V_{ON}) when carrying current (I_{ON}) on the on-state leading to 'conduction' power loss. The doping concentration and thickness of the drift region of the power device must be carefully chosen based upon the desired breakdown voltage[2]. Smaller drift region doping levels with larger thickness are needed to support larger voltages. This increases the on-state voltage drop in rectifiers with increasing blocking voltage ratings. They also have a finite leakage current (I_{OFF}) when blocking voltage (V_{OFF}) in the off-state producing power loss. In addition, power rectifiers switch from the on-state to the off-state in a finite time interval and display a large reverse recovery current before settling down to the off-state operating point. The reverse recovery process can produce power losses as great as those observed in the on-state.

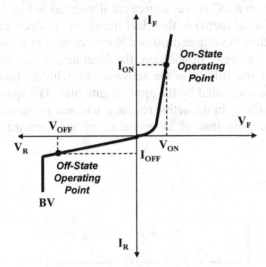

Fig. 1.4 Characteristics of a typical power rectifier.

The finite power dissipation in the power rectifier produces internal heating. This limits the maximum on-state current density to maintain an acceptable maximum junction temperature. The power dissipated in a power rectifier is given by:

$$P_D = (V_{ON} * J_{ON}) + (V_{OFF} * J_{OFF}) + P_{SW} \qquad [1.1]$$

where the first term is the on-state power loss, the second term is the off-state power loss, and the third term is the switching power loss. The on-state current density (J_{ON}) is determined by the maximum junction temperature ($T_{J,MAX}$):

$$T_{J,MAX} = \frac{P_D}{R_\theta} + T_A \qquad [1.2]$$

where R_θ is the thermal impedance and T_A is the ambient temperature. The power dissipation in power devices increases when their voltage rating is increased due to an increase in the on-state voltage drop and switching loss per cycle. The typical on-state current density for power devices ranges from 25 to 150 A/cm^2 depending up on the blocking voltage rating and the packaging technology.

The *i-v* characteristics of an ideal power switch designed for operation from a DC power source are illustrated in Fig. 1.5. As in the case of the ideal rectifier, the ideal transistor conducts current in the on-state with zero voltage drop and blocks voltage in the off-state with zero leakage current. In addition, the ideal device can operate with a high current and voltage in the active region with the forward current in this mode controlled by the applied gate bias. The spacing between the characteristics in the active region is uniform for an ideal transistor indicating a gain that is independent of the forward current and voltage.

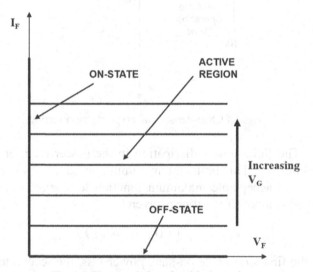

Fig. 1.5 Characteristics of an ideal transistor.

The *i-v* characteristics of a typical power switch are illustrated in Fig. 1.6. This device exhibits a finite resistance when carrying current in the on-state as well as a finite leakage current while operating in the off-state (not observable in the figure because its value is much lower than the on-state current levels). The breakdown voltage of a typical transistor is also finite as indicated in the figure with 'BV' where the current abruptly increases. The typical transistor can operate with a high current and voltage in the active region. This current is controlled by the base current for a bipolar transistor while it is determined by a gate voltage for a MOSFET or IGBT (as indicated in the figure). It is preferable to have gate voltage controlled characteristics because the drive circuit can be integrated to reduce its cost. The spacing between the characteristics in the active region is non-uniform for a typical transistor with a square-law behavior for devices operating with channel pinch-off in the current saturation mode[2]. Recently, devices operating under a new super-linear mode have been proposed and demonstrated for wireless base-station applications[3].

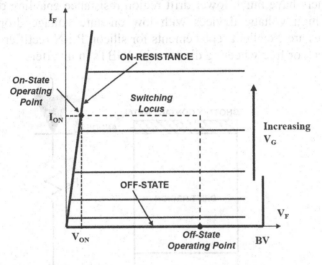

Fig. 1.6 Characteristics of a typical transistor.

1.2 Unipolar Power Rectifiers

Bipolar power devices operate with the injection of minority carriers during on-state current flow. These carriers must be removed when the switching the device from the on-state to the off-state. This is accomplished by either charge removal via the gate drive current or via the electron-hole recombination process. These processes introduce significant power losses that degrade the power management efficiency. It is therefore preferable to utilize unipolar current conduction in a power device.

The commonly used unipolar power diode structure is the Schottky rectifier that utilizes a metal-semiconductor barrier to produce current rectification[2]. The high voltage Schottky rectifier structure also contains a drift region, as show in Fig. 1.7, which is designed to support the reverse blocking voltage. The resistance of the drift region increases rapidly with increasing blocking voltage capability as discussed later in this book. Silicon Schottky rectifiers are commercially available with blocking voltages of up to 150 volts. Beyond this value, the on-state voltage drop of silicon Schottky rectifiers becomes too large for practical applications[2]. Silicon P-i-N rectifiers are favored for designs with larger breakdown voltages due to their lower on-state voltage drop despite their slower switching properties. As shown later in the book, silicon carbide Schottky rectifiers have much lower drift region resistance enabling design of very high voltage devices with low on-state voltage drop. These devices are excellent replacements for silicon P-i-N rectifiers used as fly-back or free-wheeling diodes with IGBTs in inverters.

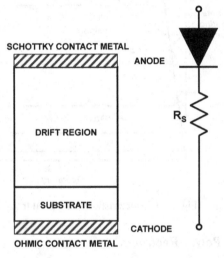

Fig. 1.7 Power Schottky rectifier structure and its equivalent circuit.

A major problem observed in Schottky rectifiers is the large increase in the reverse leakage current with increasing reverse bias voltage. An increase in reverse leakage current by more than one-order of magnitude occurs due to Schottky barrier lowering and pre-breakdown impact ionization in silicon devices[2]. A much worse increase in leakage current by six orders of magnitude due to Schottky barrier lowering and tunneling is observed for silicon carbide and gallium nitride Schottky rectifiers. This is a serious problem for high temperature operation and stability for these rectifiers.

The rapid increase in leakage current for Schottky rectifiers with increasing reverse bias voltage can be mitigated by using the

Junction-Barrier controlled Schottky (JBS) structure shown in Fig. 1.8. This structure contains P$^+$ regions surrounding the Schottky contacts. A depletion region extends from the junction and forms a potential barrier under the Schottky contact. This suppresses the electric field at the Schottky contact. The lower electric field at the Schottky contact reduces the Schottky barrier lowering and tunneling at the contact[4].

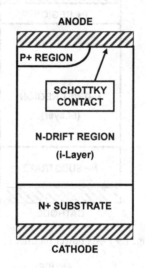

Fig. 1.8 JBS power rectifier structure.

1.3 Bipolar Power Rectifiers

The silicon P-i-N rectifier was commercialized in the 1950s. It is a bipolar device whose structure is illustrated in Fig. 1.9. Current flow in the on-state current occurs by the injection of minority (holes) from the P$^+$ region into the N-drift region. The minority carrier concentration exceeds the doping concentration of the drift region at operating on-state current levels. This is defined as *high-level injection*. The concentration of holes and electrons becomes equal in the drift region due to charge neutrality. These injected mobile carriers greatly reduce the resistance of the drift region allowing the P-i-N rectifier to carry high on-state current density with a small on-state voltage drop. The presence of the electrons and holes in the drift region is referred to as *stored charge*.

The stored charge in the drift region of the P-i-N rectifier must be removed before it is able to support a large reverse bias voltage. It undergoes a *reverse recovery process* during which a high reverse current is observed. The reverse recovery process produces considerable power dissipation not only in the P-i-N rectifier but also the transistor used to control its current flow. This results in reduced efficiency and increases the cost of the power electronics.

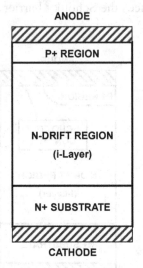

Fig. 1.9 P-i-N power rectifier structure.

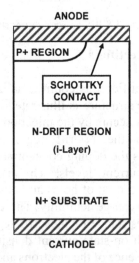

Fig. 1.10 MPS power rectifier structure.

The stored charge in the P-i-N rectifier can be greatly reduced while maintaining a low on-state voltage drop by using the MPS rectifier structure[4] shown in Fig. 1.10. Although similar in structure to the JBS rectifier, the principle of operation of the MPS rectifier is quite different. In the JBS rectifier, the on-state voltage drop is too low (~0.5 volts for silicon devices and 1.5 volts for silicon carbide devices) for the P-N junction to inject minority carriers into the drift region. In contrast, the on-state voltage drop for the MPS rectifier exceeds the built-in potential of the P-N junction. The injected minority carriers from the P-N junction produce conductivity modulation of the drift region as in the case of the P-i-N rectifiers. However, the stored charge is reduced in the MPS rectifier because the minority carrier density is zero at the Schottky contact. MPS rectifiers have a superior trade-off curve between on-state voltage drop and reverse recovery current when compared with P-i-N rectifiers.

1.4 Unipolar Power Transistors

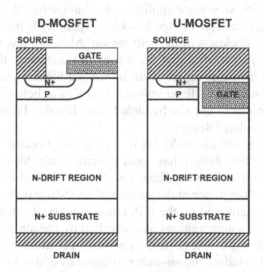

Fig. 1.11 Silicon power MOSFET structures.

Power switches can be designed as normally-on (depletion-mode) structures or as normally-off (enhancement-mode) structures. All power switches used by the industry are normally-off or enhancement mode devices. Normally-on structures create problems of large destructive shoot through currents when turning-on the power electronics because the gate drive required to keep these devices in the off-state has not yet been generated. The most commonly used unipolar power transistor is the silicon power Metal-Oxide-Semiconductor Field-Effect-Transistor or MOSFET. Although other structures, such as JFETs or SITs have been explored[5], they have not been popular for power electronic applications because of their normally-on behavior. Any proposed gallium nitride or silicon carbide device structure must be compatible with this important power circuit requirement to be acceptable for use by the industry.

The commercially available silicon power MOSFET products are based up on the structures shown in Fig. 1.11. The D-MOSFET was first commercially introduced in the 1970's and contains a 'planar-gate' structure. The P-base region and the N^+ source regions are self-aligned to the edge of the polysilicon gate electrode by using ion-implantation of boron and phosphorus with their respective drive-in thermal cycles. The n-type channel is defined by the difference in the lateral extension of the junctions under the gate electrode. The device supports positive voltage applied to the drain across the P-base/N-drift region junction. The voltage blocking capability is determined by the doping and thickness of the drift region. Although low voltage (< 100 V) silicon power MOSFET have low on-resistances, the drift region resistance increases rapidly with increasing blocking voltage limiting the performance of silicon power MOSFETs to below 200 volts. It is common-place to use the Insulated Gate Bipolar Transistors (IGBT) for higher voltage designs.

The silicon U-MOSFET structure became commercially available in the 1990s. It has a gate structure embedded within a trench etched into the silicon surface. The N-type channel is formed on the side-wall of the trench at the surface of the P-base region. The channel length is determined by the difference in vertical extension of the P-base and N^+ source regions as controlled by the ion-implant energies and drive times for the dopants. The silicon U-MOSFET structure was developed to reduce the on-state resistance by elimina-tion of the JFET component within the D-MOSFET structure[2].

An important innovation for silicon power MOSFETs is the concept of charge coupling to alter the electric field distribution in the drift region and allow it to support high voltages with large doping concentrations in the drift region[6]. The first charge coupled vertical silicon power MOSFET proposed in 1997 was the GD-MOSFET structure shown in Fig. 1.12 on the left-hand-side[7,8]. In comparison with the U-MOSFET structure, this device contains a deep trench region with a source connected electrode. A uniform electric field is generated in the drift region by using a graded doped drift region with high doping concentrations[9]. Breakdown voltages well above the parallel-plane breakdown voltage can be achieved using this idea. The specific on-resistance of these devices has been shown to be well below (5 to 25 times) that for the conventional silicon devices for blocking voltages ranging from 50 to 1000 volts[6]. Many companies have released products using this approach.

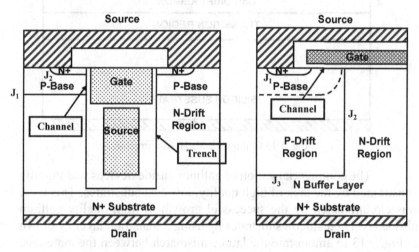

Fig. 1.12 Charge-coupled silicon power MOSFET structures.

An alternate charge-coupled silicon power MOSFET announced in 1999 was the COOLMOS[T] structure shown in Fig. 1.12 on the right hand side. Here, the charge coupling is accomplished across the vertical P-N junction formed between columns of P and N drift regions[10]. Many studies have been performed to optimize this device structure for blocking voltages of 500 to 1000 volts[6]. It has been demonstrated that the COOLMOS structure has about 3 to 10 times lower specific on-resistance than the conventional silicon power MOSFETs at a breakdown voltage of 600 volts. Many companies have commercialized this device structure under various names.

Any proposed GaN or SiC power switch technology must compete with not only the conventional silicon power MOSFET structures but also the new charge coupled silicon power MOSFETs. The charge coupled silicon devices offer much better performance but require a more expensive fabrication process. This difference must also be taken into account.

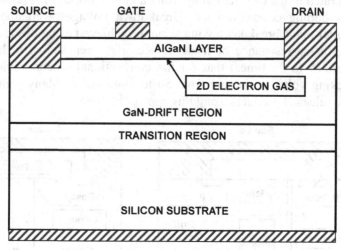

Fig. 1.13 Gallium nitride HEMT structure.

The commercialization of gallium nitride devices was initially constrained by the lack of high quality, low cost substrates. This issue was circumvented by the successful growth of high quality gallium nitride layers on silicon substrates by using a transition layer as shown in Fig. 1.13 to ameliorate the lattice mismatch between the materials. A thin aluminum-gallium-nitride layer is formed on top of the gallium nitride layer to produce the two-dimensional electron gas in the gallium nitride. The electrons in the 2D electron gas have a high mobility ($\sim$2000 cm^2/V-s) and charge ($\sim$10^{13} cm^{-2}). This creates a high conductivity current path through the GaN drift region with low specific on-resistance despite the lateral device structure.

1.5 Bipolar Power Devices

The commonly available silicon power bipolar devices were the bipolar transistor and the gate turn-off thyristor or GTO[2]. These devices were originally developed in the 1950's and widely used for power switching applications until the 1970s when the availability of silicon power MOSFETs and IGBTs supplanted them. The structures of the bipolar transistor and GTO are shown in Fig. 1.14. In both devices, injection of minority carriers into the drift region modulates its conductivity reducing the on-state voltage drop. However, this charge must be subsequently removed during switching resulting in high turn-off power losses. These devices require a large control (base or gate) current which must be implemented with discrete components leading to an expensive bulky system design.

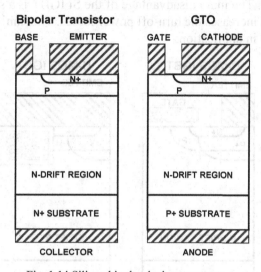

Fig. 1.14 Silicon bipolar device structures.

Several groups have worked on the development of bipolar transistors[11,12,13,14] and GTOs[15,16] using silicon carbide. The large junction potential for silicon carbide results in a relatively high on-state voltage drop for these devices when compared with commercially available silicon devices. Recently, interest has shifted from the GTO to the development of silicon carbide IGBTs.

1.6 MOS-Bipolar Power Devices

The most widely used silicon high voltage (> 300 volt) device for power switching applications is the *Insulated Gate Bipolar Transistor* (IGBT) which was developed in the 1980s by combining the physics of operation of bipolar transistors and MOSFETs[17]. The structure of the IGBT is deceptively similar to that for the power MOSFET as shown in Fig. 1.15 if viewed as a mere replacement of the N^+ substrate with a P^+ substrate. This substitution creates a four-layer parasitic thyristor which can latch up resulting in destructive failure due to loss of gate control. Fortunately the parasitic thyristor can be defeated by the addition of the P$^+$ region within the cell[2]. The benefit of the P$^+$ substrate is the injection of minority carriers into the N-drift region resulting in greatly reducing its resistance. This has enabled the development of high voltage IGBT products with high current carrying capability. The main disadvantage of the Si IGBT is a slow switching speed that increases the turn-off power losses. This limits its operating frequency in applications.

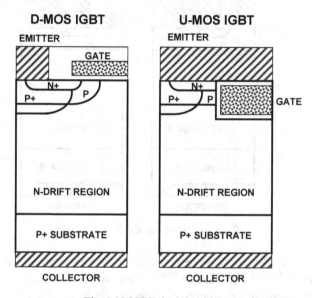

Fig. 1.15 IGBT device structures.

The development of the IGBT in silicon carbide has been analyzed and attempted by a few research groups[18,19,20]. The large junction potential and high resistance of the P$^+$ substrate for silicon carbide results in a relatively high on-state voltage drop for these devices when compared with commercially available silicon devices. Their switching speed is also compromised by the injected stored charge in the on-state. Significant improvements in the characteristics of SiC IGBTs has more recently been achieved by optimization of the buffer layer[21].

1.7 Summary

The motivation for the development of silicon carbide and gallium nitride unipolar devices has been reviewed in this chapter. Although excellent unipolar silicon Schottky rectifiers and power MOSFETs are commercially available with breakdown voltages below 200 volts, the resistance of their drift region increases rapidly at higher breakdown voltages producing significant power losses in applications. This problem can be overcome using silicon carbide and gallium nitride based unipolar devices. These devices also offer low switching losses enabling increasing the circuit operating frequency which reduces the size of passive components and filters in applications.

References

[1] B.J. Baliga, "The IGBT Device: Physics, Design, and Applications of the Insulated Gate Bipolar Transistor", Elsevier Press, 2015.

[2] B.J. Baliga, "Fundamentals of Power Semiconductor Devices", Springer-Science, 2008.

[3] B. J. Baliga, "Silicon RF Power Devices", World Scientific Press, 2005.

[4] B.J. Baliga, "Advanced Power Rectifier Concepts", Springer-Science, New York, 2010.

[5] B. J. Baliga, "Modern Power Devices", John Wiley and Sons, 1987.

[6] B.J. Baliga, "Advanced Power MOSFET Concepts", Springer-Science, New York, 2010.

[7] B.J. Baliga, "Vertical Field Effect Transistors having improved Breakdown Voltage Capability and Low On-state Resistance", U.S. Patent # 5,637,898, Issued June 10, 1997.

[8] B.J. Baliga, "Power Semiconductor Devices having improved High Frequency Switching and Breakdown Characteristics", U.S. Patent # 5,998,833, Issued December 7, 1999.

[9] B.J. Baliga, "Trends in Power Discrete Devices", IEEE International Symposium on Power Semiconductor Devices and ICs, Abstract P-2, pp. 5-10, 1997.

[10] L. Lorenz *et al.*, "COOLMOS – A New Milestone in High Voltage Power MOS" IEEE International Symposium on Power Semiconductor Devices and ICs, pp. 3-10, 1999.

[11] E. Danielsson *et al.*, "Extrinsic base design of SiC Bipolar Transistors", Silicon Carbide and Related Materials, pp. 1117-1120, 2003.

[12] I. Perez-Wurfll *et al.*, "Analysis of Power Dissipation and High Temperature Operation in 4H-SiC Bipolar Junction Transistors", Silicon Carbide and Related Materials, pp. 1121-1124, 2003.

[13] J. Zhang *et al.*, "High Power (500V-70A) and High Gain (44-47) 4H-SiC Bipolar Junction Transistors", Silicon Carbide and Related Materials, pp. 1149-1152, 2003.

[14] A. Agarwal *et al.*, "SiC BJT Technology for Power Switching and RF Applications", Silicon Carbide and Related Materials 2003, pp. 1141-1144, 2004.

[15] P. Brosselard *et al.*, "Influence of different Peripheral Protections on the Breakover Voltage of a 4H-SiC GTO Thyristor", Silicon Carbide and Related Materials, pp. 1129-1132, 2003.

[16] A.K. Agarwal *et al.*, "Dynamic Performance of 3.1 kV 4H-SiC Asymmetrical GTO Thyristors", Silicon Carbide and Related Materials, pp. 1349-1352, 2003.

[17] B. J. Baliga, "Evolution of MOS-Bipolar Power Semiconductor Technology", Proceedings of the IEEE, pp. 409-418, 1988.

[18] T.P. Chow, N. Ramaungul and M. Ghezzo, "Wide Bandgap Semiconductor Power devices", Materials Research Society Symposium, Vol. 483, pp. 89-102, 1998.

[19] R. Singh, "Silicon Carbide Bipolar Power Devices – Potentials and Limits", Materials Research Society Symposium, Vol. 640, pp. H4.2.1-12, 2001.

[20] J. Wang *et al.*, "Comparison of 5kV 4H-SiC N-channel and P-channel IGBTs", Silicon Carbide and Related Materials, pp. 1411-1414, 2000.

[21] J.W. Palmour, "Silicon Carbide Power Device development for Industrial Markets", IEEE Int. Electron Devices Meeting, Abstract 1.1.1, pp. 1-8, 2014.

Chapter 2

Material Properties

In this chapter, the measured properties for gallium nitride and silicon carbide are reviewed and compared with those for silicon. These properties are then used to obtain other parameters (such as the built-in potential) which are relevant to the analysis of the performance of devices discussed in the rest of the book.

Power devices are being developed from both gallium nitride grown on gallium nitride substrates and gallium nitride grown on silicon substrates. In the former case, it is relevant to use the measured properties of bulk GaN. In the latter case, the technology is based up on the formation of a high electron mobility layer by using an aluminum nitride layer on top of the gallium nitride layer. The properties of the two-dimensional electron gas formed at the surface of the GaN layer are of importance in determining the characteristics of power devices.

In the case of silicon carbide, the discussion is limited to the 4H poly-type of silicon carbide because its properties are superior to those reported for the 6H poly-type. The 4H-SiC material is commercially available from several sources for the development and manufacturing of power devices because of its superior properties. Manufacturers are continually reducing the micropipe density in this material while increasing the diameter of commercially available wafers.

2.1 Fundamental Properties

The basic material properties of relevance to power devices are the energy band gap, the impact ionization coefficients, the dielectric constant, the thermal conductivity, the electron affinity, and the carrier mobility. In addition, the density of states in the conduction and valence band are required for calculation of the intrinsic carrier concentration. The intrinsic carrier concentration and the built-in potential extracted by using this information have been computed and compared with those for silicon in this section. This section provides a summary of the fundamental properties of gallium nitride[1,2,3,4] and silicon carbide[5,6,7] and compares them with those for silicon[8]. The properties for the 4H poly-type have been included here.

Properties	Silicon	GaN	4H-SiC
Energy Band Gap (eV)	1.11	3.44	3.26
Relative Dielectric Constant	11.7	10.4	9.7
Thermal Conductivity (W/cm K)	1.5	1.3	3.7
Electron Affinity (eV)	4.05	4.1	3.8
Density of States Conduction Band (cm^{-3})	2.80×10^{19}	2.3×10^{18}	1.23×10^{19}
Density of States Valency Band (cm^{-3})	1.04×10^{19}	4.6×10^{18}	4.58×10^{18}

Table 2.1 Fundamental Material Properties.

2.2 Energy Band Gap

Gallium nitride and silicon carbide have an energy band gap that is much larger than that for silicon allowing their classification as a wide band gap semiconductor. The energy band gap of semiconductors decreases with increasing temperature due to thermal expansion of the lattice. A semi-empirical formula for change in band gap for silicon with temperature[9] is given by:

$$E_G(\text{Si}) = 1.169 - \frac{4.9 \times 10^{-4}\, T^2}{T + 655} \qquad [2.1]$$

where T is the absolute temperature in °K.

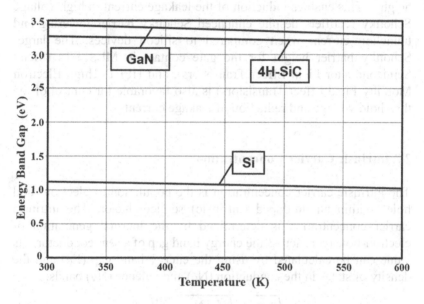

Fig. 2.1 Energy band gap of Si, GaN and 4H-SiC.

Similarly, the temperature dependence of the band gap for GaN[2] is given by:

$$E_G(\text{GaN}) = 3.51 - \frac{9.09 \times 10^{-4}\, T^2}{T + 830} \qquad [2.2]$$

and that for 4H-SiC[10] is given by:

$$E_G(4\,\text{H-SiC}) = 3.30 - \frac{8.20 \times 10^{-4}\, T^2}{T + 1800} \qquad [2.3]$$

The energy band gap for GaN and 4H-SiC can be compared with that for Si in Fig. 2.1. The energy band gaps for GaN and 4H-SiC at room temperature (300 °K) are 3.34 and 3.26 eV, respectively, about 3 times larger than that for silicon, allowing their classification as wide-band-gap semiconductors. Since a larger band gap results in a smaller generation of carriers in the depletion regions of devices, the theoretical bulk leakage current for GaN and 4H-SiC devices which utilize P-N junctions to support voltages is very small. However, in practical devices, the leakage current is much larger than predicted by space charge generation due to surface generation at the edge terminations.

The larger band gap for GaN and 4H-SiC is also favorable for producing metal-semiconductor contacts with larger Schottky barrier heights. This enables reduction of the leakage current in high voltage Schottky rectifiers despite enhanced Schottky barrier lowering and tunneling current when compared to silicon devices. The larger Schottky barrier height for the gate contact in MESFETs (Metal Semiconductor Field Effect Transistors), and HFETs (high Electron Mobility Field Effect Transistors) is also favorable for control of the threshold voltage and reduction of leakage current.

2.3 Intrinsic Carrier Concentration

The intrinsic carrier concentration is the population of electrons and holes within an un-doped (intrinsic) semiconductor. The intrinsic carrier concentration is determined by the thermal generation of electron-hole pairs across the energy band gap of a semiconductor. Its value can be calculated by using the energy band gap (E_G) and the density of states in the conduction (N_C) and valence (N_V) bands:

$$n_i = \sqrt{n.p} = \sqrt{N_C.N_V}\, e^{-[E_G(T)/2kT]} \qquad [2.4]$$

where k is Boltzmann's constant (1.38×10^{-23} J/$^{\circ}$K) and T is the absolute temperature in $^{\circ}$K. For silicon, the intrinsic carrier concentration is given by:

$$n_i = 3.87 \times 10^{16}\, T^{3/2} e^{-(7.02 \times 10^3)/T} \qquad [2.5]$$

For GaN, it is given by:

$$n_i = 1.98 \times 10^{16}\, T^{3/2} e^{-(2.143 \times 10^4)/T} \qquad [2.6]$$

while that for 4H-SiC, it is given by:

$$n_i = 1.70 \times 10^{16}\, T^{3/2} e^{-(2.041 \times 10^4)/T} \qquad [2.7]$$

Using these equations, the intrinsic carrier concentration can be calculated as a function of temperature. The results are plotted in

Fig. 2.2 using the traditional method for making this plot using an inverse temperature scale for the x-axis. An additional Fig. 2.3 is provided here because it is easier to relate the value of the intrinsic carrier concentration to the temperature with its value plotted on the x-axis. It can be clearly observed that the intrinsic carrier concentration for silicon carbide is far smaller than for silicon due to the large difference in band gap energy. The values for GaN are even smaller due to its slightly larger band gap.

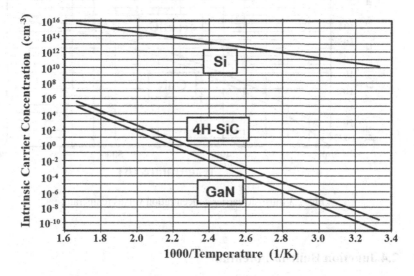

Fig. 2.2 Intrinsic carrier concentration versus inverse temperature.

At room temperature (300 °K), the intrinsic carrier concentration for silicon is 1.38×10^{10} cm^{-3} while that for 4H-SiC and GaN are 2.5×10^{-10} cm^{-3} and 9.8×10^{-12} cm^{-3}, respectively. These low value for the intrinsic concentrations in GaN and 4H-SiC make the bulk generation current negligible. Surface generation currents are usually much larger in magnitude due to the presence of states within the band gap.

For silicon, the intrinsic carrier concentration becomes equal to a typical doping concentration of 1×10^{14} cm^{-3} at a relatively low temperature of 470 °K or 197 °C. In contrast, the intrinsic carrier concentration for GaN and 4H-SiC are only 9×10^4 cm^{-3} and 4×10^5 cm^{-3} even at 600 °K or 327 °C. The development of mesoplasmas has been associated with the intrinsic carrier concentration becoming comparable to the doping concentration.[11] Mesoplasmas create current filaments that have very high current density leading to destructive failure in semiconductors. This is much less likely in gallium nitride and silicon carbide devices.

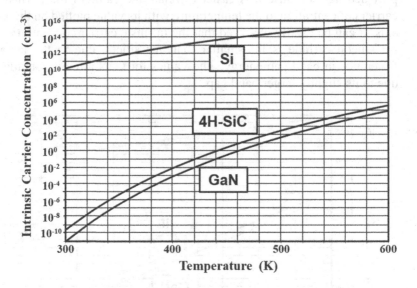

Fig. 2.3 Intrinsic carrier concentration versus temperature.

2.4 Junction Built-in Potential

The built-in potential of P-N junctions can play an important role in determining the operation and design of power semiconductor devices. As an example, the built-in potential determines the zero-bias depletion width which is important for calculation of the on-resistance of power DMOSFETs. It has a strong impact on the on-state voltage drop in JBS rectifiers, and is an important parameter used for the design of normally-off accumulation mode MOSFETs. These structures are discussed in detail later in the book.

The built-in voltage is given by:

$$V_{bi} = \frac{kT}{q} \ln\left(\frac{N_A^-.N_D^+}{n_i^2} \right)$$ [2.8]

where N_A^- and N_D^+ are the ionized impurity concentrations on the two sides of an abrupt P-N junction. For silicon, their values are equal to the doping concentration because of the small dopant ionization energy levels. This does not apply to gallium nitride and silicon carbide because of the much larger dopant ionization levels as discussed later in this chapter.

The calculated junction built-in potential for GaN and 4H-SiC P-N junctions is plotted in Fig. 2.4 as a function of temperature. In making the plots, the product $(N_A^-.N_D^+)$ was assumed to be 10^{35} cm^{-6}. This would be applicable for a typical doping concentration of 1×10^{19} cm^{-3} on the heavily doped side of the junction and 1×10^{16} cm^{-3} on the lightly doped side of the junction.

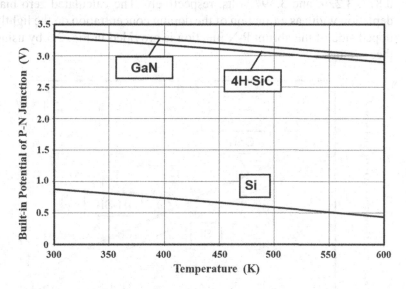

Fig. 2.4 Built-in potential for a P-N junction in GaN, SiC, and Si.

The built-in potential for GaN (3.40 V) and 4H-SiC (3.23 V) are much larger than that for silicon (0.88 V) at 27 °C due to the far smaller values for the intrinsic carrier concentration. This can be a disadvantage due to the larger zero bias depletion width for silicon carbide as shown in the next section. As an example, the larger zero bias depletion width consumes space within the D-MOSFET cell structure increasing the on-resistance by constricting the area through which current can flow. In contrast, the larger built-in potential for SiC and its associated larger zero bias depletion width can be taken advantage of in making innovative device structures, such as the ACCUFET, that are tailored to the unique properties of SiC.

2.5 Zero Bias Depletion Width

The zero-bias depletion width of a P-N junction with high doping concentration on the p-side can be obtained using[12]:

$$W_0 = \sqrt{\frac{2\varepsilon_S V_{bi}}{q N_D}}$$ [2.9]

where N_D is the doping concentration on the n-side. The junction built-in potential for silicon, 4H-SiC and GaN at room temperature are 0.877, 3.229, and 3.397 volts, respectively. The calculated zero bias depletion width as a function of the doping concentration on the lightly doped side of the abrupt P-N junction is provided in Fig. 2.5 by using these values.

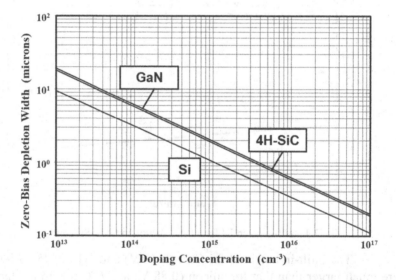

Fig. 2.5 Zero-bias depletion width in Si, SiC, and GaN P-N junctions.

The zero bias depletion width for GaN and SiC P-N junctions is much (about 2x) larger than for the silicon case for the same doping concentration. However, it is worth noting that the doping concentration in GaN and SiC devices is also much greater for a given breakdown voltage than for silicon as discussed later in the chapter. In all cases, it can be observed from the figure that the zero bias depletion width can be substantial in size at low doping concentrations making it important to take this into account during device design and analysis.

2.6 Impact Ionization Coefficients

Impact ionization occurs in semiconductors when mobile electrons and holes are accelerated by high electric fields within depletion regions of reverse blocking junctions. The mobile charges gain energy during their motion which is imparted to the lattice during scattering events. When the electric field becomes sufficiently large, the energy gained by the mobile particles can excite electron-hole pairs across the energy band gap. These new electrons and holes are in turn accelerated by the electric field producing the generation of more electron-hole pairs. This precipitates avalanche breakdown of the junction with the on-set of a large amount of current flow which limits its maximum voltage withstand capability.

The main advantage of wide band gap semiconductors for power device applications is the larger electric field required for onset of impact ionization for these materials. A larger electric field is required for a semiconductor with a greater energy band gap because the mobile charges must gain more energy to create electron-hole pairs. The impact ionization process is characterized by using impact ionization coefficients defined as the number of electron-hole pairs created by a mobile particle moving through a distance of 1 cm with the imposed electric field.

The impact ionization coefficients for semiconductors is described by Chynoweth's Law[13,14]:

$$\alpha = a.e^{-b/E} \qquad [2.10]$$

where E is the electric field component in the direction of current flow. The parameters a and b are constants that depend upon the semiconductor material and the temperature. For silicon, the impact ionization rate is much larger for electrons than for holes and been measured as a function of electric field and temperature.[15,16] The data can be modeled by using $a_n = 7 \times 10^5$ per cm and $b_n = 1.23 \times 10^6$ V/cm for electrons and $a_p = 1.6 \times 10^6$ per cm and $b_p = 2.0 \times 10^6$ V/cm for holes at room temperature.[17]

The impact ionization coefficients for 4H-SiC along the c-axis were measured as a function of temperature by using an electron beam excitation method by Raghunathan and Baliga.[18] This method allowed extraction of impact ionization rates in defect free regions of the material by isolating diodes containing defects with an EBIC (Electron Beam Induced Current) image. This was important because substantially enhanced impact ionization rates were discovered when the measurements were conducted at defect sites.[19] For defect free material, the extracted values for the impact ionization coefficient parameters for holes in 4H-SiC were found to be:

$$a_p(4H\text{-}SiC\ R/B) = 6.46 \times 10^6 - 1.07 \times 10^4 T \qquad \textbf{[2.11]}$$

with

$$b_p(4H\text{-}SiC\ R/B) = 1.75 \times 10^7 \qquad \textbf{[2.12]}$$

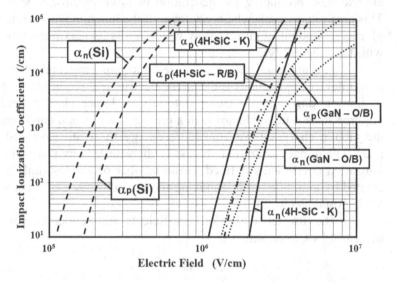

Fig. 2.6 Measured impact ionization coefficients in Si, GaN, and 4H-SiC.

The impact ionization coefficient parameters for holes and electrons in 4H-SiC have also been measured by Konstantinov and co-workers along the c-axis.[20,21] At room temperature, the measured values can be modelled using:

$$a_p(4H - SiC\ K) = 8.07 \times 10^6 \qquad \textbf{[2.13]}$$

with

$$b_p(4H - SiC\ K) = 1.5 \times 10^7 \qquad \textbf{[2.14]}$$

and

$$a_n(4H - SiC\ K) = 3.13 \times 10^8 \qquad [2.15]$$

with

$$b_n(4H - SiC\ K) = 3.45 \times 10^7 \qquad [2.16]$$

Impact ionization in GaN has been theoretically modelled by a number of researchers.[22,23,24] Experimental measurements of the impact ionization coefficients were indirectly extracted from the characteristics of AlGaN/GaN HFET characteristics with many assumptions that make the data inaccurate.[25] Impact ionization coefficients were obtained by using avalanche photodiodes but the data is influenced by the presence of defects in the large diode area.[26] Most recently, the impact ionization coefficient parameters for holes and electrons in GaN were measured by Ozbeck and Baliga[27] using electron beam excitation of diodes with reduced electric fields at the edges. The measured values at room temperature can be modelled using:

$$a_p(GaN-O/B) = 6.4 \times 10^5 \qquad [2.17]$$

with

$$b_p(GaN-O/B) = 1.46 \times 10^7 \qquad [2.18]$$

and

$$a_n(GaN-O/B) = 1.50 \times 10^5 \qquad [2.19]$$

with

$$b_n(GaN-O/B) = 1.41 \times 10^7 \qquad [2.20]$$

The impact ionization coefficients for GaN and 4H-SiC can be compared with those for silicon using Fig. 2.6. It can be seen that the onset of significant generation of carriers by impact ionization ($\alpha = 10^3$ cm^{-1}) occurs in silicon at electric fields of about 2×10^5 V/cm. An electric field of about 2×10^6 V/cm is required in GaN and in 4H-SiC to achieve the same magnitude for the impact ionization coefficient. As a consequence, breakdown in GaN and 4H-SiC devices occurs when the electric fields are in the range of 2-3 $\times 10^6$ V/cm — an order of magnitude larger than that for silicon.

In the case of GaN and 4H-SiC, the impact ionization coefficients for holes have been found to be much larger than those for electrons which is opposite that found for silicon. The measured impact ionization coefficients for holes in 4H-SiC measured by Konstaninov and co-workers is larger than that reported by Raghunathan/Baliga. The values measured by Konstantinov and co-workers is more commonly used when modelling devices to provide a conservative analysis of breakdown in devices.[28,29]

2.7 Bulk Electron Mobility

As in the case of silicon, the mobility for electrons is larger than that for holes in GaN and 4H-SiC. It is therefore favorable to make unipolar devices using N-type drift regions rather than P-type drift regions. The doping concentration for the drift region in GaN and SiC devices is substantially larger than in silicon devices for obtaining the same breakdown voltage as discussed in detail later in the book. It is useful to have models for the mobility as a function of doping concentration over a broad range of doping levels for all of these semiconductors. This behavior has also been theoretically modeled taking into account acoustic and polar optical phonon scattering as well as intervalley scattering.[30]

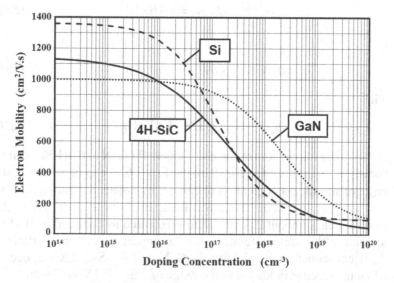

Fig. 2.7 Bulk mobility for electrons in Si, GaN, and 4H-SiC.

For n-type silicon, the measured data[31] for mobility of electrons at room temperature as a function of the doping concentration can be modeled using:

$$\mu_n(Si) = \frac{5.10 \times 10^{18} + 92 N_D^{0.91}}{3.75 \times 10^{15} + N_D^{0.91}}$$ [2.21]

At low doping levels, the electron mobility in silicon has a value of 1360 cm^2/V-s while at very high doping levels, it asymptotes to a value of 92 cm^2/V-s.

The mobility in 4H-SiC is anisotropic due to its hexagonal crystal structure. The mobility parallel to the c-axis is 20% larger than perpendicular to the c-axis. The mobility parallel to the c-axis is discussed here because it is of greatest interest for analysis of vertical discrete 4H-SiC power devices. For n-type 4H-SiC, the mobility of electrons at room temperature as a function of the doping concentration can be modeled using[6]:

$$\mu_n(4H-SiC) = \frac{4.05 \times 10^{13} + 20 N_D^{0.61}}{3.55 \times 10^{10} + N_D^{0.61}}$$ [2.22]

At low doping levels, the electron mobility in 4H-SiC has a value of 1140 cm^2/V-s, while at very high doping levels it asymptotes to a value of 20 cm^2/V-s. These values are consistent with experimental results.[32,33,34] Theoretical computation of the electron mobility in 4H-SiC has also been reported.[35]

There is considerable variation in the measured electron mobility for GaN in the literature. For n-type GaN, the mobility of electrons at room temperature as a function of the doping concentration can be best modeled using the data in the selected references[1,36,37,38]:

$$\mu_n(GaN) = \frac{2.0 \times 10^{17} + 60 N_D^{0.78}}{2.0 \times 10^{14} + N_D^{0.78}}$$ [2.23]

At low doping levels, the electron mobility in GaN has a value of 1000 cm^2/V-s, while at very high doping levels it asymptotes to a value of 60 cm^2/V-s.

The electron mobility for silicon, GaN and 4H-SiC at room temperature is plotted in Fig. 2.7 as a function of doping concentration. In all cases, the mobility decreases with increasing doping due to enhanced Coulombic scattering of electrons by the ionized donors.

This data is useful for development of power device models later in the book. It is noteworthy that the electron mobility for GaN is smaller than that for 4H-SiC at doping levels below 1×10^{16} cm^{-3} and larger than that for 4H-SiC at larger doping concentrations.

The mobility in semiconductors decreases with increasing temperature due to enhanced phonon scattering. This holds true for the electron mobility in silicon, GaN, and 4H-SiC as shown in   Fig. 2.8. For silicon, the temperature dependence of the electron mobility at low doping concentrations can be modeled using[39]:

$$\mu_n(Si) = 1360 \left(\frac{T}{300} \right)^{-2.42} \qquad [2.24]$$

For 4H-SiC, the temperature dependence of the electron mobility at low doping concentrations can be modeled using[5]:

$$\mu_n(4H-SiC) = 1140 \left(\frac{T}{300} \right)^{-2.70} \qquad [2.25]$$

For GaN, the temperature dependence of the electron mobility at low doping concentrations can be modeled using[36,37]:

$$\mu_n(GaN) = 1000 \left(\frac{T}{300} \right)^{-2} \qquad [2.26]$$

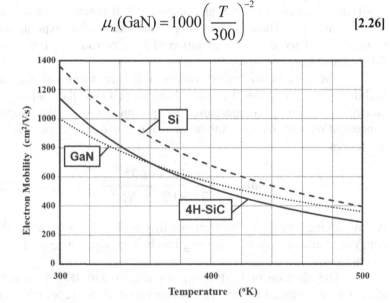

Fig. 2.8 Bulk mobility for electrons in Si, GaN, and 4H-SiC.

The variation of the mobility with temperature is shown in the above figure only up to 500 °K (227 °C) because power devices are usually not rated to operate at higher temperatures. Although there is considerable interest in the performance of silicon carbide devices at much higher temperatures due to its wide band gap structure, the operation of devices above 500 °K is problematic due to degradation of the ohmic contacts and the surface passivation. It is worth pointing out that the mobility for electrons reduces by a factor of about 2-times when the temperature rises to 125 °C and by a factor of about 3-times at 225 °C. This increases the resistance of the drift region in unipolar devices such as Schottky rectifiers and power MOSFETs. The increase in resistance reduces the maximum current handling capability. It is worth pointing out that the electron mobility for GaN becomes larger than that for 4H-SiC at 500 °K.

2.8 Bulk Hole Mobility

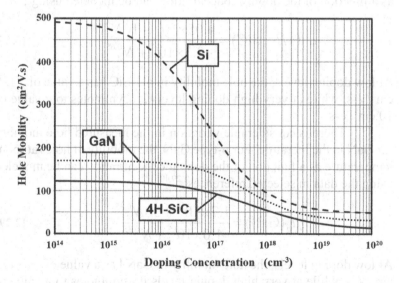

Fig. 2.9 Bulk mobility for holes in Si, GaN, and 4H-SiC.

Although power devices are invariably based on electron transport in GaN and 4H-SiC, information on the mobility for holes is required during design and analysis of devices, such as power MOSFETs. During switching transients, the capacitive charging currents flow through the p-base region of power MOSFETs. This current flow can trigger the turn-on of the parasitic bipolar transistor. Analysis of the resistance of the p-base region requires knowledge of the mobility for holes in the semiconductors.

It is useful to have models for the hole mobility as a function of doping concentration over a broad range of doping levels for silicon, GaN and 4H-SiC. For p-type silicon, the measured data[24] for mobility of holes at room temperature as a function of the doping concentration can be modeled using:

$$\mu_p(Si) = \frac{2.9 \times 10^{15} + 47.7 N_A^{0.76}}{5.86 \times 10^{12} + N_A^{0.76}}$$ [2.27]

At low doping levels, the hole mobility in silicon has a value of 495 cm^2/V-s while at very high doping levels, it asymptotes to a value of 48 cm^2/V-s.

For p-type 4H-SiC, the mobility of holes at room temperature as a function of the doping concentration can be modeled using[3]:

$$\mu_p(4H - SiC) = \frac{4.05 \times 10^{13} + 10 N_A^{0.65}}{3.3 \times 10^{11} + N_A^{0.65}}$$ [2.28]

At low doping levels, the hole mobility in 4H-SiC has a value of 120 cm^2/V-s, while at very high doping levels it asymptotes to a value of 10 cm^2/V-s.

There is considerable variation in the measured hole mobility for GaN in the literature. For p-type GaN, the mobility of holes at room temperature as a function of the doping concentration can be modeled using the data in selected references[1,29,40]:

$$\mu_p(GaN) = \frac{1.7 \times 10^{17} + 30 N_A^{0.85}}{1.0 \times 10^{15} + N_A^{0.85}}$$ [2.29]

At low doping levels, the hole mobility in GaN has a value of 170 cm^2/V-s, while at very high doping levels it asymptotes to a value of 30 cm^2/V-s.

The mobility for holes in silicon, GaN and 4H-SiC at room temperature is plotted in Fig. 2.9 as a function of doping concentration. In all cases, the mobility decreases with increasing doping due to enhanced Coulombic scattering of holes by the ionized acceptors. This data is useful for development of power device models later in the book.

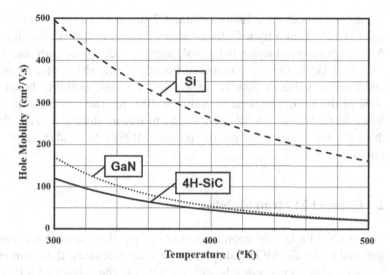

Fig. 2.10 Bulk mobility for holes in Si, GaN, and 4H-SiC.

The mobility in semiconductors decreases with increasing temperature due to enhanced phonon scattering. This holds true for the hole mobility in silicon, GaN, and 4H-SiC as shown in Fig. 2.10. For silicon, the temperature dependence of the hole mobility at low doping concentrations can be modeled using[31]:

$$\mu_p(Si) = 495\left(\frac{T}{300}\right)^{-2.2} \qquad [2.30]$$

For 4H-SiC, the temperature dependence of the hole mobility at low doping concentrations can be modeled using[28,41]:

$$\mu_p(4H\text{-}SiC) = 120\left(\frac{T}{300}\right)^{-3.4} \qquad [2.31]$$

For GaN, the temperature dependence of the hole mobility at low doping concentrations can be modeled using[37]:

$$\mu_p(GaN) = 170\left(\frac{T}{300}\right)^{-4.0} \qquad [2.32]$$

The variation of the mobility with temperature is shown in the above figure only up to 500 °K (227 °C) because power devices are usually not rated to operate at higher temperatures. The mobility for holes reduces by a factor of about 2-times when the temperature rises to 125 °C and by a factor of about 3-times at 225 °C for silicon. A much greater reduction in the hole mobility occurs for GaN and 4H-SiC. For GaN, the hole mobility becomes only 30% of the room temperature value at 400 °K. In 4H-SiC, the hole mobility becomes 40% of the room temperature value at 400 °K. The low mobility for holes in GaN and 4H-SiC can create problems during current flow through the base regions of power MOSFETs under transient operation.

2.9 Channel Electron Mobility

Power MOSFETs rely upon control of current flow through a channel induced under the MOS gate region. In silicon devices, the channel is formed using an inversion layer – electrons at the surface of a P-base region for n-channel devices and holes at the surface of an N-base region. Silicon carbide power MOSFETs of interest are n-channel devices because of the much larger mobility for electrons than holes in bulk 4H-SiC material.

An n-channel region is formed by application a positive gate bias across the gate oxide to produce a strong electric field orthogonal to the semiconductor surface. Band bending at the semiconductor surface creates an inversion layer of electrons at the interface between the P-base region and the gate oxide at sufficiently large gate bias voltages[42]. The electrons in the inversion layer are confined very close to the interface.

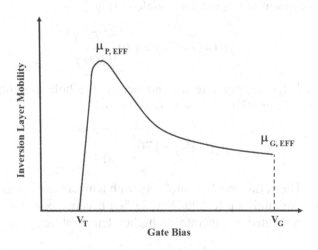

Fig. 2.11 Inversion layer effective mobility.

Many additional mechanisms influence the scattering of electrons in the inversion layer when compared with bulk transport. They include Coulombic scattering by fixed charge in the gate oxide, charges in interface states, surface phonon scattering, and surface roughness scattering. Among these, surface roughness scattering is the most significant for inversion layers in 4H-SiC[43]. The wafer surface for 4H-SiC is much rougher than for silicon due to the off-axis orientation of the substrate required by epitaxial growth and the greater difficulty in polishing this material.

The typical variation of the effective or conductivity inversion layer mobility is shown schematically in Fig. 2.11. It has a peak value ($\mu_{P,EFF}$) close to the threshold voltage and then reduces with increasing gate bias. The value of the effective inversion layer mobility ($\mu_{G,EFF}$) at the gate bias voltage (V_G) is of interest for computation of the on-resistance for power MOSFETs. This behavior can be modelled using:

$$\mu_{EFF} = \frac{\mu_{P,EFF}}{1 + K_{inv}\left(V_G - V_{TH}\right)} \qquad [2.33]$$

where $\mu_{P,EFF}$ is the peak inversion layer mobility, K_{inv} is a constant that describes the rate of decrease in mobility with increasing electric field in the oxide, V_G is the gate bias and V_{TH} is the threshold voltage.

The effective inversion layer mobility for 4H-SiC was initially reported to be very low (< 0.1 cm²/V-s). A high effective inversion layer mobility for 4H-SiC n-channel lateral MOSFETs was first reported in 1998.[44,45] Lateral MOSFETs fabricated using low temperature (410 °C) deposited SiO_2 followed by a wet oxidation cycle at 1000 °C for 400 minutes were found to exhibit peak inversion layer mobility of 176 cm²/V-s at room temperature which reduced to 140 cm²/V-s at high gate bias voltages. The effective mobility was found to increase with increasing temperature indicating presence of traps at the interface between the oxide and the semiconductor. These MOSFETs had a threshold voltage of 2 volts at room temperature. The results obtained using deposited oxide were subsequently reproduced with a peak mobility of 80 cm²/V-s.[46] 10-kV 4H-SiC SiC power MOSFETs have been successfully fabricated using a deposited gate oxide followed by NO annealing[47]. These devices had a favorable threshold voltage of 7.5 V with a channel mobility of 15 cm²/V-s.

However, it is preferable to utilize thermally grown oxide for the fabrication of power devices due to their superior breakdown strength. It was found that thermal oxidation of 4H-SiC in either dry or wet ambient produces a very high density of interface states that trap the electrons in the inversion layers. It is necessary to determine the concentration of very fast interface states.[48] The effective mobility measured with the presence of the interface states is very low due to the reduced mobile charge in the inversion layer. Hall measurements performed on inversion layers have determined that the inversion mobility is in the range of 100 cm^2/V-s.

During the last 20 years, considerable effort has been undertaken to understand the interface between SiO_2 and 4H-SiC and correlate this with the inversion layer mobility.[47] The best method for improvement of effective mobility in 4H-SiC for MOSFETs fabricated using thermally grown oxides has been reported to be by post-oxidation annealing in nitric oxide (NO) and nitrous oxide (N_2O). It has been reported that the channel mobility is inversely proportional to the interface state density. A significant reduction in the interface state density has been accomplished by performing annealing at 1175 °C in a nitric oxide (NO) ambient for 2 hours.[49,50,51,52] An effective channel mobility for electrons of 30-35 cm^2/V-s was achieved using this process at the operating gate bias. More recent work[53,54] indicates that charge trapping has been sufficiently suppressed by the nitric oxide annealing leading to effective channel mobility of about 60 cm^2/V-s.

It would be preferable to fabricate 4H-SiC power MOSFETs using ion-implanted P-base regions as in the case of silicon devices. Due to the low diffusion rate for dopants in the 4H-SiC, it is necessary to stagger the edge for the implantation of the P-base and N^+ source regions for the 4H-SiC devices[55,56] to create the DiMOSFET structure. The damage in the ion implanted regions must be removed by annealing followed by surface preparation to reduce interface states. The inversion layer mobility in n-channel lateral MOSFETs fabricated on aluminum implanted layers in 4H-SiC has been reported.[57] The ion-implant dose and energy was selected to achieve a doping concentration of 1×10^{17} cm^{-3} followed by annealing for 10 minutes at 1600 °C. The gate oxide was grown at 1200 °C and then placed in an alumina environment. Peak inversion layer mobility of 100 cm^2/V-s was observed with a high threshold voltage of about 10 volts.

High voltage power MOSFETs must be typically designed with a threshold voltage of 5 volts. This requires reducing the P-base doping concentration[58] to about 1×10^{16} cm^{-3}. Such low P-base doping concentration is comparable to the doping in the N-drift region leading to reach-through breakdown problems. To avoid this, it is necessary to include a deeper highly doped P-type barrier layer below the P-base region to create the shielded-base power MOSFET structure.[59] The shielded-base structure can also be achieved by using a retrograde ion-implantation profile to achieve a low doping concentration at the surface. This approach is widely practiced in the industry to manufacture 4H-SiC power MOSFETs as discussed later in the book. The inversion layer mobility has been measured on P-type epitaxial layers with doping concentration of 1×10^{16} cm^{-3} grown on top of aluminum ion-implanted highly doped (5×10^{18} cm^{-3}) shielding layers.[60] It was found to range from 20 to 30 cm^2/V-s when compared to 2 cm^2/V-s without the epitaxial layer.

2.10 Electron Velocity-Field Curves

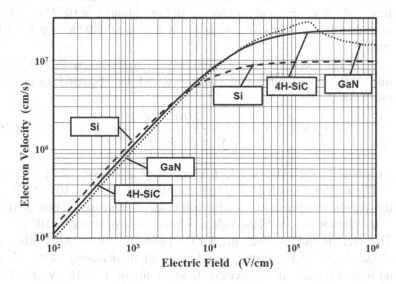

Fig. 2.12 Velocity-field curves for electrons in Si, GaN, and 4H-SiC at 300 °K.

Power devices must operate with high electric fields in the drift region in order to support high voltages over short distances. The velocity for electrons and holes in semiconductors increases in proportion to the electric field at low values for the field. The proportionality constant is defined as the mobility which was described in the previous sections. At larger electric field values, the mobile carriers gain enough energy to generate optical phonon scattering in the semiconductors. This results in a saturation of the drift velocity at high electric fields. In addition, inter-valley scattering in Gallium Nitride produces a peak in the velocity-field curve. The velocity-field curves for electrons in the three semiconductors can be compared using Fig. 2.12.

The electron velocity-field curve at room temperature for Si with low doping concentrations can be modeled using the measured data[61]:

$$v_n(\text{Si}) = \frac{9.85 \times 10^6 \, E}{\left[1.04 \times 10^5 + \left(E^{1.3}\right)\right]^{0.77}} \qquad [2.34]$$

At low electric fields, the electron mobility in Si has a value of 1360 cm^2/V-s at small doping concentrations (below 1×10^{15} cm^{-3}). In this regime of operation, the electron velocity increases in proportion to the electric field until the field reaches a value of about 1×10^3 V/cm. The electron velocity then increases more slowly with increasing electric field until it saturates at an electric field of about 1×10^5 V/cm. The saturated drift velocity for electrons in silicon at 300 °K is 9.7×10^6 cm/s.

The electron velocity-field curve at room temperature for 4H-SiC with low doping concentrations can be modeled using the measured data[62,63]:

$$v_n(4\text{H-SiC}) = \frac{2.20 \times 10^7 \, E}{\left[2.27 \times 10^5 + \left(E^{1.25}\right)\right]^{0.8}} \qquad [2.35]$$

At low electric fields, the electron mobility in 4H-SiC has a value of 1140 cm^2/V-s at small doping concentrations (below 1×10^{15} cm^{-3}). In this regime of operation, the electron velocity increases in propor-tion to the electric field until the field reaches a value of about 1×10^4 V/cm. This electric field is much greater than that for silicon. After this, the electron velocity increases more slowly with increase-ing electric field until it saturates at an electric field of about 1×10^5 V/cm. The saturated drift velocity for electrons in 4H-SiC at 300 °K is 2.2×10^7 cm/s. This values is twice as large as that observed for silicon.

The electron velocity-field curve for GaN has been theoretically modelled with non-parabolic multi-valley balance equations.[64] The model predicts a peak electron velocity of 3.4×10^7 cm/s at an electric field of about 2×10^5 V/cm. Similar results have been reported using Monte-Carlo models but with lower peak electron velocity of about 2.5 to 3.0×10^7 cm/s.[65,66,67,68] From the theoretical predictions, the electron velocity-field curve at room temperature for GaN with low doping concentrations can be fitted using:

$$v_n(GaN) = \frac{2.70 \times 10^7 \ E}{\left[3.46 \times 10^5 + \left(E^{1.25}\right)\right]^{0.8}} \qquad [2.36]$$

At low electric fields, the electron mobility in GaN has a value of 1000 cm^2/V-s at small doping concentrations (below 1×10^{16} cm^{-3}). In this regime of operation, the electron velocity increases in proportion to the electric field until the field reaches a value of about 1×10^5 V/cm. This electric field is much greater than that for silicon. After this, the electron velocity increases more slowly with increasing electric field until it reaches a peak value of 2.7 $\times$ 10^7 V/cm at an electric field of about 1.5 $\times$ 10^5 V/cm. The velocity then reduces with increases electric field until it saturates. The saturated drift velocity for electrons in GaN at 300 °K is 1.5 $\times$ 10^7 cm/s. This values is larger than that observed for silicon but smaller than that observed in 4H-SiC.

2.11 Hole Velocity-Field Curves

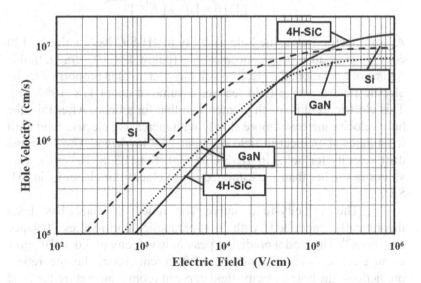

Fig. 2.13 Velocity-field curves for holes in Si, GaN, and 4H-SiC at 300 °K.

The velocity-field curves for holes in the three semiconductors can be compared using Fig. 2.13. The velocity-field curve for holes at room temperature for Si with low doping concentrations can be modeled using the measured data[42]:

$$v_p(\text{Si}) = \frac{8.91 \times 10^6 \ E}{\left[1.41 \times 10^5 + \left(E^{1.2}\right)\right]^{0.83}} \qquad [2.37]$$

At low electric fields, the hole mobility in Si has a value of 495 cm^2/V-s at small doping concentrations (below 1×10^{15} cm^{-3}). In this regime of operation, the hole velocity increases in proportion to the electric field until the field reaches a value of about 1×10^4 V/cm. The hole velocity then increases more slowly with increasing electric field until it saturates at an electric field of about 1×10^5 V/cm. The saturated drift velocity for holes in silicon at 300 °K is 8.9×10^6 cm/s.

There is no reliable data for the velocity-field curve in p-type 4H-SiC. The saturated drift velocity for holes is about 1.3×10^7 cm^2/V-s. The velocity-field curve for holes at room temperature for 4H-SiC with low doping concentrations can be modeled using:

$$v_p(4\text{H-SiC}) = \frac{1.3 \times 10^7 \ E}{\left[1.16 \times 10^6 + \left(E^{1.2}\right)\right]^{0.83}} \qquad [2.38]$$

At low electric fields, the hole mobility in 4H-SiC has a value of 120 cm^2/V-s at small doping concentrations (below 1×10^{15} cm^{-3}). In this regime of operation, the electron velocity increases in proportion to the electric field until the field reaches a value of about 1×10^5 V/cm. This electric field is much greater than that for silicon. After this, the hole velocity increases more slowly with increasing electric field until it saturates at an electric field of about 1×10^6 V/cm. The saturated drift velocity for holes in 4H-SiC at 300 °K is 1.3×10^7 cm/s. This values is smaller than the saturated drift velocity for electrons in 4H-SiC.

The velocity-field curve for holes in GaN has been theoretically modelled with non-parabolic multi-valley balance equations[64]. The model predicts a peak hole velocity of 7.0×10^6 cm/s at an electric field of about 1×10^7 V/cm. From the theoretical predictions, the hole velocity-field curve at room temperature for GaN with low doping concentrations can be fitted using:

$$v_p(\text{GaN}) = \frac{7.0 \times 10^6 \, E}{\left[3.63 \times 10^5 + \left(E^{1.2} \right) \right]^{0.83}} \qquad [2.39]$$

At low electric fields, the hole mobility in GaN has a value of 170 cm^2/V-s at small doping concentrations (below 1×10^{16} cm^{-3}). In this regime of operation, the hole velocity increases in proportion to the electric field until the field reaches a value of about 5×10^4 V/cm. After this, the hole velocity increases more slowly with increasing electric field until it reaches a peak value of 7×10^6 V/cm at an electric field of about 5×10^5 V/cm. The saturated drift velocity for holes in GaN at 300 °K is 7×10^6 cm/s. This values is smaller as that observed for silicon.

2.12 GaN 2DEG Mobility

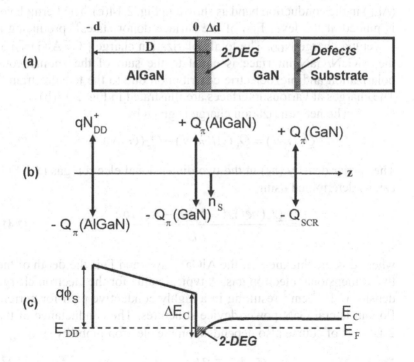

Fig. 2.14 Typical AlGaN/GaN system: (a) heterostructure; (b) charge distribution; and (c) band diagram.

The AlGaN/GaN *high electron mobility transistor (HEMT)* has evolved as an important candidate for power electronic applications. Such structures have also been referred to as *modulation doped field effect transistors (MODFETs)* and *2-dimensional gate field effect transistors (TEGFETs)*. In these types of devices, the basic idea is to separate the mobile electrons from the Coulombic scattering due to the charge of the donor atoms. In the case of the AlGaN/GaN heterostructure, the mobile electrons in the channel originate from the ionized surface donor states.[69]

A cross-section of the basic AlGaN/GaN heterostructure is illustrated in Fig. 2.14(a). It consists of a GaN layer grown on a substrate, which is usually silicon for the case of power devices to reduce the cost. A highly defective layer if formed at the interface between the GaN and the substrate due to the lattice mismatch. An AlGaN layer is grown on top of the GaN layer. The energy band gap of the AlGaN layer is larger than that of GaN creating a discontinuity (ΔE_C) in the conduction band as shown in Fig. 2.14(c). The Fermi level is pinned at the level E_{DD} of the surface donor states[69] producing a screening charge N_{DD}. The total polarization charge [$-Q_\pi(AlGaN)$] at the AlGaN/GaN interface is equal to the sum of the spontaneous polarization and piezoelectric contribution due to the tensile strain.[70] The charges at various interfaces are illustrated in Fig. 2.14(b).

The net polarization charge is given by:

$$Q_\pi(net) = Q_\pi(AlGaN) - Q_\pi(GaN) \qquad \textbf{[2.40]}$$

The carrier density (n_S) in the two-dimensional electron gas (2-DEG) can be determined using:

$$n_S = \frac{Q_\pi(net).d - \varepsilon_S(\phi_S - \Delta E_C/q)}{qD} \qquad \textbf{[2.41]}$$

where d is the thickness of the AlGaN layer and D is the depth of the two-dimensional electron gas. A typical value for the electron charge density is 10^{13} cm^{-2} resulting in a highly conductive path for current flow in lateral GaN power device structures. The conductance of the 2-DEG is of course also dependent on the electron mobility ($\mu_{2\text{-DEG}}$):

$$\sigma_{2-DEG} = q\mu_{2-DEG}n_S \qquad \textbf{[2.42]}$$

Considerable effort has been put into improving the conductivity of the 2-DEG layer over the years by optimization of the growth conditions for the layers, the thickness of the AlGaN layer, and the aluminum composition of the AlGaN layer. In 2002, 2-DEG layers with sheet carrier concentrations above 1×10^{13} cm^{-2} were reported[71] with electron mobility of up to 2000 cm^2/V-s. The GaN and AlGaN layers were grown on 100-mm diameter silicon substrates using an MOCVD process. The HFET structure consisted of an undoped AlGaN layer with thickness of 50 Angstroms, an AlGaN donor layer with 2×10^{18} cm^{-3} silicon atoms, an undoped AlGaN layer, and a GaN cap layer. The aluminum mole fraction in the AlGaN layers was 25%. The average sheet resistance across the wafers was measured at 274 Ω/sq. The typical electron mobility in the 2-DEG was 1590 cm^2/V-s.

In 2010, it was reported[72] that the 2-DEG electron sheet carrier density could be manipulated from 1×10^{13} to 2×10^{14} cm^{-2} by adjusting the tensile stress in the GaN layer. The electron mobility was found to decrease with increasing tensile stress from 1000 to 100 cm^2/V-s. The smallest sheet resistance of 290 Ω/sq was observed at the largest stress values.

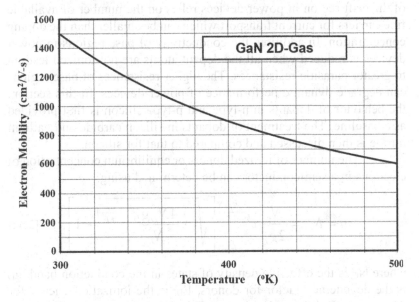

Fig. 2.15 Electron mobility in GaN 2D-gas layer.

The temperature dependence of the electron mobility in the 2-DEG layer was reported[73] in 1999. For the 2-DEG layer with a sheet carrier density of 2.6×10^{12} cm^{-3}, the temperature dependence of the electron mobility is given by:

$$\mu_{2-DEG}(\text{GaN}) = \mu_{2-DEG}(300K)\left(\frac{T}{300}\right)^{-1.8} \qquad [2.43]$$

The change in the electron mobility with temperature is shown in Fig. 2.15. The reduction in the mobility with increasing temperature is attributed to increased phonon scattering.

2.13 Dopant Ionization

At doping concentrations below 10^{19} cm^{-3}, the donor and acceptor level can be treated at a discrete position away from the band edges. The electron and hole carrier concentrations are not equal to the doping concentration because all of the dopant atoms are not ionized. The number of ionized dopants becomes smaller when the temperature is reduced and the doping concentration is increased. The conductivity of the drift region in power devices relies on the number of available free carriers for current transport which can be smaller than the doping concentration. In addition, the conductivity of base regions in power devices is reduced when all the dopant atoms are not ionized leading to greater parasitic resistances. The higher resistances of base regions can degrade dynamic performance of power devices. In this section, the behavior of dopants in n-type and p-type silicon is first provided as a baseline. The treatment of dopants in silicon carbide and gallium nitride is then discussed and compared to that for silicon.

The number of ionized donors or equilibrium concentration of electrons in a semiconductor can be determined using:

$$n = N_D^+ = \frac{N_C}{2g_D}e^{\left(-\frac{E_{DI}}{kT}\right)}\left[\sqrt{1 + \frac{4N_D g_D}{N_C}e^{\left(\frac{E_{DI}}{kT}\right)}} - 1\right] \qquad [2.44]$$

where N_C is the effective density of states in the conduction band, g_D is the degeneracy factor for donors, E_{DI} is the ionization energy for donors, k is Boltzmann's constant, T is the absolute temperature, and N_D is the donor doping concentration. Similarly, the number of ionized acceptors or equilibrium concentration of holes in a semiconductor can be determined using:

$$p = N_A^- = \frac{N_V}{2g_A} e^{\left(-\frac{E_{Al}}{kT}\right)} \left[\sqrt{1 + \frac{4N_A g_A}{N_V} e^{\left(\frac{E_{Al}}{kT}\right)}} - 1 \right] \qquad [2.45]$$

where N_V is the effective density of states in the valence band, g_A is the degeneracy factor for acceptors, E_{Al} is the ionization energy for acceptors, k is Boltzmann's constant, T is the absolute temperature, and N_A is the acceptor doping concentration. The degeneracy factor for donors is typically 2 and that for acceptors is typically 4.

2.13.1 Dopant Ionization in Silicon

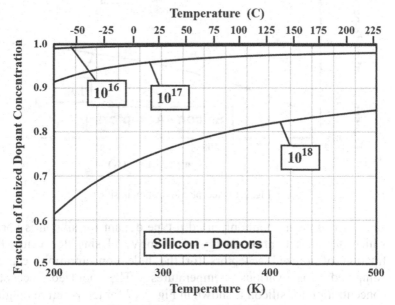

Fig. 2.16 Donor ionization in silicon.

The most commonly used n-type dopant for silicon is phosphorus which has an activation energy of 46 meV.[74] Using this value, the fraction of ionized donors (or the electron concentration) can be computed for various temperatures. The ionized donor concentration for silicon is shown in Fig. 2.16 for temperature ranging from 200 to 500 °K. The corresponding temperature in °C is also shown on the chart for convenience. Typical drift region for silicon power devices have low doping concentrations below 10^{16} cm^{-3}. From Fig. 2.16, it can be concluded that all the donors are ionized for the entire temperature range in the drift region of silicon power devices. For higher doping concentrations, such as the n-type base regions of p-channel power MOSFETs, all the donors cannot be assumed to completely ionized especially at below room temperature (300 °K).

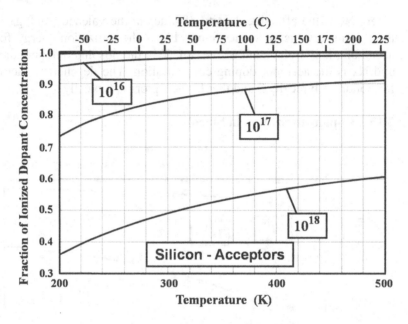

Fig. 2.17 Acceptor ionization in silicon.

The most commonly used p-type dopant for silicon is boron which has an activation energy of 44 meV.[74] Using this value, the fraction of ionized acceptors (or the hole concentration) can be computed for various temperatures. The ionized acceptor concentration for silicon is shown in Fig. 2.17 for temperature ranging from 200 to 500 °K. The fraction of ionized acceptors is smaller than donors for the same temperature and dopant concentration. Typical drift region for silicon power devices have low doping concentrations below 10^{16} cm^{-3}. From Fig. 2.17, it can be concluded that all the acceptors are ionized for the entire temperature range in the drift region of silicon power devices. For higher doping concentrations, such as the p-type base regions of n-channel power MOSFETs, all the donors cannot be assumed to completely ionized especially at below room temperature (300 °K).

2.13.2 Dopant Ionization in 4H-SiC

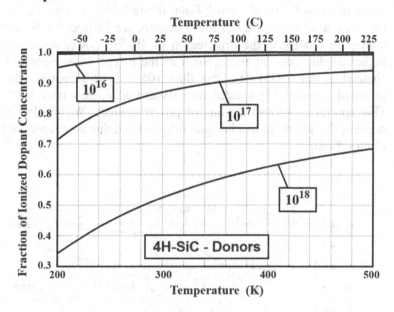

Fig. 2.18 Donor ionization in 4H-SiC.

The most commonly used n-type dopants for 4H-SiC are nitrogen and phosphorus with an activation energy of 61 meV.[75] Using this value, the fraction of ionized donors (or the electron concentration) can be computed for various temperatures. The ionized donor concentration for 4H-SiC is shown in Fig. 2.18 for temperature ranging from 200 to 500 °K. The corresponding temperature in °C is also shown on the chart for convenience. Typical drift region for 4H-SiC power devices have low doping concentrations below 10^{16} cm^{-3}. From Fig. 2.18, it can be concluded that all the donors are ionized for the entire temperature range in the drift region of 4H-SiC power devices. For higher doping concentrations, such as the n-type base regions of p-channel power MOSFETs, all the donors cannot be assumed to completely ionized especially at below room temperature (300 °K).

The most commonly used p-type dopant for 4H-SiC is aluminum which has an activation energy of 200 meV.[75] Using this value, the fraction of ionized acceptors (or the hole concentration) can be computed for various temperatures. The ionized acceptor concentration for 4H-SiC is shown in Fig. 2.19 for temperature ranging from 200 to 500 °K. The fraction of ionized acceptors is much smaller than donors for the same temperature and dopant concentration.

Typical drift region for 4H-SiC power devices have low doping concentrations below 10^{16} cm^{-3}. From figure 2.19, it can be concluded that only a small fraction of the acceptors are ionized for the entire temperature range in the drift region of 4H-SiC power devices. For higher doping concentrations, such as the p-type base regions of n-channel power MOSFETs, less than 10% of the acceptors can be assumed to completely ionized even at elevated temperatures (500 °K). This applies to neutral regions of power devices. Within depletion regions, the acceptors will become fully ionized due to presence of the electric field.

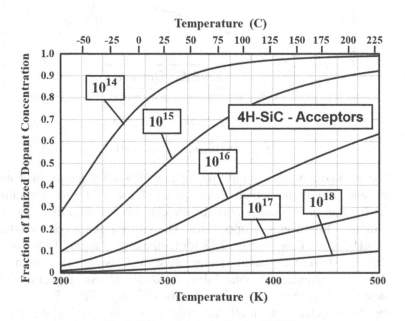

Fig. 2.19 Acceptor ionization in 4H-SiC.

2.13.3 Dopant Ionization in GaN

The most commonly used n-type dopants for GaN are oxygen and silicon with an activation energy of 30 meV.[76] Using this value, the fraction of ionized donors (or the electron concentration) can be computed for various temperatures. The ionized donor concentration for GaN is shown in Fig. 2.20 for temperature ranging from 200 to 500°K. The corresponding temperature in °C is also shown on the chart for convenience. Typical drift region for GaN power devices have low doping concentrations below 10^{16} cm^{-3}. From Fig. 2.20, it can be concluded that all the donors are ionized for the entire temperature range in the drift region of GaN power devices.

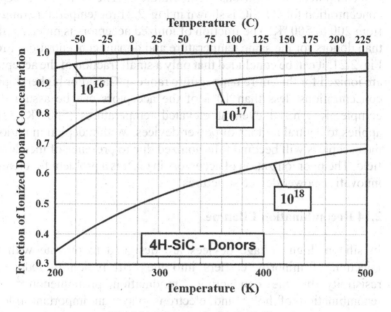

Fig. 2.20 Donor ionization in GaN.

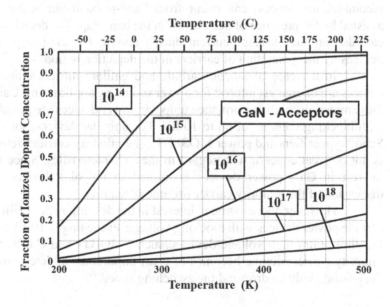

Fig. 2.21 Acceptor ionization in GaN.

The most commonly used p-type dopant for GaN is Magnesium which has an activation energy of 220 meV[76]. Using this value, the fraction of ionized acceptors (or the hole concentration) can be computed for various temperatures. The ionized acceptor concentration for 4H-SiC is shown in Fig. 2.21 for temperature ranging from 200 to 500 °K. The fraction of ionized acceptors is much smaller than donors for the same temperature and dopant concentration. From Fig. 2.21, it can be concluded that only a small fraction of the acceptors are ionized for the entire temperature range in GaN. For higher doping concentrations, less than 10% of the acceptors can be assumed to completely ionized even at elevated temperatures (500 °K). This applies to neutral regions of power devices. Within depletion regions, the acceptors will become fully ionized due to presence of the electric field. The poor ionization of acceptors in GaN is a problem for creating innovative power device structures.

2.14 Recombination Lifetime

In silicon, high voltage devices are designed to operate with the injection of minority carriers into the drift region to reduce its resistivity by the conductivity modulation phenomenon.[12] The recombination of holes and electrons plays an important role in determining the switching speed of these structures. The recombination process can occur from band-to-band but is usually assisted by the presence of deep levels in the band gap. The deep levels can also produce leakage current during reverse blocking in power devices by the generation of carriers in the depletion region.

In the case of silicon carbide and gallium nitride, unipolar devices are of greatest interest for power switching applications. Using these materials, high performance unipolar power devices are possible with blocking voltages of up to 10-kV. For unipolar devices, such as Schottky rectifiers and power MOSFETs, the minority carrier lifetime is not of great concern with regard to their characteristics. Since the interest in GaN power devices is limited to unipolar devices, no discussion of the minority carrier lifetime is required.

However, there has been interest in the development of silicon carbide power devices with blocking voltages exceeding 10-kV.[77,78] For these ultra-high voltage devices, such as IGBTs and GTOs, the minority carrier lifetime becomes of great importance in determining the on-state voltage drop and the switching speed.[79]

Before 2000, the minority carrier lifetime values reported for 4H-SiC material were relatively low (10^{-9} to 10^{-7} seconds)[7]. This is consistent with diffusion lengths of between 1 and 2 microns measured by using the EBIC technique.[80] It is likely that these low values for the lifetime and diffusion lengths were associated with surface recombination. In 1999, Kimoto *et al.*, demonstrated that surface recombination plays an important role in determining the switching speed of p-n junction diodes.[81] From their measurements, a low bulk lifetime of 0.33 μs and a surface recombination velocity of 5×10^4 cm/s were extracted. In 2003, Galeckas *et al.*, pointed out that the lifetime in the N^+ substrates of 4H-SiC has a low lifetime of   10-100 ns.[82] Consequently, the recombination of minority carriers in the epitaxial layers was accelerated by diffusion into the substrates.

A major breakthrough in improving the lifetime in the epitaxial drift layers for 4H-SiC power devices was accomplished by identification of the $Z_{1/2}$ defect as the dominant recombination center.[83] The reduction of the concentration of the $Z_{1/2}$ center has been accomplished by two methods — namely, carbon ion-implanta-tion with annealing at 1700 °C and thermal oxidation at 1300 °C for long periods (e.g. 5 hours) followed by argon annealing at 1550 °C for 30 minutes.[84] This has allowed increasing the lifetime from 0.68 μs in as-grown layers to 6.6 μs after the oxidation treatment. Surface passivation using a plasma-enhanced chemical vapor deposited oxide (PECVD) layer that is annealed in nitric oxide at 1300 °C for 30 minutes was demonstrated to reduce the interface defect density leading to a lifetime of 13 μs.

In general, the lifetime (τ) in 4H-SiC drift regions is given by:

$$\frac{1}{\tau} = \frac{1}{\tau_{SRH}} + \frac{1}{\tau_{BB}} + \frac{1}{\tau_{AU}} \qquad [2.46]$$

where τ_{SRH} is the Shockley-Read-Hall (SRH) recombination lifetime, τ_{BB} is the band-to-band recombination lifetime, and τ_{AU} is the Auger lifetime.[85] In this expression, the SRH lifetime can be computed using[12]:

$$\tau_{SRH} = \tau_{p0}\left\{\left[1 + e^{(E_i - E_F)/kT}\right] + \zeta\left[e^{(2E_i - E_r - E_F)/kT}\right]\right\} \qquad [2.47]$$

where τ_{p0} is the minority carrier lifetime in heavily doped n-type material, E_i is the intrinsic level position, E_F is the Fermi level position, E_r is the recombination center position, ζ is the capture cross-section ratio, k is Boltzmann's constant, and T is the absolute temperature.

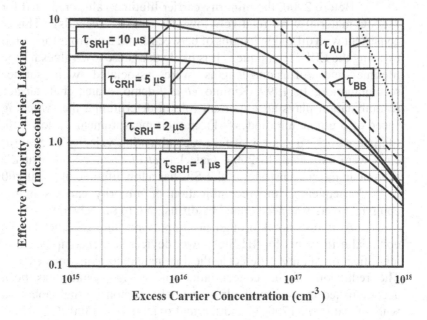

Fig. 2.22 Effective minority carrier lifetime in 4H-SiC.

The band-to-band lifetime is given by:

$$\tau_{BB} = \frac{1}{R_{BB}\left(n_0 + p_0 + \Delta n\right)} \quad [2.48]$$

where R_{BB} is the band-to-band recombination coefficient, n_0 and p_0 are the equilibrium electron and hole concentrations and Δn is the excess carrier concentration. The band-to-band recombination coefficient[86] for 4H-SiC has a value of 1.5×10^{-12} cm^3/s.

The lifetime determined by the Auger recombination process is given by:

$$\tau_{AU} = \frac{1}{C_n\left(n_0^2 + 2n_0\Delta n + \Delta n^2\right) + C_p\left(p_0^2 + 2p_0\Delta n + \Delta n^2\right)} \quad [2.49]$$

where C_n and C_p are the Auger recombination coefficients. For 4H-SiC, the Auger recombination coefficients[87] have a value of 5.0×10^{-31} cm^6/s for C_n and 2.0×10^{-31} cm^6/s for C_p.

The effective minority carrier lifetime in 4H-SiC obtained by using the above models is depicted in Fig. 2.22. The influence of band-to-band and Auger recombination begins to occur only when the excess (or injected) carrier concentration exceeds 10^{16} cm^{-3} and becomes pronounced at injected concentration of 10^{17} cm^{-3}. Such high injected carrier concentrations are observed in bipolar power devices such as PiN rectifiers, gate turn-off thyristors and IGBTs during on-state and surge current flow.

2.15 Lifetime Control

In the case of silicon power devices, unipolar devices become inefficient at blocking voltages above 200 volts. Bipolar power devices such as PiN rectifiers and IGBTs are required for applications with higher blocking voltages. The switching speed of the bipolar devices depends on the minority carrier lifetime in the drift region. The lifetime in the drift region can be controlled most conveniently by using electron irradiation.[88,89,90] This process has also been applied successfully for controlling the minority carrier lifetime in 4H-SiC devices.

It has been demonstrated that low-energy electron irradiation produces the $Z_{1/2}$ centers in 4H-SiC leading to a reduction of the minority carrier lifetime[91,92]. Although other defects are also generated by the electron irradiation, they can be removed by annealing at 900-1000 °C. The density of the $Z_{1/2}$ centers produced by electron irradiation is given by:

$$N_{Z_{1/2}} = K_{Z1/2} \cdot \phi_e \qquad \text{[2.50]}$$

where $K_{Z1/2}$ is the electron irradiation coefficient (cm^{-1}) that depends on the electron energy and ϕ_e is the electron irradiation fluence (cm^{-2}). The values for $K_{Z1/2}$ increase from 1.5×10^{-4} to 2×10^{-3} to 5×10^{-3} to 3×10^{-2} when the electron irradiation energy is increased from 116 to 160 to 200 to 250 keV. These results need to be extended to electron energy levels of 1 MeV and greater (as used for silicon power devices) to ensure that the radiation can penetrate through the entire drift region in power devices.

2.16 Contacts for Power Devices

In this section discusses contacts used for 4H-SiC and GaN devices. Ohmic contacts are required for the N$^+$ source and drain regions of the field effect transistors and the cathode region of Schottky rectifiers. Rectifying contacts are required for making Schottky rectifiers from 4H-SiC and for the gate region in GaN HFET devices.

2.16.1 Metal-Semiconductor Contacts

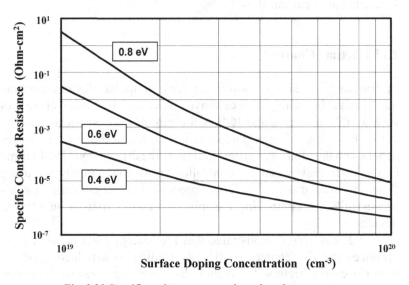

Fig. 2.23 Specific resistance at metal-semiconductor contacts.

Contacts to power devices are made by using metal films deposited on the surface of its semiconductor layers. A common requirement is to make ohmic contacts to n-type and p-type regions. This can be achieved by using a metal-semiconductor contact with low barrier height and a high doping concentration in the semiconductor to promote tunneling current across the contact. For metal-semiconductor contacts with high doping level in the semiconductor, the contact resistance determined by the tunneling process is dependent upon the barrier height and the doping level[93]:

$$R_c = \exp\left[\frac{2\sqrt{\varepsilon_S m^*}}{h} \left(\frac{\phi_{bn}}{\sqrt{N_D}} \right) \right] \qquad \text{[2.51]}$$

In order to take full advantage of the low specific on-resistance of the drift region in silicon carbide devices, it is important to obtain a specific contact resistance that is several orders of magnitude smaller than that of the drift region. This is also necessary because the contact areas are often only a small fraction (less than 10 percent) of the active area of most power device structures. Typically, specific contact resistances of less than 1×10^{-5} ohm-cm^2 are desirable to n-type regions.

The specific contact resistance calculated using the above formula is plotted in Fig. 2.23 as a function of the doping concentration using the barrier height as a parameter. Unfortunately, the barrier heights of metal contacts to silicon carbide tend to be large due to its wide band gap. From the figure, it can be concluded that a specific contact resistance of 1×10^{-5} ohm-cm^2 can be obtained for a doping concentration of 5×10^{19} cm^{-3} if the barrier height is 0.6 eV.

2.16.2 Ohmic Contacts to n-Type 4H-SiC

High surface doping concentrations can be achieved for N-type 4H-SiC by using hot-implantation of nitrogen, phosphorus or arsenic followed by appropriate high temperature annealing.[94] Ohmic contacts with specific resistances of less than 10^{-5} Ohm-cm^2 have been reported to ion-implanted layers by using nickel and titanium[95,96] after annealing at 950-1000 °C for a few minutes. Contacts to highly doped n-type 4H-SiC have also been achieved by using titanium carbide[97] with specific contact resistance of 1×10^{-5} ohm-cm^2.

2.16.3 Ohmic Contacts to p-Type 4H-SiC

Ohmic contacts with low specific resistance can also be obtained for p-type 4H-SiC by using high doping concentrations with aluminum ion-implants. The best results are obtained using either titanium[98] annealed at 800 °C for a few minutes or by using titanium carbide[99] to yield a resistance of $2-4 \times 10^{-5}$ Ohm-cm^2. Ohmic contacts to p-type 4H-SiC have also been reported by using nickel annealed at 1000 °C for a few minutes but the contact resistance is considerably larger (7×10^{-3} Ohm-cm^2).

2.16.4 Schottky Barrier Contacts to n-Type 4H-SiC

Metal-semiconductor contacts can also be used to make Schottky barrier rectifiers with high breakdown voltages by using silicon carbide. In this case, it is advantageous to have a relatively large barrier height to reduce the leakage current. The commonly used metals for formation of Schottky barriers to 4H-SiC are Titanium and Nickel. The barrier heights measured for these contacts range from 1.10 to 1.25 eV for Titanium and 1.30 to 1.60 eV for Nickel.[100,101,102] The use of these metals to fabricate high voltage device structures is discussed in subsequent chapters of the book.

2.16.5 Ohmic Contacts to n-Type GaN

GaN-on-silicon power FETs are of interest due to the low sheet resistance of the 2D-gas produced at the AlGaN/GaN interface. In order to take advantage of this high conductivity of the resulting drift region, it is important to achieve very low contact resistances for the source and drain regions. Ohmic contacts in the GaN HFET structure must contact the 2D-gas to enable efficient current transport between the source and drain regions. The most widely used metal system[103,104,105,106,107,108] for the Ohmic contacts to GaN HFETs is a Ti/Al/Ni/Au stack with a typical thickness of 25nm/150nm/50nm/100 nm. The stack is typically annealed at 850 °C for 30 seconds to form the Ohmic contacts. It has been shown that the metal spikes through the top AlN layer into the GaN layer forming the connection with the 2D-gas layer.[109] The specific contact resistance achieved by this approach is typically 2×10^{-6} Ohm-cm^2. For a lateral device, the contact resistance is often expressed as about 0.1 Ohm per cm of gate width.

During recent years, there has been interest in created gold free contacts to make the device process compatible with silicon foundries. A Ti/AlSi/Mo stack annealed at 800 °C has been demonstrated[110] to produce a contact resistance of 0.1 Ohm/cm. In addition, a Ti/Al/W stack annealed at 875 °C for 30 seconds has been demonstrated[111] to produce a contact resistance of 0.1 Ohm/cm.

2.16.6 Schottky Contacts for GaN HEMTs

The GaN HEMT structure was originally created using a Schottky gate contact. The Schottky gate contact is typically made using Ni/Au stack[106,107,109] with a thickness of 50 nm/250 nm. The Schottky barrier height for the Ni contact has been reported[112] to be 0.71 eV. This is relatively low value that leads to high leakage currents in the devices.

2.17 Fabrication Technology for Silicon Carbide Devices

Due to the substantial infrastructure available to fabricate silicon devices, it is advantageous to manufacture silicon carbide devices using the same technology platform. During the last fifteen years of research, it has been established that silicon carbide devices can be fabricated using the same equipment used for silicon devices in most instances. However, it has been found that much higher temperatures are needed for the annealing of ion implanted regions in silicon carbide to activate the dopants and remove the damage. In addition, the design of device structures in silicon carbide requires giving special consideration to the low diffusion coefficients for impurities. These unique issues are discussed in this section of the chapter.

2.17.1 Diffusion Coefficients and Solubility of Dopants

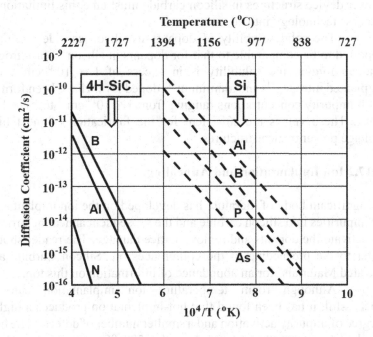

Fig. 2.24 Diffusion rates for dopants in 4H-SiC.

In silicon power devices, it is common-place to drive the dopants after ion-implantation to increase the junction depth. The widely used double-diffused (DMOS) process for silicon power MOSFETs utilizes the difference in the junction depth of the P-base and N^+ source regions to define the channel length without relying upon high resolution lithography to achieve sub-micron dimensions.[12]

The diffusion coefficients for dopants in 4H-SiC can be compared with those in silicon using Fig. 2.24. In silicon large junction depths can be achieved by diffusion even at 1200 °C by using aluminum and gallium as dopants[113]. In contrast, for silicon carbide the diffusion coefficients are very small even at relatively high temperatures. For nitrogen — the commonly used N-type dopant, the effective diffusion coefficient is reported as 5×10^{-12} cm^2/s even at 2450 °C. For aluminum — the commonly used P-type dopant, the effective diffusion coefficient is reported as 3×10^{-14} to 6×10^{-12} cm^2/s even at 1800-2000 °C. Due to the on-set of dissociation at these temperatures with attendant generation of defects, it is not practical to drive dopants in silicon carbide after ion implantation. The design of power device structures in silicon carbide must take this limitation in process technology into account.

The solid solubility of dopants in silicon carbide has been reported to be comparable to that for dopants in silicon[5]. For nitrogen and aluminum, the solubility is in excess of 1×10^{20} cm^{-3}. Ion implanted nitrogen[114] and aluminum[115] doped layers have been formed with impurity concentrations ranging from 1×10^{19} cm^{-3} to 1×10^{20} cm^{-3}. These values are sufficient for the fabrication of most high voltage power device structures.

2.17.2 Ion Implantation and Annealing

A significant body of literature has developed on the ion implantation of impurities into silicon carbide and the subsequent annealing process to activate the dopants and remove lattice damage. The reader should refer to the proceeding of the conferences on 'Silicon Carbide and Related Materials' for an abundance of information on this topic.

Although room temperature ion implantation can be successful, it has been found that hot-implantation produces a higher degree of impurity activation and a smaller number of defects. The hot-implants are usually performed in the 500 °C range followed by anneals performed at between 1200 and 1600 °C. At above 1600 °C, degradation of the silicon carbide surface is observed due to sublimation or evaporation.[94,114] The surface degradation can be mitigated by using a graphite boat to host the implanted wafer with another silicon carbide wafer placed on top (proximity anneals).[116]

2.17.3 Gate Oxide Formation

Silicon bipolar power devices, such as the bipolar power transistor and the gate turn-off thyristor (GTO), were supplanted with MOS-gated devices in the 1980s. The silicon power MOSFET replaced the bipolar transistor for lower voltage (< 200 V) applications while the IGBT replaced bipolar transistors and GTOs for high voltage (> 200 V) applications. The main advantage of these MOS-gated device architectures was voltage controlled operation which greatly simplified the control circuit making it amenable to integration.[12] This feature must be extended to silicon carbide power devices to make them attractive from an applications perspective.

The issues that must be considered for the gate dielectric in MOS-gated silicon carbide devices are the quality of the oxide-semiconductor interface and the ability of the oxide to withstand higher electric fields than within silicon devices.[117] The quality of the oxide-semiconductor interface determines not only the channel mobility but impacts the threshold voltage of the devices. The electric field in the oxide is related to the electric field in the semiconductor by Gauss's Law:

$$E_{ox} = \frac{\varepsilon_{semi}}{\varepsilon_{ox}} E_{semi} \approx 3 E_{semi} \qquad [2.52]$$

In silicon, the maximum electric field at breakdown is in the 3×10^5 V/cm range. Consequently, the maximum electric field in the oxide remains well below its breakdown field strength of 10^7 V/cm. In contrast, the maximum electric field for breakdown in silicon carbide is in the 3×10^6 V/cm range. Consequently, the electric field in the oxide can approach its breakdown strength and easily exceed a field of 3×10^6 V/cm, which is considered to be the threshold for reliable operation. One approach to overcome this problem is to use gate dielectric material with a larger permittivity.[118,119] In these references, it has been theoretically shown that the specific on-resistance can be reduced by an order of magnitude by using high dielectric constant gate material. A permittivity of about 15 (versus 3.85 for silicon dioxide) was found to be adequate for allowing full use of the high electric field strength for breakdown in silicon carbide without problems with unacceptably high oxide electric fields. One example of such a dielectric is zirconium oxide.[120]

A second approach utilizes improved device structural architecture to screen the gate dielectric from the high electric fields within the silicon carbide. This is discussed in detail in the chapters on shielded planar and shielded trench gate MOSFETs in this book. Such devices can then be made using silicon dioxide whose properties are well understood. In this case, the silicon dioxide can be either grown by the thermal oxidation of silicon dioxide or by the formation of the oxide using chemical vapor deposition processes. Many studies on both of these techniques are available in the literature.

The thermal oxidation of SiC has been reviewed in the literature and the growth rate of the oxide has been compared with that on silicon surfaces.[121] Some selected data is shown in Fig. 2.25 to illustrate the much lower growth rate of thermal oxide on SiC when compared with silicon for both dry and wet oxidation conditions. The data shown in this figure are for 6H-SiC but similar values are applicable to the 4H-SiC polytype.[122] It is obvious that a much higher temperature and longer time duration is required for the growth of oxides on silicon carbide. This oxide has been shown to be essentially silicon dioxide with no carbon incorporated in the film. The small amount of aluminum incorporated into the oxide grown on p-type SiC[123] has been found to have no adverse effect[57] on the MOS properties.

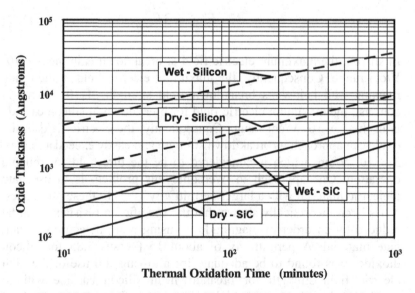

Fig. 2.25 Comparison of thermal oxidation of SiC with silicon at 1200 °C.

The gate dielectric produced with thermal oxidation of 6H-SiC has been found to be satisfactory for making MOSFETs with reasonable inversion layer mobility.[124] However, this method did not produce sufficiently high quality interfaces on 4H-SiC to make MOSFETs. It was discovered at PSRC that a low temperature oxide subjected to a wet nitrogen anneal enabled fabrication of n-channel inversion layer MOSFETs in 4H-SiC with an effective inversion layer mobility of 165 cm^2/V-s.[44,45] Although the gate oxide for these MOSFETs was unusually large (9000 angstroms), high performance MOSFETs with similar mobility values, were subsequently fabricated at PSRC using a gate oxide thickness of 850 angstroms.[125] At PSRC, an extensive evaluation of process conditions was performed to determine their impact on the inversion layer mobility. This work has been validated by other groups.[46] In addition, inversion layer mobility in the range of 50 cm^2/V-s has been reported on thermally grown oxides subjected to annealing in nitric oxide.[50,51,52,53,54] These experiments indicate that power MOSFETs with low specific on-resistance can be developed from 4H-SiC.

2.17.4 Reactive Ion Etching of Trenches

The first silicon power MOSFETs were manufactured using the Double-diffused or DMOS process to form a planar gate architecture.[126] Although these devices offered excellent input impedance and fast switching performance for low voltage (< 100 volt) applications, their on-resistance was found to be limited by the presence of a JFET region. The JFET region could be eliminated by adopting a trench gate or UMOS process.[12] The trench gate structure in the silicon power MOSFET is fabricated by using reactive ion etching. An equivalent process for silicon carbide was not developed until the mid 1990s. Until then, alternative methods such as etching with molten KOH solutions or the use of amorphization of selective SiC regions was explored.[127]

Reactive ion etching is the most convenient process for formation of trenches in silicon carbide enabling the fabrication of devices with processes compatible with silicon device manufacturing technology. Many gas compositions have been tried for the formation of trenches in silicon carbide. For application to power devices, the trenches should have a nearly vertical profile with a slightly rounded bottom to reduce electric field enhancement. The trench surface must be smooth and free of damage in order to reduce interface states and obtain high mobility along the inversion channel formed on the trench sidewalls. Reactive ion etching using a SF_6/O_2 mixture has been found to produce trenches that meet these requirements[128]. Other reactive ion etching chemistries using fluorinated gases[129] have also been reported to produce trenches with vertical profiles and smooth surfaces.

2.18 Summary

The properties of gallium nitride and silicon carbide relevant to power devices have been reviewed in this chapter. An improved knowledge of the impact ionization coefficients has allowed projection of very high performance for GaN and SiC based unipolar devices. The status of process technology for the fabrication of these unipolar devices has also been summarized here.

References

[1] S.J. Pearson, C.R. Abernathy and F. Ren, "Gallium Nitride Processing for Electronics, Sensors, and Spintronics", Springer-Science, New York, 2006.

[2] T. Hanada, "Basic Properties of ZnO, GaN, and Related Materials", in 'Oxide and Nitride Semiconductors: Processing, Properties, and Applications', Ed. T. Yao and S-K. Hong, Springer-Science, New York, 2009.

[3] "GaN – Gallium Nitride", www.ioffe.rssi.ru.

[4] J.F. Muth *et al.*, "Absorption Coefficient, Energy gap, Exciton Binding Energy, and Recombination Lifetime of GaN obtained from Transmission Measurements", Applied Physics Letters, Vol. 71, pp. 2572-2574, 1997.

[5] G. L. Harris, "Properties of Silicon Carbide", IEE Inspec, 1995

[6] M. Ruff, H. Mitlehner and R. Helbig, "SiC Devices: Physics and Numerical Simulations", IEEE Transactions on Electron Devices, Vol. ED-41, pp. 1040-1054, 1994.

[7] N.G. Wright *et al.*, "Electrothermal Simulation of 4H-SiC Power Devices", Silicon Carbide, III-Nitrides, and Related Materials - 1997, Material Science Forum, Vol. 264, pp. 917-920, 1998.

[8] S.M. Sze, "Physics of Semiconductor Devices", John Wiley and Sons, 1981

[9] S.M. Sze and K.K. Ng, "Physics of Semiconductor Devices", Third Edition, pp. 15-16, John Wiley, New York, 2007.

[10] T. Kimoto and J.A. Cooper, "Fundamentals of Silicon Carbide Technology", pp. 17-18, John Wiley, New York, 2014.

[11] S. K. Ghandhi, "Semiconductor Power Devices", John Wiley and Sons, 1977.

[12] B.J. Baliga, "Fundamentals of Power Semiconductor Devices", Springer-Science, New York, 2008.

[13] A.G. Chynoweth, "Ionization Rates for Electrons and Holes in Silicon, Physical Review, Vol. 109, pp.1537-1545, 1958.

[14] A. G. Chynoweth, "Uniform Silicon P-N Junctions II. Ionization rates for Electrons", J. Applied Physics, Vol. 31, pp 1161-1165, 1960.

[15] C. R. Crowell and S. M. Sze, "Temperature dependence of Avalanche Multiplication in Semiconductors", Applied Physics Letters, Vol. 9, pp 242-244, 1966.

[16] R. Van Overstraeten and H. De Man, "Measurement of the Ionization Rates in Diffused Silicon P-N Junctions", Solid State Electronics, Vol. 13, pp. 583-590, 1970.

[17] B.J. Baliga, "Fundamentals of Power Semiconductor Devices", Section 2.1.5, Springer-Science, New York, 2008.

[18] R. Raghunathan and B. J. Baliga, "Temperature dependence of Hole Impact Ionization Coefficients in 4H and 6H SiC", Solid State Electronics, Vol. 43, pp. 199-211, 1999.

[19] R. Raghunathan and B. J. Baliga, "Role of Defects in producing Negative Temperature Dependence of Breakdown Voltage in SiC", Applied Physics Letters, Vol. 72, pp. 3196-3198, 1998.

[20] A.O. Konstantinov et al., "Ionization Rates and Critical Electric Fields in 4H-SiC", Applied Physics Letters, Vol. 71, pp. 90, 1997.

[21] A.O. Konstantinov et al., "Study of Avalanche Breakdown and Impact Ionization in 4H Silicon Carbide", Journal of Electronic Materials, Vol. 27, pp. 335-341, 1998.

[22] J. Kolnik et al., "Monte Carlo calculation of Electron Initiated Impact Ionization in Bulk Zinc-Blende and Wurtzite GaN", Journal of Applied Physics, Vol. 81, pp. 726-733, 1997.

[23] I.H. Oguzman et al., "Theory of Hole Initiated Impact Ionization in Bulk Zincblende and Wurtzite GaN", Journal of Applied Physics, Vol. 81, pp. 7827-7834, 1997.

[24] C. Bulutay, "Electron Initiated Impact Ionization in AlGaN Alloys", Semiconductor Science and Technology, Vol. 17, pp. L59-L62, 2002.

[25] K. Kunihiro *et al.*, "Experimental Evaluation of Impact Ionization Coefficients in GaN", IEEE Electron Device Letters, Vol. 20,   pp. 608-610, 1999.

[26] R. McClintock *et al.*, "Hole-initiated Multiplication in Back-illuminated GaN Avalanche Photodiodes", Applied Physics Letters, Vol. 90, pp. 141112-1 – 141112-3, 2007.

[27] B.J. Baliga, "Gallium Nitride Devices for Power Electronic Applications", Semiconductor Science and Technology, Vol. 28, pp. 1-8, 2013.

[28] T. Kimoto and J.A. Cooper, "Fundamentals of Silicon Carbide Technology", pp. 29, John Wiley, New York, 2014.

[29] A. Akturk *et al.*, "Comparison of 4H-SiC Impact Ionization Models using Experiments and Self-Consistent Simulations", Journal of Applied Physics, Vol. 104, pp. 026101 1-3, 2008.

[30] H. Iwata and K. M. Itoh, "Theoretical Calculation of the Electron Hall Mobility in n-type 4H- and 6H-SiC", Silicon Carbide and Related Materials – 1999, Materials Science Forum, Vol. 338-342, pp. 879-884, 2000.

[31] C. Jacobini *et al.*, "A Review of some Charge Transport Properties of Silicon", Solid State Electronics, Vol. 20, pp. 77-89, 1977.

[32] J. Pernot *et al.*, "Electrical Transport in n-type Silicon Carbide", Journal of Applied Physics, Vol. 90, pp. 1869-1875, 2001.

[33] S. Kagamuhara *et al.*, "Parameters required to Simulate Electrical Characteristics of SiC Devices for n-type 4H-SiC", Journal of Applied Physics, Vol. 96, pp. 5601-5607, 2004.

[34] T. Kimoto and J.A. Cooper, "Fundamentals of Silicon Carbide Technology", pp. 25, John Wiley, New York, 2014.

[35] H. Iwata and K.M. Itah, "Theoretical Calculation of the Electron Hall Mobility in n-type 4H- and 6H-SiC", Material Science Forum, Vol. 338-342, pp. 729-732, 2000.

[36] S.J. Pearton *et al.*, "Fabrication and Performance of GaN Electronic Devices", Material Science and Engineering Research, Vol. 30, pp. 55-212, 2000.

[37] T.T. Mnatsakanov *et al.*, "Carrier Mobility Model for GaN", Solid-State Electronics, Vol. 47, pp. 111-115, 2003.

[38] V.W.L. Chin, T.L. Tansley and T. Osotchan, "Electron Mobilities in Gallium, Indium, and Aluminum Nitrides", Journal of Applied Physics, Vol. 75, pp. 7365-7372, 1994.

[39] C. Canali *et al.*, "Electron Drift Velocity in Silicon", Phys. Rev. Vol. B12, pp. 2265-2284, 1975

[40] www.iue.tuwien.ac.at/phd/vitanov/node61.html.

[41] A. Koizumi *et al.*, "Temperature and Doping Dependencies of Electrical Properties in Al-Doped 4H-SiC Epitaxial Layers", Journal of Applied Physics, Vol. 106, 013716, 2009.

[42] B.J. Baliga, "Fundamentals of Power Semiconductor Devices", Chapter 6, Springer-Science, New York, 2008.

[43] T. Kimoto and J.A. Cooper, "Fundamentals of Silicon Carbide Technology", pp. 333, John Wiley, New York, 2014.

[44] S. Sridevan and B.J. Baliga, "Inversion Layer Mobility in SiC MOSFETs", Material Science Forum, Vol. 264-268, pp. 997-1000, 1998.

[45] S. Sridevan and B.J. Baliga, "Lateral n-Channel Inversion Mode 4H-SiC MOSFETs", IEEE Electron Device Letters, Vol. 19, pp. 228-230, 1998.

[46] D. Alok, E. Arnold and R. Egloff, "Process dependence of Inversion-layer Mobility in 4H-SiC Devices", Material Science Forum, Vol. 338-342, pp. 1077-1080, 2000.

[47] S-H. Ryu *et al.*, "Critical Issues for MOS Based Power Devices in 4H-SiC", Materials Science Forum, Vol. 615-617, pp. 743-748, 2009.

[48] T. Kimoto and J.A. Cooper, "Fundamentals of Silicon Carbide Technology", pp. 247, John Wiley, New York, 2014.

[49] G.Y. Chung *et al.*, "Improved Inversion Channel Mobility for 4H-SiC MOSFETs following High Temperature Anneals in Nitric Oxide", IEEE Electron Device Letters, Vol. 22, pp. 176-178, 2001.

[50] J.R. Williams *et al.*, "Passivation of the 4H-SiC/SiO2 Interface with Nitric Oxide", Material Science Forum, Vol. 389-393, pp. 967-972, 2002.

[51] S. Dhar *et al.*, "Effect of Nitric Oxide Annealing on the Interface Trap Density near the Conduction Band Edge of 4H-SiC at the oxide/(1120) 4H-SiC Interface", Applied Physics Letters, Vl. 84, pp. 1498-1500, 2004.

[52] C-Y Lu *et al.*, "Effect of Process Variations and Ambient Temperature on Electron Mobility at the SiO2/4H-SiC Interface", IEEE Transactions on Electron Devices, Vol. 50, pp. 1582-1588, 2003.

[53] S. Dhar *et al.*, "Inversion Layer Carrier Concentration and Mobility in 4H-SiC MOSFETs", Journal of Applied Physics, Vol. 108, pp. 054509, 2010.

[54] S. Dhar *et al.*, "Temperature Dependence of Inversion Layer Carrier Concentration and Hall Mobility in 4H-SiC MOSFETs", Material Science Forum, Vol. 717-720, pp. 713-716, 2012.

[55] B.J. Baliga and M. Bhatnagar, "Method of Fabricating Silicon Carbide Field Effect Transistor", U.S. Patent 5,322,802, Issued June 21, 1994.

[56] S-H Ryu *et al.*, "Design and Process Issues for Silicon Carbide Power DiMOSFETs", Material Research Society Symposium Proceedings, Vol. 640, pp. H4.5.1-H4.5.6, 2001.

[57] G. Gudjonsson *et al.*, "High Field-Effect Mobility in n-Channel Si Face 4H-SiC MOSFETs with Gate Oxide Grown on Aluminum Ion-Implanted Material", IEEE Electron Device Letters, Vol. 26, pp. 96-98, 2005.

[58] B.J. Baliga, "Fundamentals of Power Semiconductor Devices", Chapter 6, pp. 476-479, Springer-Science, New York, 2008.

[59] B.J. Baliga, "Silicon Carbide Semiconductor Devices having Buried Silicon Carbide Conduction Barrier Layers Therein", U.S. Patent 5,543,637, Issued August 6, 1996.

[60] S. Haney and A. Agarwal, "The Effects of Impant Activation Anneal on the Effective Inversion Layer Mobility of 4H-SiC MOSFETs", Journal of Electronic Materials, Vol. 27, pp. 666-671, 2008.

[61] C. Canali *et al.*, "Electron and Hole Drift Velocity Measurements in Silicon", IEEE Transactions on Electron Devices, Vol. ED-22, pp. 1045-1047, 1975.

[62] I.A. Khan and J.A. Cooper, "Measurements of High-Field Transport in Silicon Carbide", IEEE Transactions on Electron Devices, Vol. 47, pp. 269-273, 2000.

[63] T. Kimoto and J.A. Cooper, "Fundamentals of Silicon Carbide Technology", pp. 28, John Wiley, New York, 2014.

[64] J.C. Cao and X.L. Lei, "Non-parabolic Multi-valley Balance-equation approach to Impact Ionization: Application to Wurtzite GaN", European Physics Journal, Vol. B7, pp. 79-83, 1999.

[65] A.F.M. Anwar, S. Wu and R.T. Webster, "Temperature Dependent Transport Properties in GaN, $Al_xGa_{1-x}N$, and $In_xGa_{1-x}N$ Semiconductors", IEEE Transactions on Electron Devices, Vol. 48, pp. 567-572, 2001.

[66] M. Farahmand *et al.*, "Monte Carlo Simulation of Electron Transport in the III-Nitride Wurtzite Phase Materials System: Binary and Ternaries", IEEE Transactions on Electron Devices, Vol. 48, pp. 535-342, 2001.

[67] B. Benbakhti *et al.*, "Electron Transport Properties of Gallium Nitride for Microscopic Power Device Modelling", Journal of Physics, Vol. 193, pp. 1-4, 2009.

[68] S. Chen and G. Wang, "High-Field Properties of Carrier Trnsport in Bulk Wurtzite GaN: A Monte Carlo Perspective", Journal of Applied Physics, Vol. 103, pp. 023703-1 – 023703-6, 2008.

[69] U.K. Misra and J. Singh, "Semiconductor Device Physics and Design", Springer, New York, 2008.

[70] F. Sacconi *et al.*, "Spontaneous and Piezoelectric Polarization Effects on the Output Characteristics of AlGaN/GaN Heterojunction Modulation Doped FETs", IEEE Transactions on Electron Devices, Vol. 48, pp. 450-457, 2001.

[71] J.D. Brown *et al.*, "AlGaN/GaN HFETs Fabricated on 1000-mm GaN on Silicon (111) Substrates", Solid State Electronics, Vol. 46, pp. 1535-1539, 2002.

[72] K-S. Im *et al.*, "Normally-Off GaN MOSFET based on AlGaN/GaN Heterostructure with extremely high 2DEG Density grown on Silicon Substrate", IEEE Electron Device Letters, Vol. 31, pp. 192-194, 2010.

[73] X.Z. Dang *et al.*, "Measurement of Drift Mobility in AlGaN/GaN Heterostructure Field-Effect Transistors", Applied Physics Letters, Vol. 74, pp. 3890-3892, 1999.

[74] S.M. Sze and K.K. Ng, "Physics of Semiconductor Devices", Third Edition, Wiley, New York, 2007.

[75] T. Kimoto and J.A. Cooper, "Fundamentals of Silicon Carbide Technology", pp. 513, John Wiley, New York, 2014.

[76] T.A.G. Eberlein *et al.*, "Shallow Acceptors in GaN", Applied Physics Letters, Vol. 91, pp. 132105-1 – 132105-2, 2007.

[77] Q. Zhang *et al.*, "SiC Power Devices for Microgrids", IEEE Transactions on Power Electronics, Vol. 25, pp. 2889-2896, 2010.

[78] L. Cheng *et al.*, "Strategic Overview of High-Voltage SiC Power Device Development aiming at Global Energy Savings", Material Science Forum, Vols. 778-780, pp. 1089-1094, 2014.

[79] J.A. Cooper *et al.*, "Power MOSFETs, IGBTs, and Thyristors in SiC: Optimization, Experimental Results, and Theoretical Performance", IEEE International Electron Devices Meeting, Abstract 7.2.1, pp. 149-152, 2009.

[80] R. Raghunathan and B.J. Baliga, "EBIC Measurements of Diffusion Lengths in Silicon Carbide", 38th Electronic Materials Conference, Abstr. I-6, 1996.

[81] T. Kimoto *et al.*, "Performance Limiting Surface Defects in SiC Epitaxial p-n Junction Diodes", IEEE Transactions on Electron Devices, Vol. 46, pp. 471-477, 1999.

[82] A. Galeckas, J. Linnros and M. Lindstedt, "Characterization of Carrier Lifetime and Diffusivity in 4H-SiC using Time-resolved Imaging Spectroscopy of Electroluminescence", Material Science and Engineering, Vol. B102, pp. 304-307, 2003.

[83] T. Kimoto and J.A. Cooper, "Fundamentals of Silicon Carbide Technology", pp. 174, John Wiley, New York, 2014.

[84] T. Kimoto *et al.*, "Enhancement of Carrier Lifetime in n-type 4H-SiC Epitaxial Layers by improving Surface Passivation", Applied Physics Express, Vol. 3, pp. 121201-1 – 121201-3, 2010.

[85] B.J. Baliga, "Fundamentals of Power Semiconductor Devices", pp. 59-83, Springer-Science, New York, 2008.

[86] N. Kaji *et al.*, "Ultrahigh-Voltage SiC p-i-n Diodes with Improved Forward Characteristics", IEEE Transactions on Electron Devices, Vol. 62, pp. 374-381, 2015.

[87] A. Galeckas *et al.*, "Auger Recombination in 4H-SiC: Unusual Temperature Behavior", Applied Physics Letters, Vol. 71, pp. 3269-3271, 1997.

[88] B.J. Baliga, "Fundamentals of Power Semiconductor Devices", pp. 75-80, Springer-Science, New York, 2008.

[89] B.J. Baliga, "Fast Switching Insulated Gate Transistors", IEEE Electron Device Letters, Vol. EDL-4, pp. 452-454, 1983.

[90] B.J. Baliga, "Switching Speed Enhancement in Insulated Gate Transistors by Electron Irradiation", IEEE Transactions on Electron Devices, Vol. ED-31, pp. 1790-1795, 1984.

[91] T. Kimoto and J.A. Cooper, "Fundamentals of Silicon Carbide Technology", pp. 177-179, John Wiley, New York, 2014.

[92] K. Danno, D. Nakamura and T. Kimoto, "Investigation of Carrier Lifetime in Epilayers and Lifetime Control by Electron Irradiation", Applied Physics Letters, Vol. 90, pp. 202109-1 – 202109-3, 2007.

[93] S.M. Sze, "Physics of Semiconductor Devices", page 304, John Wiley and Sons, 1981.

[94] S. Imai *et al.*, "Hot-Implantation of Phosphorus Ions into 4H-SiC", Silicon Carbide and Related Materials – 1999, Materials Science Forum, Vol. 338-342, pp. 861-864, 2000.

[95] S. Tanimoto *et al.*, "Ohmic Contact Structure and Fabrication Process Applicable to Practical SiC Devices", Silicon Carbide and

Related Materials – 2001, Materials Science Forum, Vol. 389-393, pp. 879-884, 2002.

[96] T. Marinova *et al.*, "Nickel based Ohmic Contacts on SiC", Material Science and Engineering, Vol. B46, pp. 223-226, 1997.

[97] S.K. Lee *et al.*, "Low Resistivity Ohmic Titanium Carbide Contacts to n- and p-type 4H-SiC", Solid State Electronics, Vol. 44, pp. 1179-1186, 2000.

[98] J. Crofton *et al.*, "Titanium and Aluminum-Titanium Ohmic Contacts to p-type SiC", Solid State Electronics, Vol. 41, pp. 1725-1729, 1997.

[99] S.K. Lee *et al.*, "Electrical Characterization of TiC Contacts to Aluminum Implanted 4H-SiC", Applied Physics Letters, Vol. 77, pp. 1478-1480, 2000.

[100] A. Kestle *et al.*, "A UHV Study of Ni/SiC Schottky Barrier and Ohmic Contact Formation", Silicon Carbide and Related Materials – 1999, Materials Science Forum, Vol. 338-342, pp. 1025-1028, 2000.

[101] K.V. Vassilevski *et al.*, "4H-SiC Schottky Diodes with high On/Off Current Ratio", Silicon Carbide and Related Materials – 2001, Materials Science Forum, Vol. 389-393, pp. 1145-1148, 2002.

[102] R. Raghunathan, D. Alok and B.J. Baliga, "High Voltage 4H-SiC Schottky Barrier Diodes", IEEE Electron Device letters, Vol. EDL-16, pp. 226-227, 1995.

[103] B.T. Hughes *et al.*, "Fabrication and Characterization of AlGaN/GaN HFETs on MOVPE Layers", Symposium on High Performance Electron Devices for Microwave and Optoelectronic Applications, pp. 59-64, 1999.

[104] S. Joblot *et al.*, "AlGaN/GaN HEMTs on (001) silicon Substrates", Electronics Letters, Vol. 42, No. 2, 2006.

[105] T.J. Anderson *et al.*, "An AlN/Ultrathin AlGaN/GaN HEMT Structure for Enhancement-Mode Operation using Selective Etching", IEEE Electron Device Letters, Vol. 30, pp. 1251-1253, 2009.

[106] G. Li *et al.*, "Threshold Voltage Control in AlGaN/AlN/GaN HEMTs by Work Function Engineering", IEEE Electron Device letters, Vol. EDL-31, pp. 954-956, 2010.

[107] A.D. Koehler *et al.*, "Atomic Layer Epitaxy AlN for Enhanced AlGaN/GaN MEMT Passivation", IEEE Electron Device letters, Vol. EDL-34, pp. 1115-1117, 2013.

[108] Y-H Wang *et al.*, "6.5 V High Threshold Voltage AlGaN/GaN Power MIS HEMT using Mutlilayer Fluorinated Gate Stack", IEEE Electron Device letters, Vol. EDL-36, pp. 381-383, 2015.

[109] Y. Dora *et al.*, "Effect of Ohmic Contacts on Buffer Leakage of GaN Transistors", IEEE Electron Device letters, Vol. EDL-27, pp. 529-531, 2006.

[110] N. Ikeda *et al.*, "Over 55A, 800V High Power AlGaN/GaN HFETs for Power Switching Application", Physica Status Solidi (a), Vol. 284, pp. 2028-2031, 2007.

[111] H. Huang *et al.*, "Au-Free Normally-Off AlGaN/GaN-on-Si MIS-HEMTs using combined partially Recessed and Fluorinated Trap-Charge Gate Structures", IEEE Electron Device letters, Vol. EDL-35, pp. 569-571, 2014.

[112] G.H. Jessen *et al.*, "Gate Optimization of AlGaN/GaN HEMTs using WSi, Ir, Pd, and Ni Schottky Contacts", IEEE Gallium Arsenide Integrated Circuits Symposium, pp. 277-279, 2003.

[113] B.J. Baliga, "Deep Planar Gallium and Aluminum Diffusions in Silicon", Journal of the Electrochemical Society, Vol. 126, pp. 292-296, 1979.

[114] S. Blanque *et al.*, "Room Temperature Implantation and Activation Kinetics of Nitrogen and Phosphorus in 4H-SiC crystals", Silicon Carbide and Related Materials – 2003, Materials Science Forum, Vol. 457-460, pp. 893-896, 2004.

[115] Y. Negoro *et al.*, "Low Sheet Resistance of High Dose Aluminum Implanted 4H-SiC using (1120) Face", Silicon Carbide and Related Materials – 2003, Materials Science Forum, Vol. 457-460, pp. 913-916, 2004.

[116] R.K. Chilukuri, P. Ananthanarayanan, V. Nagapudi and B.J. Baliga, "High Voltage P-N Junction Diodes in Silicon Carbide using Field Plate Edge Termination", Material Research Society Proceedings, Vol. 572, pp. 81-86, 1999.

[117] B. J. Baliga, "Critical Nature of Oxide/Interface Quality for SiC Power Devices", Microelectronics Engineering, Vol. 28, pp. 177-184, 1995.

[118] S. Sridevan, P.K. McLarty and B.J. Baliga, "Silicon Carbide Switching Devices having Nearly Ideal Breakdown Voltage Capability and Ultra-Low On-State Resistance", U.S. Patent # 5,742,076, Issued April 21, 1998.

[119] S. Sridevan, P.K. McLarty and B.J. Baliga, "Analysis of Gate Dielectrics for SiC Power UMOSFETs", IEEE Int. Symposium on Power Devices and ICs, pp. 153-156, 1997.

[120] V.V. Afanas'ev *et al.*, "Oxidation of Silicon Carbide: Problems and Solutions", Silicon Carbide and Related Materials – 2001, Materials Science Forum, Vol. 389-393, pp. 961-966, 2002.

[121] J. A. Cooper, "Silicon Carbide MOSFETs", in 'Wide Energy Bandgap Electronic Devices', Edited by F. Ren and J. C. Zolper, World Scientific Press, 2003.

[122] A. Golz *et al.*, "Oxidation Kinetics of 3C, 4H, and 6H silicon carbide", Silicon Carbide and Related Materials – 1995, Institute of Physics Conference Series, Vol. 142, pp. 633-636, 1996.

[123] S. Sridevan, P.K. McLarty and B.J. Baliga, "On the Presence of Aluminum in Thermally Grown Oxides on 6H-SiC", IEEE Electron Device Letters, Vol. 17, pp. 136-138, 1996.

[124] L.A. Lipkin and J.W. Palmour, "Improved Oxidation Procedures for reduced SiO_2/SiC Defects". J. Electronic Materials, Vol. 25, pp. 909-915, 1996.

[125] S. Sridevan and B.J. Baliga, "Phonon Scattering limited Mobility in SiC Inversion Layers", PSRC Technical Report, TR-98-03.

[126] D.A. Grant and J. Gowar, "Power MOSFETs", John Wiley and Sons, 1989.

[127] D. Alok and B.J. Baliga, "A Novel Method for Etching Trenches in Silicon Carbide", J. Electronic Materials, Vol. 24, pp. 311-314, 1995.

[128] M. Kothandaraman, D. Alok and B.J. Baliga, "Reactive Ion Etching of Trenches in 6H-SiC", J. Electronic Materials, Vol. 25, pp. 875-878, 1996.

[129] V. Saxena and A.J. Steckl, "Fast and Anisotropic Reactive Ion Etching of 4H and 6H SiC in NF_3", Silicon Carbide and Related Materials – 1997, Materials Science Forum, Vol. 264-268, pp. 829-832, 1998.

Chapter 3

Breakdown Voltage

It is preferable to utilize a unipolar power device structure, such as the power MOSFET, than a bipolar power device, such as an IGBT, in power electronic applications due to the much smaller switching power losses. In the case of silicon power devices, the on-state voltage drop for unipolar devices becomes very high when their drift regions are designed to support more than 200 volts. This motived the development of the IGBT which is now widely used for applications where the circuit voltages exceed 200 volts[1].

The main advantage of a wide band gap semiconductor for power device applications stems from the very low resistance of the drift region even when it is designed to support large voltages. The highest voltage that can be supported by a drift region is determined by the onset of impact ionization in the semiconductor with increasing electric field within the region. In the previous chapter, it was shown that the onset of impact ionization occurs at much larger electric fields in gallium nitride and silicon carbide when compared with silicon. In this chapter, the design of the drift region is discussed. It is demonstrated that doping concentration of the drift region is much larger in wide band gap semiconductors when compared with silicon devices designed to support the same voltage. Furthermore, the width of the drift region for wide band gap devices is much smaller than that required for silicon devices.

Although the parallel-plane junction analysis is representative of the active region of power devices where current flow transpires, the maximum blocking voltage (highest value the device can support) is limited by breakdown at its periphery. The electric field in discrete power devices is invariably enhanced at its edges. All discrete power devices require an edge termination around the active area that is designed to improve the blocking voltage capability and make it approach the ideal value of the parallel-plane junction. Consequently, the edge termination design limits the maximum voltage that can be supported by the device. The design of edge terminations for GaN and SiC discrete power devices is described in this chapter.

3.1 Parallel-Plane Breakdown

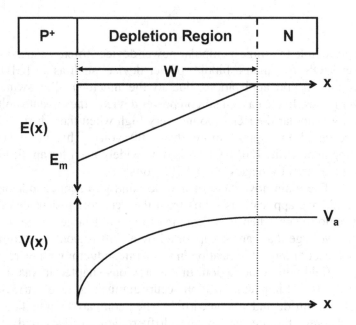

Fig. 3.1 Electric field and potential distribution for an abrupt parallel-plane P⁺N junction.

The analysis of a one-dimensional abrupt junction can be used to understand the design of the drift region within power devices[2]. The case of a P⁺/N junction is illustrated in Fig. 3.1 where the P⁺ side is assumed to be very highly doped so that the electric field supported within it can be neglected. When this junction is reverse-biased by the application of a positive bias to the N-region, a depletion region is formed in the N-region together with the generation of a strong electric field within it that supports the voltage. The Poisson's equation for the N-region is then given by:

$$\frac{d^2V}{dx^2} = -\frac{dE}{dx} = -\frac{Q(x)}{\varepsilon_S} = -\frac{qN_D}{\varepsilon_S}$$ [3.1]

where Q(x) is the charge within the depletion region due to the presence of ionized donors, ε_S is the dielectric constant for the semiconductor, q is the electron charge, and N_D is the donor concentration in the uniformly doped N-region.

Integration of the above equation with the boundary condition that the electric field must go to zero at the edge of the dep-letion region (i.e. at x = W) provides the electric field distribution:

$$E(x) = -\frac{qN_D}{\varepsilon_S}(W - x) \qquad [3.2]$$

The electric field has a maximum value of E_m at the P$^+$/N junction (x = 0) and decreases linearly to zero at x = W.

Integration of the electric field distribution through the depletion region provides the potential distribution:

$$V(x) = -\frac{qN_D}{\varepsilon_S}(Wx - \frac{x^2}{2}) \qquad [3.3]$$

This equation is obtained by using the boundary condition that the potential is zero at x = 0 within the P$^+$ region. The potential varies quadratically as illustrated in the figure. The thickness of the depletion region (W) is related to the applied reverse bias (V_a):

$$W = \sqrt{\frac{2\varepsilon_S V_a}{qN_D}} \qquad [3.4]$$

Using these equations, the maximum electric field at the junction can be obtained:

$$E_m = \sqrt{\frac{2qN_D V_a}{\varepsilon_S}} \qquad [3.5]$$

When the applied bias increases, the maximum electric field approaches values at which significant impact ionization begins to occur. The impact ionization coefficients for GaN and 4H-SiC were compared in chapter 2 (see Fig. 2.6).

The breakdown voltage is determined by the ionization integral becoming equal to unity[2]:

$$\int_0^W \alpha.dx = 1 \qquad [3.6]$$

where α is the impact ionization coefficient discussed in chapter 2. The exponential format for the impact ionization coefficients (see Eq. [2.10]) is not amenable to deriving simple analytical solutions for the breakdown voltage.

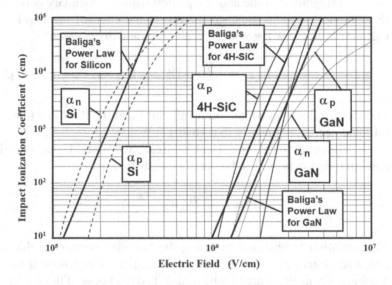

Fig. 3.2 Impact ionization coefficients for Si, GaN and 4H-SiC using a power law.

In order to obtain a closed form solution for the breakdown voltage, it is convenient to use a power law behavior for the impact ionization coefficient in place of Chynoweth's law[2]. For silicon, this was originally performed using the Fulop's power law formula:

$$\alpha_F(Si) = 1.8 \times 10^{-35} E^7 \qquad [3.7]$$

However, this formula has been found to overestimate the breakdown voltages of actual devices[3]. The Baliga's power law for impact ionization coefficients in silicon, given by:

$$\alpha_B(Si) = 3.507 \times 10^{-35} E^7 \qquad [3.8]$$

provides a good match between the analytically calculated breakdown voltage and those measured on silicon devices[3]. The impact ionization coefficient computed using Baliga's power law for silicon is compared with the measured impact ionization coefficients for holes and electrons in Fig. 3.2. The values obtained using the power law fall between the values for electrons and holes up to impact ionization coefficient values of 1×10^4 cm^{-1}.

Similarly, a Baliga's power law for the impact ionization coefficient in the case of 4H-SiC is given by:

$$\alpha_B(4H\text{-}SiC) = 1.0 \times 10^{-41} E^7 \tag{3.9}$$

This expression provides a good fit with the Konstantinov *et al.*,[4] data as shown in Fig. 3.2 for holes. The breakdown voltages calculated for 4H-SiC devices using this power law are consistent with experimental results[5].

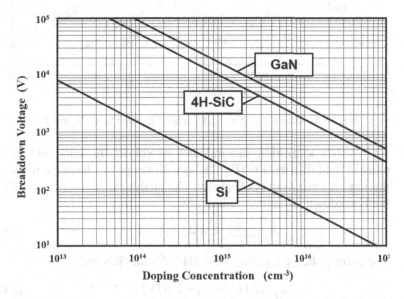

Fig. 3.3 Breakdown voltages for abrupt parallel-plane junctions in Si, GaN, and 4H-SiC.

A Baliga's power law for the impact ionization coefficient in the case of GaN is given by:

$$\alpha_B(\text{GaN}) = 1.5 \times 10^{-42} E^7 \tag{3.10}$$

This expression provides a good fit with the Ozbek and Baliga[6] data as shown in Fig. 3.2. The values obtained using the power law fall between the values for electrons and holes. The breakdown voltages calculated for GaN devices using this power law are consistent with experimental results[7].

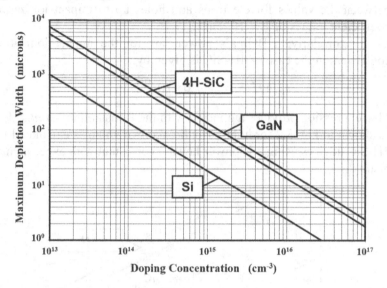

Fig. 3.4 Maximum depletion width at breakdown in Si, GaN, and 4H-SiC.

The ionization integral can be solved by using the linear electric field distribution (Eq. [3.2]) with the power law equations for the impact ionization coefficients. The analytical solution derived for the breakdown voltage for silicon[2] is:

$$BV_{PP}(\text{Si}) = 4.45 \times 10^{13} N_D^{-3/4} \qquad [3.11]$$

The corresponding equations for 4H-SiC and GaN are:

$$BV_{PP}(4\text{H}-\text{SiC}) = 1.67 \times 10^{15} N_D^{-3/4} \qquad [3.12]$$

and

$$BV_{PP}(\text{GaN}) = 2.82 \times 10^{15} N_D^{-3/4} \qquad [3.13]$$

The breakdown voltages for one-dimensional parallel-plane junctions obtained by using the analytical solutions are plotted in Fig. 3.3. For the same doping concentration, the breakdown voltage for 4H-SiC is 36.3 times larger than for silicon and the breakdown voltage for GaN is 61.5 times larger than for silicon. For example, with a doping concentration of 1×10^{15} cm^{-3}, the breakdown voltage from the silicon case is 258 volts while that for 4H-SiC and GaN are 9382 and 15,885 volts, respectively. From the design point of view, it is more important to compare the doping concentration in the drift region to achieve the same breakdown voltage. For the same breakdown voltage of 1000 volts, the doping concentrations in the drift region for the silicon, 4H-SiC, and GaN one-dimensional parallel-plane junctions are 1.65×10^{14} cm^{-3}, 1.98×10^{16} cm^{-3}, and 3.99×10^{16} cm^{-3}, respectively. In general, the doping concentrations in the drift region for 4H-SiC, and GaN one-dimensional parallel-plane junctions are 120 and 243 times greater than that for silicon devices. This increase in doping concentration greatly decreases the resistance for the drift region in 4H-SiC and GaN discrete vertical devices as discussed in a later chapter.

The depletion width in the drift region reaches its largest value when it is supporting the breakdown voltage. An analytical expression for the maximum depletion can be derived by combining Equ. [3.4] with the equations for the breakdown voltage. The analytical equations for the case of silicon, 4H-SiC and GaN are:

$$W_{PP}(\text{Si}) = 2.404 \times 10^{10} N_D^{-7/8}$$ [3.14]

$$W_{PP}(4\,\text{H-SiC}) = 1.34 \times 10^{11} N_D^{-7/8}$$ [3.15]

and

$$W_{PP}(\text{GaN}) = 1.80 \times 10^{11} N_D^{-7/8}$$ [3.16]

The maximum depletion layer width, reached at the onset of breakdown, predicted by these equations is shown in Fig. 3.4. As an example, with a doping concentration of 1×10^{15} cm^{-3}, the maximum depletion width for the silicon case is 18.3 microns while that for 4H-SiC and GaN are 100 and 135 microns, respectively. In general, for the same doping concentration, the maximum depletion width in 4H-SiC and GaN are 5.5x and 7.4x larger than that in silicon because they can sustain a much larger electric field. More importantly, for the same breakdown voltage, the depletion width in 4H-SiC and GaN are smaller than for a silicon device by a factor of 12 and 16.5 times, respectively, because of the higher doping concentrations in the drift region. As an example, for the same breakdown voltage of 1000 volts, the maximum depletion width in the drift region for the silicon, 4H-SiC, and GaN one-dimensional parallel-plane junctions are 18.3 microns, 7.37 microns, and 5.37 microns, respectively. The smaller thickness of the drift region, in conjunction with the far larger doping concentration, results in an enormous reduction in the specific on-resistance of the drift region in 4H-SiC and GaN when compared with silicon as discussed in a later chapter.

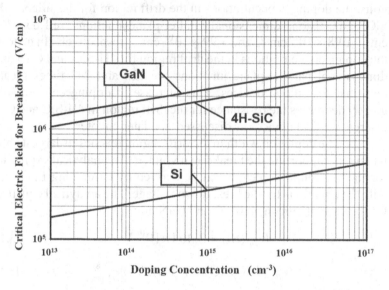

Fig. 3.5 Critical electric field for breakdown in Si, GaN and 4H-SiC.

Breakdown occurs when the ionization integral becomes equal to unity. There is a corresponding peak electric field at the junction for each doping concentration in the drift region. These values are defined as the critical electric field for breakdown. An analytical solution for the critical electric field can be derived by using Equ. [3.5] with the equations for the breakdown voltage. The critical electric fields are given by:

$$E_C(Si) = 3.70 \times 10^3 \, N_D^{1/8} \qquad [3.17]$$

$$E_C(4\text{H-}SiC) = 2.49 \times 10^4 \, N_D^{1/8} \qquad [3.18]$$

and

$$E_C(\text{GaN}) = 3.13 \times 10^4 \, N_D^{1/8} \qquad [3.19]$$

The critical electric field for 4H-SiC and GaN can be compared with that for silicon using Fig. 3.5. For the same doping concentration, the critical electric field in 4H-SiC and GaN are 6.62x and 8.32x larger than for silicon. The larger critical electric field in 4H-SiC allows supporting high voltages with larger doping concentration in the drift region with a smaller depletion layer width.

3.2 Punch-Through Breakdown

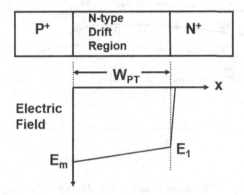

Fig. 3.6 Electric field distribution for the punch-through structure.

In the case of bipolar power devices, such as PiN rectifiers and asymmetric IGBTs, it is preferable to employ a punch-through drift region design to reduce its thickness[2]. The punch-through structure is illustrated in Fig. 3.6 with its electric field distribution at high reverse bias voltages. The electric field has a shallow slope in the drift region because it is designed with a low doping concentration. The electric field is truncated when the depletion region reaches the N[+] substrate due to its high doping concentration.

The breakdown voltage for the punch-through diode structure is give by[2]:

$$BV_{PT} = E_C W_{PT} - \frac{qN_D W_{PT}^2}{2\varepsilon_S}$$ [3.20]

The breakdown voltage for the case of silicon punch-through diodes is provided in Fig. 3.7 for comparison with those for 4H-SiC and GaN. A 5 micron thick drift region can support about 100 volts for the silicon case. The breakdown voltage increases to 1000 volts when the drift region thickness for the silicon diode is increased to 50 microns. The reduction in breakdown voltage at lower doping levels does not occur in practice[8].

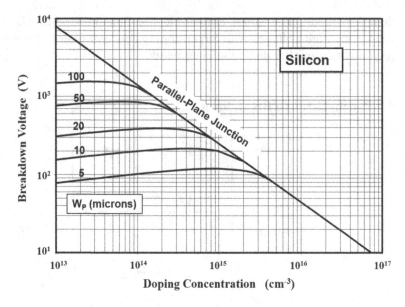

Fig. 3.7 Breakdown voltage for silicon punch-through diodes.

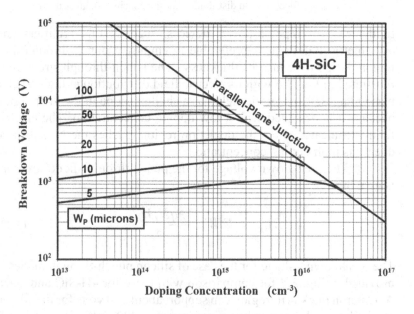

Fig. 3.8 Breakdown voltage for 4H-SiC punch-through diodes.

The breakdown voltage for the case of 4H-SiC punch-through diodes is provided in Fig. 3.8 for various values of drift region thickness. A 5 micron thick drift region can support about 800 volts in the case of 4H-SiC. The breakdown voltage increases to about 7000 volts when the drift region thickness for the 4H-SiC diode is increased to 50 microns. The much larger values than in the case of silicon are due to the greater critical electric field for breakdown in this material.

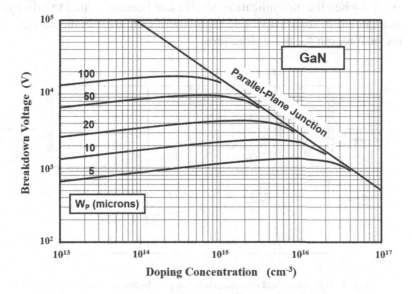

Fig. 3.9 Breakdown voltage for GaN punch-through diodes.

The breakdown voltage for punch-through GaN diodes is provided in Fig. 3.9 for various values of drift region thickness. A 5 micron thick drift region can support about 1000 volts in the case of GaN. The breakdown voltage increases to about 10,000 volts when the drift region thickness for the GaN diode is increased to 50 microns. The much larger values than in the case of silicon are due to the greater critical electric field for breakdown in this material. The GaN diodes support slightly larger voltages than 4H-SiC diodes.

3.3 Open-Base Transistor Breakdown

Silicon carbide bipolar power devices like GTOs and IGBTs are usually fabricated using the asymmetrical structure[2] with the depletion layer designed to punch-through to a buffer layer located adjacent to the collector junction as illustrated in Fig. 3.10. The forward blocking capability of these devices is determined by the open-base transistor breakdown phenomenon. The maximum blocking voltage occurs when the common base current gain of the wide base transistor becomes equal to unity. The common base current gain is determined by the product of the emitter injection efficiency, the base transport factor, and the multiplication factor. Using the avalanche breakdown criteria when the multiplication co-efficient becomes equal to infinity, as assumed in some papers, leads to significant error in the design of the drift region for these structures.

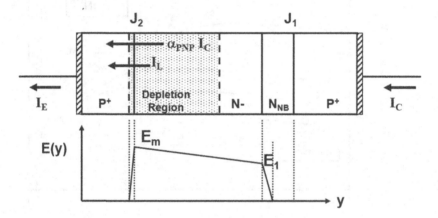

Fig. 3.10 Electric field distribution in the punch-through transistor structure.

The emitter injection efficiency is smaller than unity due to the high doping concentration of the N-buffer layer. The emitter injection efficiency for the P^+ collector/N-buffer junction (J_1) can be obtained by using[2]:

$$\gamma_E = \frac{D_{pNBL} L_{nC} N_{AC}}{D_{pNBL} L_{nC} N_{AC} + D_{nC} W_{NBL} N_{DBL}}$$

[3.21]

where D_{pNBL} and D_{nC} are the diffusion coefficients for minority carriers in the N-buffer and P^+ collector regions; N_{AC} and L_{nC} are the doping concentration and diffusion length for minority carriers in the P^+ collector region; N_{DBL} and W_{NBL} are the doping concentration and width of the N-buffer layer. In determining the diffusion coefficients and the diffusion length, it is necessary to account for impact of the high doping concentrations in the P^+ collector region and N-buffer layer on the mobility. In addition, the lifetime within the highly doped P^+ collector region is reduced due to heavy doping effects, which shortens the diffusion length.

As shown in the figure, the electric field is truncated by the high doping concentration of the N-buffer layer making the un-depleted width of the NPN transistor base region equal to the width of the N-buffer layer. The base transport factor is then given by:

$$\alpha_T = \frac{1}{\cosh\left(W_{NBL} / L_{pNB}\right)} \quad [3.22]$$

which is independent of the collector bias. Here, $L_{p,NB}$ is the diffusion length for holes in the N-buffer layer. This analysis neglects the depletion region extension within the N-buffer layer. The diffusion length for holes ($L_{p,NB}$) in the N-buffer layer depends upon the diffusion coefficient and the minority carrier lifetime in the N-buffer layer. The diffusion coefficient varies with the doping concentration in the N-buffer layer based upon the concentration dependence of the mobility. In addition, the minority carrier lifetime has been found to be dependent upon the doping concentration[9] in the case of silicon devices. Although this phenomenon has not been verified for silicon carbide, it is commonly used when performing numerical analysis of silicon carbide devices. The effect can be modeled by using the relationship:

$$\frac{\tau_{LL}}{\tau_{p0}} = \frac{1}{1 + \left(N_D / N_{REF}\right)} \quad [3.23]$$

where N_{REF} is a reference doping concentration whose value will be assumed to be 5×10^{16} cm^{-3}.

The multiplication factor for a P-N junction is given by:

$$M = \frac{1}{1 - \left(V_A / BV_{PP}\right)^n} \quad [3.24]$$

with a value of n = 6 for the case of a P^+/N junction and the avalan-che breakdown voltage of the P-base/N-base junction (BV_{PP}) *without the punch-through phenomenon*. In order to apply this equation to the punch-through case relevant to the asymmetric silicon carbide structure, it is necessary to relate the maximum electric field at the junction for the two cases. The electric field at the interface between the lightly doped portion of the N-base region and the N-buffer layer is given by:

$$E_1 = E_m - \frac{qN_D W_N}{\varepsilon_S}$$

[3.25]

The voltage supported by the device is given by:

$$V_C = \left(\frac{E_m + E_1}{2}\right) W_N = E_m W_N - \frac{qN_D}{2\varepsilon_S} W_N^2$$

[3.26]

From this expression, the maximum electric field is given by:

$$E_m = \frac{V_C}{W_N} + \frac{qN_D W_N}{2\varepsilon_S}$$

[3.27]

The corresponding equation for the non-punch-through case is:

$$E_m = \sqrt{\frac{2qN_D V_{NPT}}{\varepsilon_S}}$$

[3.28]

Consequently, the non-punch-through voltage that determines the multiplication coefficient 'M' corresponding to the applied collector bias 'V$_C$' for the punch-through case is given by:

$$V_{NPT} = \frac{\varepsilon_S E_m^2}{2qN_D} = \frac{\varepsilon_S}{2qN_D}\left(\frac{V_C}{W_N} + \frac{qN_D W_N}{2\varepsilon_S}\right)^2$$

[3.29]

The multiplication coefficient for the asymmetric silicon carbide IGBT structure can be computed by using this non-punch-through voltage:

$$M = \frac{1}{1 - (V_{NPT}/BV_{PP})^n}$$

[3.30]

The multiplication coefficient increases with increasing collector bias. The open-base transistor breakdown voltage (and the forward blocking capability of the punch-through transistor structure) is determined by the collector voltage at which the multiplication factor becomes equal to the reciprocal of the product of the base transport factor and the emitter injection efficiency.

The doping concentration of the N-buffer layer must be sufficiently large to prevent reach-through of the electric field to the P^+ collector region. Although the electric field at the interface between the N-base region and the N-buffer layer is slightly smaller than at the blocking junction (J_2), a worse case analysis can be done by assuming that the electric field at this interface is close to the critical electric field for breakdown in the drift region. The minimum charge in the N-buffer layer to prevent reach-through can be then obtained using:

$$N_{DBL}W_{NBL} = \frac{\varepsilon_S E_C}{q} \qquad [3.31]$$

Using a critical electric for breakdown in silicon carbide of 2×10^6 V/cm for a doping concentration of 1.5×10^{14} cm^{-3} in the N-base region, the minimum charge in the N-buffer layer to prevent reach-through for a silicon carbide punch-through transistor structure is found to be 1.07×10^{13} cm^{-2}. An N-buffer layer with doping concentration of 5×10^{16} cm^{-3} and thickness of 5 microns has a charge of 2.5×10^{13} cm^{-2} that satisfies this requirement.

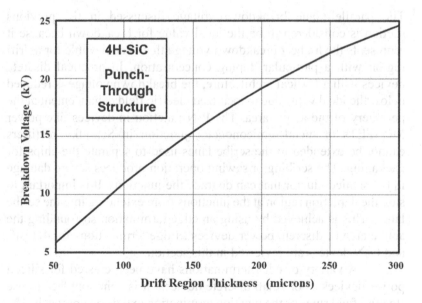

Fig. 3.11 Open base breakdown voltages for the 4H-SiC punch through transistor structures.

The drift region thickness required to achieve various breakdown voltages for the 4H-SiC transistor punch-through structure can be obtained using the above equations. From Fig. 3.11, it can concluded that the drift region thickness must be increased from 50 microns to 300 microns to achieve an increase in breakdown voltage from 5 kV to 25 kV. For this analysis, the structure was assumed to have a P^+ collector region with doping concentration of 1×10^{19} cm^{-3} and the buffer layer had a thickness of 5 microns with a doping concentration of 5×10^{16} cm^{-3}. The lifetime in the drift region was assumed to be 1 microsecond. Using these parameters, the emitter injection efficiency computed using Eq. [3.21] is 0.971. When the device is close to breakdown, the entire N-base region is depleted and the base transport factor computed by using Eq. [3.22] in this case is 0.903. Based up on Eq. [7.3], open-base transistor breakdown will then occur when the multiplication coefficient becomes equal to 1.14 for the above values for the injection efficiency and base transport factor.

3.4 Silicon Carbide Device Edge Terminations

The parallel-plane breakdown voltage discussed in the previous section is considered to be the ideal value for breakdown because it represents the highest breakdown voltage that is achievable for a drift region with a particular doping concentration. In practical discrete devices with a vertical architecture, the breakdown voltage is reduced below the ideal value due to localized electric field enhancement at the periphery of the active area. The P-N junction in devices like power MOSFETs or metal-semiconductor junction in Schottky rectifiers cannot be extended to the scribe lanes used to separate the chips for packaging. The scribing or sawing operation produces severe damage in the semiconductor that can degrade the junctions. It is important to stop the depletion region at the junctions from extending into the scribe lanes. This is achieved by using an edge termination surrounding the active area of discrete power devices[2]. Edge terminations for 4H-SiC and GaN devices are reviewed in this section.

A variety of edge terminations have been created for silicon power devices[2]. For small discrete devices, it is commonplace to use floating field rings or the junction termination extension approach. The utilization of this concept to 4H-SiC devices was initially hampered by the lack of a technology for making P-N junctions by ion-implantation.

Consequently, a unique approach was proposed and demonstrated for silicon carbide and gallium nitride by amorphisation of the surface using argon ion-implantation to produce a highly resistive ring. After improved technology for activation of ion-implants in 4H-SiC was developed, the formation of floating field rings and JTE terminations has been widely practiced.

For large discrete devices made from an entire silicon wafer, it is commonplace to use bevel edge terminations[2]. The bevel edge termination has recently been extended to small discrete devices fabricated from 4H-SiC by using a V-shaped dicing blade.

3.4.1 Unterminated Planar SiC Schottky Diode

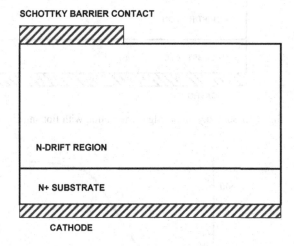

Fig. 3.12 The un-terminated Schottky diode structure.

The first high voltage vertical Schottky barrier rectifiers were demonstrated at PSRC/NCSU from 6H-SiC in 1992 using a remarkably simple process of evaporating metal through a 'dot' mask onto a silicon carbide surface after appropriate cleaning[10]. This approach was later successfully demonstrated at PSRC for the fabrication of 1000-V Schottky rectifiers from 4H-SiC in 1995[11].

The unterminated Schottky barrier diode structure, illustrated in Fig. 3.12, contains a sharp edge at the periphery of the diode leading to electric field enhancement and reduction of the breakdown voltage. This phenomenon can be observed using the results of two-dimensional numerical simulations on a Schottky diode[12]. The diode was found to exhibit a breakdown voltage of 650 volts which is significantly smaller (40%) than the parallel-plane breakdown voltage of 1670 volts. It is necessary to include an edge termination at the periphery of the Schottky contact to improve the breakdown voltage.

3.4.2 SiC Schottky Diode with Floating Metal Field Rings

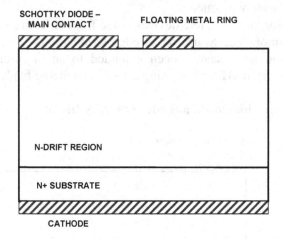

Fig. 3.13 Schottky diode edge termination with floating metal ring.

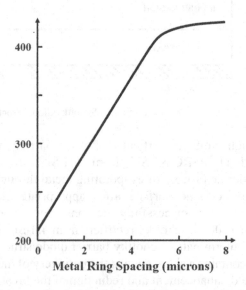

Fig. 3.14 Breakdown voltage of the Schottky diode edge termination with a single floating metal ring.

A simple method for reducing the electric field enhancement at the edge of Schottky diode without adding processing steps is by placement of floating metal field rings around the periphery of the diode[13] as illustrated in Fig. 3.13. This concept is similar to floating field rings produced in silicon devices by creating p-type regions in the n-drift region around the periphery of P-N junction diodes[2]. An optimum spacing of one-quarter of the maximum parallel-plane depletion width (W_{PP}) is predicted by analytical theory for a single floating field ring in agreement with empirical evidence in silicon devices[14].

A typical 4H-SiC Schottky rectifier with breakdown voltage of 1000 volts has a drift region with doping concentration of 2×10^{16} cm^{-3} (see Fig. 3.3). The maximum depletion width for this doping concentration is 7.3 microns. Consequently, the optimum spacing for the metal field ring should be about 2 microns. The measured breakdown voltage as a function of metal ring spacing[15] is shown in Fig. 3.14. The breakdown voltage increases when the spacing is made much larger and saturates at a value of 5 microns. This indicates that there is substantial charge at the surface of silicon carbide that spreads the depletion region.

3.4.3 SiC Schottky Diode with Resistive Schottky Extension

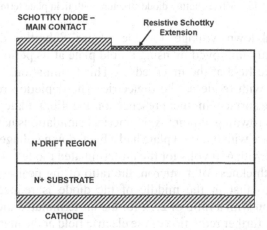

Fig. 3.15 Schottky diode edge termination with resistive Schottky extension.

The electric field at the edge of the Schottky contact can be reduced by spreading the potential along the surface. One of the methods that has been demonstrated to achieve this is by creating a resistive

Schottky extension[12] as illustrated in Fig. 3.15. The resistive region must have a high sheet resistance in the range of 10^8 ohms/sq. This has been achieved by oxidation of a thin (50 angstrom) titanium film in air at 300 °C. An increase in the breakdown voltage from 150 to 500 volts was reported using this approach.

3.4.4 SiC Schottky Diode with Field Plate Edge Termination

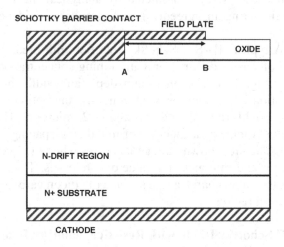

Fig. 3.16 Schottky diode structure with field plate termination.

The breakdown voltage for the planar Schottky diode can be significantly improved by using a field plate at its periphery to reduce the electric field at the metal edge[2]. This termination is illustrated in Fig. 3.16 with oxide as the dielectric. The depletion region spreads along the surface in the presence of the field plate reducing the potential crowding at point A. Numerical simulations have shown that the structure with the field plate had a breakdown voltage of 1300 volts compared with 650 volts for the un-terminated diode[12]. For the case of an oxide thickness of 1 micron, the ratio of the peak electric field at point A to that in the middle of the diode is reduced to 2.3x in comparison with a ratio of 2.7x for the un-terminated diode.

A further reduction of the electric field at the metal edge (point A) can be obtained by reducing the oxide thickness. For the case of an oxide thickness of 0.6 microns, the peak electric field at the metal corner is 2x of that in the middle. This allowed the breakdown voltage to increase to 1430 volts. A further reduction in oxide thickness to 0.2

microns leads to an even lower electric field at the metal corner. Unfortunately, the electric field at the field plate edge (point B) now becomes larger than at point A leading to a reduction of the breakdown voltage to 840 volts. This occurs because the field plate behaves like a cylindrical junction with a radius of curvature given by:

$$x_J = \left(\frac{\varepsilon_{SiC}}{\varepsilon_{OX}} \right) t_{OX} \simeq 3 t_{OX} \qquad [3.32]$$

The smaller effective radius of curvature at the edge of the field plate reduces the breakdown voltage in accordance with breakdown voltage of cylindrical junctions[16].

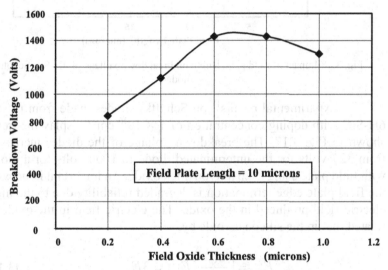

Fig. 3.17 Breakdown voltage of 4H-SiC Schottky diodes in with field plates.

The variation of the breakdown voltage with field oxide thickness obtained using numerical solutions is plotted in Fig. 3.17. It is obvious that there is an optimum oxide thickness at which the breakdown voltage reaches a maximum value. It is also important to use a sufficient length (L) for the field plate as shown in Fig. 3.18. A length of 10 microns is adequate for the simulated case. In general, the field plate length must be made larger than the maximum depletion layer width for the underlying drift region doping concentration.

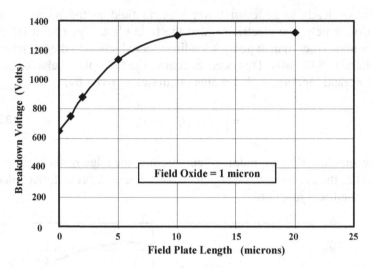

Fig. 3.18 Impact of field plate length on breakdown voltage of 4H-SiC Schottky diodes.

Experimental results[17] on Schottky diodes made from p-type 6H-SiC with doping concentration of 1×10^{16} cm^{-3} display the trend shown in Fig. 3.17. The breakdown voltage of the diodes increased from 327 volts for the unterminated diode to 1000 volts for a diode with field plate using a 0.6 micron thick oxide. An important issue with the field plate edge termination is degraded reliability due to the high electric field produced in the oxide. The electric field in the oxide is related that in the silicon carbide by:

$$E_{OX} = \left(\frac{\varepsilon_{SiC}}{\varepsilon_{OX}} \right) E_{SiC} \simeq 3 E_{SiC} \qquad [3.33]$$

This equation predicts an electric field in the oxide of close to 10 MV/cm when the electric field in the silicon carbide reaches the critical electric field for breakdown (see Fig. 3.5). Such high electric field have been known to produce a reliability problem in devices and operating the field plate edge termination with an electric field of less than 3 MV/cm severely reduces the blocking voltage for the diode[18].

One variation of the basic field plate structure is to use a ramp oxide whose thickness increases away from the Schottky contact as shown in Fig. 3.19. This type of ramp profile can be created by using phosphosilicate glass and over etching the oxide with a photoresist mask[19]. Breakdown voltages close to the ideal parallel-plane value were reported for Schottky diodes fabricated from 6H-SiC. However, the electric field in the oxide is still well above the reliability limit[19]. Similar results have been obtained for Schottky diodes fabricated using 4H-SiC[20].

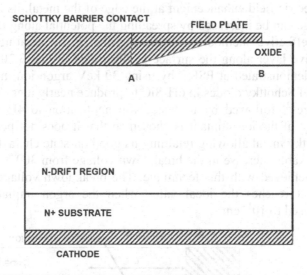

Fig. 3.19 Schottky diode structure with ramp oxide field plate termination.

Based up on Equ. [3.33], the electric field in the oxide can be reduced by using dielectrics with larger permittivity. This was first proposed for reducing the electric field in the gate oxide of power MOSFETs[21]. A field plate edge termination with aluminum nitride as the high-*k* dielectric with permittivity of 8.5 has been studied[22]. A thin silicon dioxide film was used below the AlN to improve the interface characteristics. The breakdown voltage was found to increase with increasing high-*k* dielectric thickness up to 0.6 microns similar to silicon dioxide case shown in Fig. 3.15 and then reduce beyond this thickness. The electric field in the high-*k* dielectric was calculated to be 2 MV/cm improving the oxide reliability. The breakdown voltage for the field plate termination with AlN as the dielectric was observed to be 1700 volts compared to 1100 volts with silicon dioxide as the dielectric. This is attributed to a reduced electric field at the edge of the Schottky metal (point A in Fig. 3.19) with the larger permittivity. Numerical simulations have also been performed for field plate structures with various high-k dielectrics[23]. It was found that increasing the permittivity to 25 by using Hafnium oxide improves the breakdown voltage.

3.4.5 SiC Schottky Diode Termination with Argon Implant

Since the breakdown voltage of the planar Schottky diode is limited by electric field enhancement at the edge of the metal, its breakdown voltage can be increased by spreading the potential along the surface. One effective method for achieving this is by creating a highly resistive layer along the surface as illustrated in Fig. 3.20. This was first demonstrated at PSRC by using 30 keV argon ion implantation around Schottky diodes in 6H-SiC to produce nearly ideal breakdown voltages[24] followed by its successful application to 4H-SiC[25]. The energy of the ion implant is chosen so that it does not penetrate the Schottky metal allowing retaining its good on-state characteristics. A remarkable increase in the breakdown voltage from 300 V to 1000 V was achieved with this technique. The breakdown voltage increases until it reaches the ideal value when the argon implant dose is increased to 10^{16} cm^{-2}.

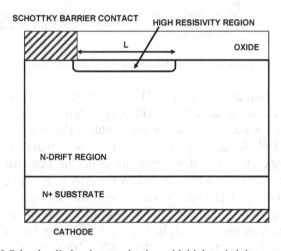

Fig. 3.20 Schottky diode edge termination with high resistivity extension.

The principle behind the approach is to utilize ion implantation to create damage in the silicon carbide lattice to produce deep levels in its band gap. This moves the Fermi level close to the center of the band gap within the implanted zone producing a high resistivity layer due to the large band gap of the semiconductor[26]. The presence of the deep level traps has been confirmed using Deep Level

Transient Spectroscopy (DLTS) measurements. The damage can be created by using any implantation species (including dopants, such as Aluminum and Boron, without sufficient annealing to remove all the deep level defects) as subsequently reported by other groups[27,28].

The design of the argon implanted edge termination has been studied to determine its optimum length[29]. It was found that a length of 100 microns is adequate to achieve a breakdown voltage of 800 V for a 6H-SiC diode fabricated using an epitaxial layer doping concentration of 2×10^{16} cm^{-3} as shown in Fig. 3.21. An argon implant dose of 1×10^{15} cm^{-2} was used at the periphery of the Schottky diodes. The length of the implanted zone is about ten times the width of the depletion region at breakdown. The leakage current was found to increase linearly with the length of the implanted zone. It is therefore important to use a length just sufficient to achieve the parallel plane breakdown voltage. It has been found that the leakage current can be reduced by two-orders of magnitude with post implantation annealing at 600 °C[30].

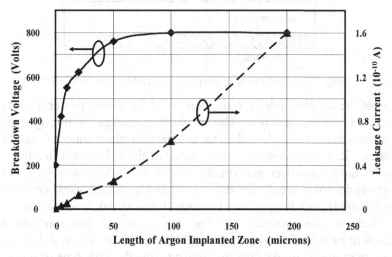

Fig. 3.21 Design of the high resistivity extension for the argon implanted edge termination.

3.5 Silicon Carbide P-N Junction Edge Terminations

The P-N junction is commonly used for the formation of a variety of power device structures. Although the emphasis in this book is on unipolar devices due to their high performance characteristics when fabricated from 4H-SiC, devices such as power MOSFETs require P-N junctions to support the blocking voltage within the active region. The P-N junction can then also be used at the edges of the devices to reduce electric field crowding. Device edge terminations that utilize P-N junctions are discussed in this section.

3.5.1 Unterminated Planar Junction

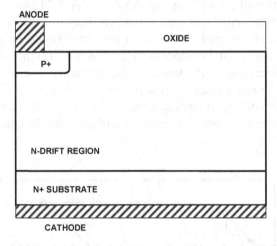

Fig. 3.22 Ion-implanted planar unterminated junction in 4H-SiC.

One of the simplest edge terminations used for silicon power devices is the cylindrical junction. The breakdown voltage of this type of termination is well below that for the parallel-plane junction due to electric field enhancement at the corner of the cylindrical junction[2]. This behavior has been analytically modeled[16]. In silicon carbide, the dopants do not move during the ion implant annealing cycles despite the relatively high temperature because of the extremely low diffusion coefficients (see chapter 2). The ion-implanted junctions must therefore be modeled with very small characteristic diffusion lengths that are representative of the ion implant straggle. This type of planar edge termination is illustrated in Fig. 3.22 with an oxide passivation.

The unterminated P-N junction structure in 4H-SiC was analyzed by performing two-dimensional numerical simulations using a drift region doping concentration of 1×10^{16} cm^{-3} and thickness of 30 microns[12]. The depth of the P$^+$ region was varied to examine its impact on the breakdown voltage. The breakdown voltage of the structure was found to increase with increasing junction depth as shown in Fig. 3.23. For a shallow junction depth of 0.2 microns, the breakdown occurred at 400 volts. This is due to a significant enhancement in the electric field at the corner of the junction. This field enhancement was found to be reduced when the junction depth was increased to 0.9 microns leading to an increase in the breakdown voltage to 900 volts. These values are well below the breakdown voltage (1670 volts) for the ideal parallel plane junction. They are consistent with the measured breakdown voltages for nitrogen implanted N^+/P diodes formed in boron implanted 6H-SiC diodes[31].

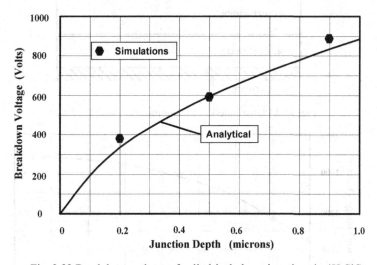

Fig. 3.23 Breakdown voltage of cylindrical planar junctions in 4H-SiC.

The breakdown voltage of cylindrical junctions can be predicted by using the following analytical relationship[2]:

$$\frac{BV_{CYL}}{BV_{PP}} = \frac{1}{2}\left[\left(\frac{r_J}{W_{PP}}\right)^2 + 2\left(\frac{r_J}{W_{PP}}\right)^{6/7}\right].\ln\left[1+2\left(\frac{W_{PP}}{r_J}\right)^{8/7}\right] - \left(\frac{r_J}{W_{PP}}\right)^{6/7}$$

[3.34]

where r_J is the radius of curvature of the junction. The breakdown voltage calculated using this equation is shown in Fig. 3.23 by the solid line for the case of 4H-SiC with a doping concentration of 1×10^{16} cm^{-3}. The breakdown voltage obtained by the two dimensional numerical simulations are in reasonable agreement with the analytical formula as shown by the symbols.

3.5.2 Planar Junction with Field Plate

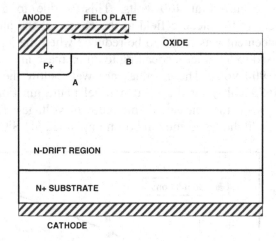

Fig. 3.24 Planar junction edge termination with field plate.

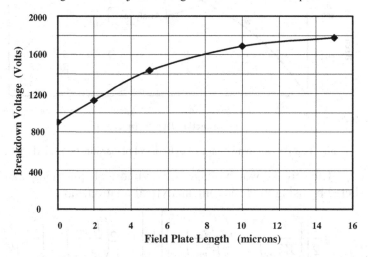

Fig. 3.25 Cylindrical planar junctions in 4H-SiC with field plate.

In the previous section, it was demonstrated that the breakdown voltage of a planar junction is limited by the enhanced electric field at the corner of the P-N junction (location A in Fig. 3.24 above). This electric field enhancement can be ameliorated by forming a field plate by extending the anode metal over the edge of the junction on top of the oxide. For silicon devices, the improvement in the breakdown voltage depends upon the length (L) of the field plate and the thickness of the oxide underneath the field plate[2]. Although this is also true in principle for the case of silicon carbide structures, an additional consideration is the much higher electric field within the oxide as discussed in section 3.4.3.

Two-dimensional numerical simulations of the planar junction with field plate were performed[12] using an N-type drift region with doping concentration of 1×10^{16} cm^{-3}. The junction depth was chosen as 0.9 microns. The addition of the field plate reduces the potential crowding at the junction by spreading the depletion region along the surface resulting in larger breakdown voltages. The increase in the breakdown voltage is shown in Fig. 3.25 with increasing length of the field plate. In these simulations, the oxide thickness was kept at 1 micron. The breakdown voltage can be increased two-fold with sufficient field plate length. A field plate length of 15 microns is adequate for this particular N-region doping concentration. The electric field at the edge of the field plate can become greater than that at the junction if the oxide thickness is reduced.

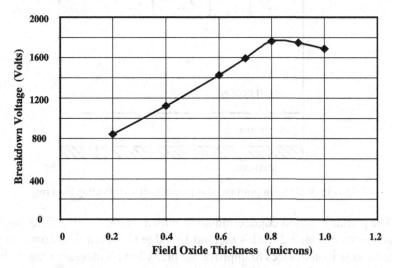

Fig. 3.26 Cylindrical planar junctions in 4H-SiC with field plate.

The impact of changing the field oxide thickness on the breakdown voltage is shown in Fig. 3.26. It can be seen that the breakdown voltage reaches a maximum value at a field oxide thickness of 0.8 microns. However, the change in breakdown voltage is small for larger oxide thickness values. These results indicate that a simple edge termination for 4H-SiC devices can be constructed using a P-N junction with a field plate extending over an oxide at the edges. The main advantage of this approach is that no additional masking or processing steps are required during device fabrication to prepare the edge termination.

Excellent breakdown voltages with low leakage currents were experimentally observed on 6H-SiC and 4H-SiC with both n-type and p-type drift regions by using field plates with either silicon dioxide or silicon nitride as the dielectric[32]. Breakdown voltages for 4H-SiC P-N diodes comparable to those for floating field rings were also reported using field plates with silicon dioxide[33].

3.5.3 Planar Junction Termination with Floating Field Rings

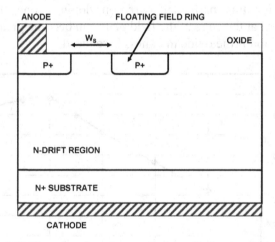

Fig. 3.27 Planar junction edge termination with floating field ring.

The planar junction edge termination with a single floating field ring[34] is shown in Fig. 3.27. This method for edge termination is commonly used in silicon devices to improve the breakdown voltages[2]. One of the advantages of this approach is that the floating field ring can be fabricated simultaneously with the main junction without adding process steps. The breakdown voltage of the termination depends upon the spacing (W_S) of the floating field ring from the main junction[2]. If the spacing is too large, the electric field at the edge of the main junction remains high leading to a breakdown voltage similar to the planar junction without the field ring. If the spacing is too small, a high electric field develops at the outer edge of the floating field ring leading to a breakdown voltage equal to that for a planar junction without the field ring. An optimum spacing is required to achieve an increase in the breakdown voltage. The breakdown voltage of a planar junction with a single floating field ring can be calculated using an analytical method[14]:

$$\frac{BV_{FFR} - BV_{CYL}}{BV_{PP}} = \left[0.5\left(\frac{r_J}{W_{PP}}\right)^2 - 0.96\left(\frac{r_J}{W_{PP}}\right)^{6/7} \right]$$

$$+ 1.92\left(\frac{r_J}{W_{PP}}\right)^{6/7} . \ln\left[1.386\left(\frac{W_{PP}}{r_J}\right)^{4/7} \right]$$

[3.35]

with the optimum floating field ring spacing given by:

$$\frac{W_S}{W_{PP}} = \sqrt{\frac{BV_{FFR}}{BV_{PP}}} - \sqrt{\frac{BV_{FFR}}{BV_{PP}} - \frac{BV_{CY}}{BV_{PP}}}$$

[3.36]

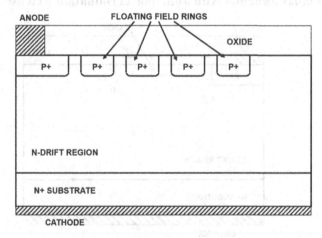

Fig. 3.28 Planar junction edge termination with multiple floating field rings.

For the case of a drift region doping concentration of 1×10^{16} cm^{-3} and a junction depth of 0.9 microns, these equations predict an increase in the breakdown voltage from about 800 V to 1200 V for an optimum field ring spacing of 4.5 microns. An increase in the breakdown voltage has been experimentally observed for junctions fabricated in 4H-SiC with field ring spacing of 2-4 microns[33]. An optimum field ring spacing of 5 microns has also been reported for diodes with a breakdown voltage of 1600 volts[35], consistent with the predictions of the analytical model.

The planar junction edge termination with multiple floating field rings is shown in Fig. 3.28. The use of more floating rings allows the depletion region to spread further along the surface than with a single ring. The results of numerical simulations and experiments on 4H-SiC P-N junction diodes with up to 4 rings have been reported[36]. It was demonstrated that an optimized design with a distance between the rings increasing from 1.5 to 2.0 to 2.5 microns produces 85% of the ideal breakdown voltage. A breakdown voltage of 1.2 kV was achieved in a 4H-SiC JBS diode by using an edge termination having 5 rings with an equal spacing of 3 microns[37]. Breakdown voltages of 70% of the ideal value of 1800 V have been achieved by using 12 guard rings[38]. 10-kV diodes have been reported by using 102 floating field rings[39] distributed over a space of 900 microns for a drift region doped at 8×10^{14} cm^{-3}.

3.5.4 Planar Junction with Junction Termination Extension

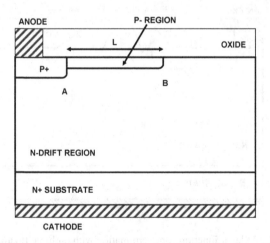

Fig. 3.29 Planar diode with single zone junction termination extension.

For silicon devices, the breakdown voltage has been shown to be greatly improved by using a 'Junction Termination Extension (JTE)'[40]. This concept has also been applied to silicon carbide devices. The JTE structure, illustrated in Fig. 3.29, contains a P-type region formed at the periphery of a P$^+$/N junction. This lightly doped P-type region is usually formed by using ion implantation to precisely control the dopant charge within the layer. If the doping concentration in the

P-type region is too high, the breakdown occurs at its edge (point B) at a lower breakdown voltage than the main junction due to its smaller radius of curvature. If the doping concentration is too low in the P-type region, it becomes completely depleted at low reverse bias voltages resulting in breakdown at the main junction (point A) at the same voltage as the un-terminated junction.

The optimum charge in the P-type region is:

$$Q = \varepsilon_S E_C \qquad [3.37]$$

which amounts to 1.72×10^{-6} C/cm^2 for a critical electric field of 2×10^6 V/cm for 4H-SiC. This is equivalent to a dopant dose of 1.1×10^{13} cm^{-2}, which is about 10 times larger than that used in silicon devices[2]. A more precise optimization of the dose can be performed by two-dimensional numerical simulations of the structure.

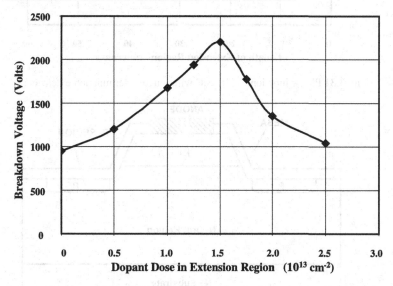

Fig. 3.30 Planar 4H-SiC diodes with single zone junction termination extension.

The results of the simulations[12], performed using an extension length of 10 microns, are summarized in Fig. 3.30 where the variation in breakdown voltage is plotted as a function of the dopant dose in the P-region. As in the case of silicon devices, there is an optimum dose at which the breakdown voltage has a maximum value. At lower doses, the breakdown voltage continues to occur at the main junction due to the peak in the electric field at point A. At higher doses, the breakdown voltage occurs at the edge of the extension due to the peak in the electric field at point B.

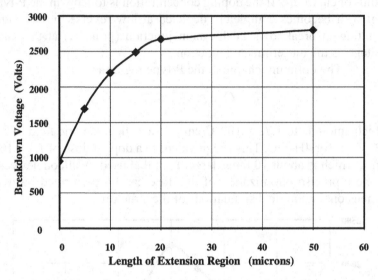

Fig. 3.31 Planar junctions in 4H-SiC with junction termination extension.

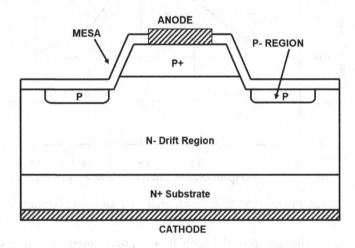

Fig. 3.32 Mesa junction in 4H-SiC with junction termination extension.

The optimum dose obtained by using the numerical simulations is in agreement with that predicted by using the simple formula in Eq. [3.37]. The breakdown voltage obtained at the optimum dose is twice that for the un-terminated junction for this extension length. The dependence of the breakdown voltage on the extension length is shown in Fig. 3.31 at the optimum dose in the extension. A breakdown voltage close to the parallel plane junction can be obtained by using an extension length of 50 microns.

This method for edge terminations has gained popularity due to its effectiveness in suppressing the edge breakdown for P-N junctions[41,42,43]. The optimum dose of the single zone JTE region is just above 1×10^{13} cm^{-2} based on this work. In order to implement this technique it was necessary to develop methods for precisely controlling the P-type dopant dose in the extension. Although the dose can be controlled very well by using ion implantation, processes for the annealing and activation of P-type dopants (Boron and Aluminum) needed to be developed. In addition, the performance has been found to be sensitive to surface charge in the passivation as also reported for silicon devices. In the case of P-N diodes made by the growth of a P$^+$ epitaxial layer on an N- drift region, the JTE termination can be created by first producing a mesa as shown in Fig. 3.32[44,45]. Diodes with a breakdown voltage of 10 kV were fabricated using an edge termination length of 200 microns and an optimum JTE dose of 1×10^{13} cm^{-2}.

3.5.5 Planar Junction with Two-Zone JTE

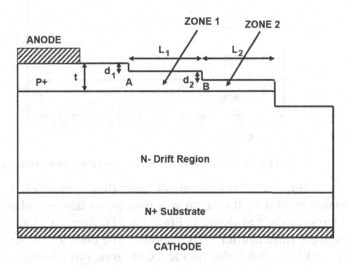

Fig. 3.33 Planar diode with two zone mesa junction termination extension.

The fabrication of the JTE region requires precise control of the dose of the dopants. In silicon devices, this can be performed by using ion-implantation and annealing to achieve 100% activation. The precise control of dopant concentration in 4H-SiC was initially difficult due to poor activation of dopants in ion implanted layers. Early work on creating JTE regions relied up on the epitaxial growth of the P-type region on the N-type drift region followed by etching mesas as illustrated in Fig. 3.33. Enhancements to the breakdown voltage were achieved using multiple zones with different dopant dose levels[38,46] achieved by etching steps in an epitaxially grown P-type layer with aluminum doping. The P-type layer was 0.7 microns thick with an aluminum concentration of 2×10^{18} cm^{-3}. A breakdown voltage of over 90% of the ideal value was achieved by using etch depths of $d_1 = 0.4$ microns and $d_2 = 0.04$ microns. This method requires precise control and uniformity for the etch rate for 4H-SiC. The dose of the second zone is 10% smaller than the first zone.

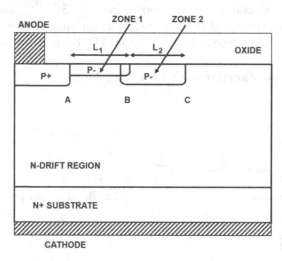

Fig. 3.34 Planar diode with two zone junction termination extension.

After the development of annealing processes for ion implanted regions in 4H-SiC, it was possible to fabricate planar JTE edge terminations[47] as shown in Fig. 3.34. JTE zones of 65 microns in length were sufficient for 1.7 kV diodes. The dose for the first zone was 1×10^{14} cm^{-2} while that for the second zone was 10-times smaller. A breakdown voltage of over 90% of the ideal value was obtained by using this approach. The leakage current was reduced significantly by using silicon dioxide as passivation on the surface.

3.5.6 Planar Junction with Multiple Floating Zone JTE

The multiple floating zone junction termination extension[48] concept utilizes floating rings with low doping concentrations as illustrated in Fig. 3.35. The zones are designed with decreasing widths (W_1, W_2, W_3,...) and increasing spacings (S_1, S_2, S_3,...). The widths of the zones are reduced by a factor of 1.02 while the spacing is increased by a factor of 1.02 to create a very gradual change in the net P-type charge along the surface. A breakdown voltage of 10-kV was achieved by using 36 and 72 zones with a dose of 2×10^{13} cm^{-2}. The advantages of this method are that a single photolithography step and ion-implant steps can be used to form the edge termination and that it has a wide tolerance in ion implant dose for the P-region. This concept has been analyzed using numerical simulations with a mesa etched P-N junction[49] and applied to fabrication of 21-kV 4H-SiC bipolar transistors[50].

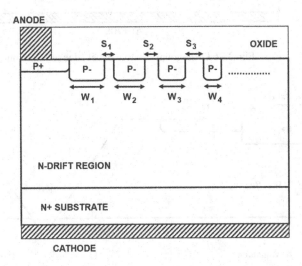

Fig. 3.35 Planar diode with multiple floating zone junction termination extension.

3.5.7 Orthogonal Bevel Edge Terminations

The use of bevels to terminate the edges of high voltage silicon power rectifiers and thyristors has been extensively used since the 1960s[2].

The positive bevel approach is based on removal of more semiconductor material from the lightly doped side of the P-N junction to spread the depletion layer over a larger distance. The resulting reduction in surface electric field has been shown to move the breakdown to the interior parallel-plane junctions allowing achieving 100% of the ideal breakdown voltage. This approach was restricted to single devices made from an entire wafer because the terminations require tapering the wafer edges.

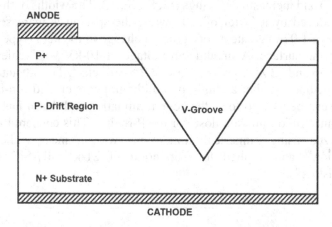

Fig. 3.36 Orthogonal positive bevel edge termination.

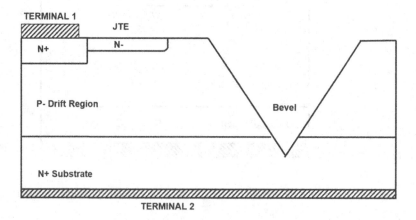

Fig. 3.37 Symmetric blocking bevel edge termination.

The application of bevel edges to silicon carbide devices requires producing bevels for multiple devices with small areas on a single wafer. This has been achieved by performing orthogonal saw cuts on 4H-SiC wafers with a V-shaped blade[51]. The saw cut produces a positive bevel at the N^+ substrate-P drift region junction as illustrated in Fig. 3.36. The damage produced by the sawing operation must be removed by reactive ion etching. Three dimensional numerical simulations have demonstrated that the orthogonal positive bevels reduce the surface electric field at the corners to even below the low electric field along the edges. This allows obtaining close to ideal breakdown voltages.

The orthogonal positive bevel is particularly suitable for making silicon carbide power devices with symmetric blocking voltage capability such as GTOs and IGBTs. The symmetric blocking capability can be accomplished by using the structure shown in Fig. 3.37. Here, the forward blocking is obtained by using the JTE edge termination at the upper N^+/P drift region junction and the reverse blocking capability is achieved with the orthogonal positive bevel[52] for the lower N^+ substrate/P drift region junction.

3.5.8 Bevel Junction Termination Extension

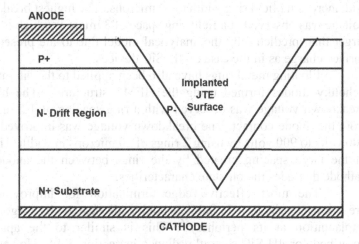

Fig. 3.38 Bevel junction termination extension.

The orthogonal positive bevel and JTE concepts have been combined to achieve the edge termination shown in Fig. 3.38. This bevel-JTE termination[53] is created by ion-implantation of aluminum to form the P- JTE region. It was found that a breakdown voltage of over 90% of the ideal value can be achieved when the aluminum dose is 1 to 2 × 10^{13} cm^{-2}. Numerical simulations have shown that the breakdown voltage is reduced when the JTE dose is too high because of a high electric field at the junction between the P- region and the N^+ substrate. The breakdown voltage is also reduced when the JTE dose is too low because of a high electric field at the junction between the P^+ region and the N drift region. The termination works well for a broad range of bevel angles ranging from 30 to 60 degrees.

3.6 Gallium Nitride Edge Terminations

Most of the recent interest has been focused on lateral GaN power transistors fabricated using GaN-on-Si wafers. These devices do not require an edge termination region. However, some effort has been undertaken to develop edge terminations for vertical discrete GaN diodes.

The use of floating metal field rings, similar to that shown in Fig. 3.13, has been applied to vertical GaN Schottky rectifiers[54]. With a single floating ring spaced at 2 microns from the main Schottky contact, the breakdown voltage improved from 150 volts to 300 volts with increasing field ring width to 20 microns. The highest breakdown voltage was observed at a field ring space of 3 microns which is much larger than predicted by the analytical model due to the presence of surface charge as in the case of 4H-SiC devices.

Floating metal rings have also been applied to the lateral GaN Schottky diodes formed using the HEMT structure[55]. The highest breakdown voltage was observed with a ring spacing of 4-5 microns from the anode contact. The breakdown voltage was improved from 700 volts to 900 volts by using 2 rings of 20 microns in width. The use of the large spacing created by the rings between the anode and cathode degrades the on-state characteritics.

The most effective edge termination for improving the breakdown voltage of GaN Schottky diodes is by argon ion implantation at its periphery[56]. This is similar to the approach described for 4H-SiC Schottky diodes in section 3.4.5. However, it was found that the minimum ion implant dose to achieve nearly ideal breakdown voltage was 2×10^{16} cm^{-2} which is two-times larger than that required for 4H-SiC. The breakdown voltage of Schottky diodes was improved from 300 volts to 1600 volts by using this approach. The minimum length of the ion implant zone for GaN devices[57] to achieve a nearly ideal breakdown voltage of 1600-V was found to be 50 microns.

3.7 Summary

The design issues pertinent to obtaining a high breakdown voltage in 4H-SiC and GaN vertical discrete device structures have been reviewed in this chapter. As a bench mark, the breakdown voltage of the abrupt parallel plane junction has been analyzed for 4H-SiC and GaN using Baliga's power law equations for the impact ionization coefficients. Theses analytical formulae allow development of closed form analytical equations for the depletion layer width and critical electric field at breakdown. The breakdown voltage of open-base transistor structures with buffer layers is provided for the design of 4H-SiC GTOs and IGBTs.

The breakdown voltage of practical devices is limited by the electric field crowding at their edges. Various edge termination methods for 4H-SiC diodes have been reviewed to determine their performance relative to the ideal parallel plane junction. The argon implanted high resistance zone and the junction extension using a P-region with a dose of 1-2×10^{13} cm^{-2} have been found to provide the best performance. Bevel edge terminations have been created to achieve high symmetric blocking voltages.

References

[1] B.J. Baliga, "The IGBT Device: Physics, Design, and Applications of the Insulated Gate Bipolar Transistor", Elsevier Press, Amsterdam, 2015.

[2] B.J. Baliga, "Fundamentals of Power Semiconductor Devices", Springer-Science, New York, 2008.

[3] B.J. Baliga, "Advanced Power MOSFET Concepts", Springer-Science, New York, 2010.

[4] A.O. Konstantinov *et al.*, "Ionization Rates and Critical Electric Fields in 4H-SiC", Applied Physics Letters, Vol. 71, pp. 90, 1997.

[5] J.W. Palmour *et al.*, "Silicon Carbide for Power Devices", IEEE International Symposium on Power Semiconductor Devices and ICs, pp. 25-32, 1997.

[6] B.J. Baliga, "Gallium Nitride Devices for Power Electronic Applications", Semiconductor Science and Technology, Vol. 28, pp. 1-8, 2013.

[7] A.M. Ozbek and B.J. Baliga, "Planar Nearly Ideal Edge-Termination Technique for GaN Devices", IEEE Electron Device Letters, Vol. 32, pp. 300-302, 2011.

[8] T. Kimoto and J.A. Cooper, "Fundamentals of Silicon Carbide Technology", p. 279, IEEE Press, 2014.

[9] B.J. Baliga and M.S. Adler, "Measurement of Carrier Lifetime Profiles in Diffused Layers of Semiconductors", IEEE Transactions on Electron Devices, Vol. ED-25, pp. 472-477, 1978.

[10] M. Bhatnagar, P.K. McLarty and B.J. Baliga, "Silicon Carbide High Voltage (400V) Schottky Barrier Diodes", IEEE Electron Device Letters, Vol. 13, pp. 501-503, 1992.

[11] R. Raghunathan, D. Alok and B.J. Baliga, "High Voltage 4H-SiC Schottky Barrier Diodes", IEEE Electron Device Letters, Vol. 16, pp. 226-227, 1995.

[12] B.J. Baliga, "Silicon Carbide Power Devices", World Scientific Press, Singapore, 2006.

[13] M. Bhatnagar et al., "Edge Terminations for SiC High Voltage Schottky Rectifiers", IEEE International Symposium on Power Semiconductor Devices and ICs, pp. 89-94, 1993.

[14] B.J. Baliga, "Closed Form Analytical Solutions for the Breakdown Voltage of Planar Junctions Terminated with a Single Floating Field Ring", Solid State Electronics, Vol. 33, pp. 485-488, 1990.

[15] M. Bhatnagar et al., "Edge Terminations for SiC High Voltage Schottky Rectifiers", International Symposium on Power Semiconductor Devices and ICs, pp. 89-94, 1993.

[16] B.J. Baliga and S.K. Ghandhi, "Analytical Solutions for the Breakdown Voltage of Abrupt Cylindrical and Spherical Junctions", Solid State Electronics, Vol. 19, pp. 739-744, 1976.

[17] M.C. Tarplee et al., "Design Rules for Field Plate Edge Termination in SiC Schottky Diodes", IEEE Transactions on Electron Devices, Vol. 48, pp. 2659-2664, 2001.

[18] S. Hu and K. Sheng, "A Study of Oxide Reliability Limitation on different Field Plate based Termination Techniques for SiC Power Devices", International Power Electronics and Motion Control Conference, Vol. 2, pp. 868-872, 2004.

[19] G. Brezeanu et al., "A Nearly Ideal SiC Schottky Barrier Device Edge Termination", International Semiconductor Conference, Vol. 1, pp. 183-186, 1999.

[20] G. Brezeanu *et al.*, "High Performance SiC Diodes based on an Efficient Planar Termination", International Semiconductor Conference, Vol. 1, pp. 27-36, 2003.

[21] S. Sridevan, P.K. McLarty and B.J. Baliga, "Analysis of Gate Dielectrics for SiC Power UMOSFETs", IEEE International Symposium on Power Semiconductor Devices and ICs, pp. 153-156, 1997.

[22] A.S. Kumta, Rusli and X. Jinghua, "Field-Plate-Terminated 4H-SiC Schottky Diodes using Al-Based High-k Dielectrics", IEEE Transactions on Electron Devices, Vol. 56, pp. 2925-2934, 2009.

[23] Y. Huang *et al.*, "Effects of Edge Termination using Dielectric Field Plates with Different Dielectric Constants, Thicknesses, and Bevel Angles", IEEE Applied Power Electronics Conference, pp. 2902-2906, 2014.

[24] D. Alok, B.J. Baliga and P.K. McLarty, "A Simple Edge Termination for Silicon Carbide with Nearly Ideal Breakdown Voltage", IEEE Electron Device Letters, Vol. ED15, pp. 394-395, 1994.

[25] D. Alok, R. Raghunathan and B.J. Baliga, "Planar Edge Termination for 4H-SiC Devices", IEEE Transactions on Electron Devices, Vol. 43, pp. 1315-1317, 1996.

[26] D. Alok, B.J. Baliga, M. Kothandaraman and P.K. McLarty, "Argon Implanted SiC Device Edge Termination: Modeling, Analysis, and Experimental Results", Institute of Physics Conference Series, Vol. 142, pp. 565-568, 1996.

[27] A. Itoh, T. Kimoto and H. Matsunami, "Excellent Reverse Blocking Characteristics of High Voltage 4H-SiC Schottky Rectifiers with Boron Implanted Edge Termination", IEEE Electron Device Letters, Vol. ED17, pp. 139-141, 1996.

[28] R. Weiss, L. Frey and H. Ryssel, "Different Ion Implanted Edge Terminations for Schottky Diodes on SiC", IEEE International Ion Implantation Technology Conference, pp. 139-142, 2002.

[29] D. Alok and B.J. Baliga, "SiC Device Edge Termination using Finite Area Argon Implantation", IEEE Transactions on Electron Devices, Vol. 44, pp. 1013-1017, 1997.

[30] A.P. Knights *et al.*, "The Effect of Annealing on Argon Implanted Edge Terminations for 4H-SiC Schottky Diodes", MRS Symposium Proceedings, Vol. 572, pp. 129-134, 1999.

[31] P.M. Shenoy and B.J. Baliga, "Planar, High Voltage, Boron Implanted 6H-SiC P-N Junction Diodes", Institute of Physics Conference Series, Vol. 142, pp. 717-720, 1996.

[32] R.K. Chilukuri, P. Ananthanarayanan, V. Nagapudi and B.J. Baliga, "High Voltage P-N Junction Diodes in Silicon Carbide using Field Plate Edge Termination", MRS Symposium Proceedings, Vol. 572, pp. 81-86, 1999.

[33] R. Singh and J. W. Palmour, "Planar Terminations in 4H-SiC Schottky Diodes with Low Leakage and High Yields", IEEE International Symposium on Power Semiconductor Devices and ICs, pp. 157-160, 1997.

[34] Y.C. Kao and E.D. Wolley, "High Voltage Planar P-N Junctions", Proceedings of the IEEE, Vol. 55, pp. 1409-1414, 1967.

[35] W. Bahng et al., "Fabrication and Characterization of 4H-SiC pn Diode with Field Limiting Ring", Silicon Carbide and Related Materials – 2003, Materials Science Forum, Vol. 457-460, pp. 1013-1016, 2004.

[36] D.C. Sheridan et al., "Simulation and Fabrication of High Voltage 4H-SiC Diodes with Multiple Floating Guard Ring Termination", Silicon Carbide and Related Materials – 2000, Materials Science Forum, Vol. 338-342, pp. 1339-1342, 2000.

[37] S-C. Kim et al., "Fabrication Characteristics of 1.2kV SiC JBS Diode", IEEE International Conference on Microelectronics, pp. 181-184, 2008.

[38] X. Li et al., "Theoretical and Experimental Study of 4H-SiC Junction Edge Termination", Silicon Carbide and Related Materials – 1999, Materials Science Forum, Vol. 338-342, pp. 1375-1378, 2000.

[39] S-H Ryu et al., "10 kV, 5A 4H-SiC Power DMOSFET", IEEE International Symposium on Power Semiconductor Devices and ICs, pp. 1-4, 2006.

[40] V.A.K. Temple, "Junction Termination Extension: a New Technique for increasing Avalanche Breakdown Voltage and controlling Surface Electric Fields in P-N Junctions", IEEE International Electron Devices Meeting, Abstract 20.4, pp. 423-426, 1977.

[41] R. Rupp et al., "Performance and Reliability Issues of SiC Schottky Diodes", Silicon Carbide and Related Materials – 1999, Materials Science Forum, Vol. 338-342, pp. 1167-1170, 2000.

[42] H.P. Felsl and G. Wachutka, "Performance of 4H-SiC Schottky Diodes with Al-Doped p-Guard-Ring Junction Termination at

Reverse Bias", Silicon Carbide and Related Materials – 2001, Materials Science Forum, Vol. 389-393, pp. 1153-1156, 2002.

[43] C. Raynaud *et al.*, "Design, Fabrication, and Characterization of 5kV 4H-SiC P+ N Planar Bipolar Diodes protected by Junction Termination Extension", Silicon Carbide and Related Materials – 2004, Materials Science Forum, Vol. 457-460, pp. 1033-1036, 2004.

[44] R. Singh *et al.*, "SiC Power Schottky and PiN Diodes", IEEE Transactions on Electron Devices, Vol. ED49, pp. 665-672, 2002.

[45] T. Hiyoshi *et al.*, "Simulation and Experimental Study on the Junction Termination Extension Structure for High-Voltage 4H-SiC PiN Diodes", IEEE Transactions on Electron Devices, Vol. 55, pp. 1841-1846, 2008.

[46] P. Alexandrov *et al.*, "High Performance C plus Al Co-implanted 500 V 4H-SiC PiN Diode", Electronics Letters, Vol. 37, pp. 531-533, 2001.

[47] R. Perez *et al.*, "Planar Edge Termination Design and Technology Considerations for 1.7kV 4H-SiC PiN Diodes", IEEE Transactions on Electron Devices, Vol. 52, pp. 2309-2316, 2005.

[48] W. Sung *et al.*, "A New Edge Termination Technique for High Voltage Devices in 4H-SiC – Multiple Floating Zone Junction Termination Extension", IEEE Electron Device Letters, Vol. 32, pp. 880-882, 2011.

[49] G. Feng, J. Suda and T. Kimoto, "Space-Modulated Junction Termination Extension for Ultrahigh-Voltage p-i-n Diodes in 4H-SiC", IEEE Transactions on Electron Devices, Vol. 59, pp. 414-418, 2012.

[50] H. Miyake *et al.*, "21-kV SiC BJTs with Space-Modulated Junction Termination Extension", IEEE Electron Device Letters, Vol. 33, pp. 1598-1600, 2012.

[51] X. Huang *et al.*, "Orthogonal Positive-Bevel Termination for Chip-Size SiC Reverse Blocking Devices", IEEE Electron Device Letters, Vol. 33, pp. 1592-1594, 2012.

[52] X. Huang *et al.*, "SiC Symmetric Blocking Terminations using Orthogonal Positive Bevel Termination and Junction Termination Extension", IEEE International Symposium on Power Semiconductor Devices and ICs, pp. 179-182, 2013.

[53] W. Sung *et al.*, "Bevel Junction Termination Extension – A New Edge Termination Technique for 4H-SiC High-Voltage Devices", IEEE Electron Device Letters, Vol. 36, pp. 594-596, 2015.

[54] S-C. Lee *et al.*, "A New Vertical GaN Schottky barrier Diode with Floating Metal Ring for High Breakdown Voltage", IEEE

International Symposium on Power Semiconductor Devices and ICs, Paper P-40, pp. 319-322, 2004.

[55] S-C Lee *et al.*, "High Breakdown Voltage GaN Schottky Barrier Diode employing Floating Metal Rings on AlGaN/GaN Heterojunction", IEEE International Symposium on Power Semiconductor Devices and ICs, pp. 247-250, 2005.

[56] A.M. Ozbek and B.J. Baliga, "Planar Nearly Ideal Edge-Termination Technique for GaN Devices", IEEE Electron Device Letters, Vol. 32, pp. 300-302, 2011.

[57] A.M. Ozbek and B.J. Baliga, "Finite Zone Argon Implant Edge-Termination for High-Voltage GaN Schottky Rectifiers", IEEE Electron Device Letters, Vol. 32, pp. 1361-1363, 2011.

Chapter 4
Ideal Specific On-Resistance

All unipolar power device structures, such as Schottky rectifiers and power MOSFETs, contain a drift region designed to support the high blocking voltages. The drift region is an essential component of the device structure without which the power device could not sustain a high voltage in power electronic circuits. The rest of the device structure is required for control of the operating points of the device such as the on-state and blocking mode. This includes the gate structure, source and drain contacts, and substrates used for handling the device wafers during processing. The resistance of these components in the device structure must be minimized by design optimization. In addition, the current flow through the drift region is not uniform making its resistance larger than desired. The ultimate goal is to create a device whose on-resistance becomes close to that of the drift region. The drift region resistance with uniform current flow represents the lowest value that is achievable with traditional one-dimensional potential distribution. This parameter is consequently called the *ideal specific on-resistance ($R_{on,sp}$)*.

Until the 1990s, power devices relied up on parallel-plane junctions with voltage supported across one-dimensional depletion regions. The solution of Poisson's equation for these drift regions indicates a triangular electric field distribution. In the 1990s, the concept of two-dimensional charge coupling was proposed to alter the electric field profile in the drift region to a more uniform distribution. It was demonstrated that the doping concentration in the drift region could be substantially increased with this approach leading to much lower specific on-resistance than possible with the one-dimensional case.

Subsequently, the high electron mobility transistor (HEMT) concept was proposed for gallium nitride power devices. In this approach, a two-dimensional electron gas is created in the gallium nitride drift region by transfer of electrons from an aluminum nitride layer. The very high electron mobility in this structure allows reducing the specific on-resistance despite using a lateral device topology.

This chapter describes the electric field distribution for the conventional one-dimensional and charge-coupled two-dimensional cases. Using this information, the specific on-resistance for the conventional one-dimensional and charge-coupled two-dimensional cases is derived. The analytical solution for the ideal specific on-resistance for the gallium nitride HEMT devices is also derived. These analytical solutions are used to define the lowest possible specific on-resistance in silicon, silicon carbide, and gallium arsenide unipolar power devices. This information is important for a proper understanding of the limits of performance of power devices made from these materials. Actual power device structures can approach but not exceed this performance barrier.

4.1 Ideal Specific On-Resistance for One-Dimensional Case

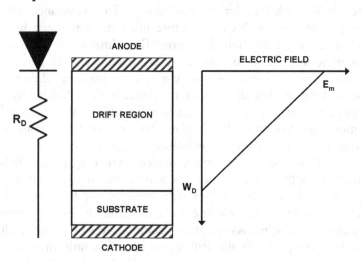

Fig. 4.1 The ideal drift region and its electric field distribution.

Most unipolar power devices contain a drift region which is designed to support the blocking voltage using a one-dimensional potential distribution. The doping concentration and thickness of the *ideal drift region* can be analyzed by assuming an abrupt junction profile with high doping concentration on one side and a low uniform doping concentration on the other side, while neglecting any junction curvature effects by assuming a parallel-plane configuration.

The solution of Poisson's equation leads to a triangular electric field distribution within a uniformly doped drift region as previously discussed in chapter 3 with the slope of the electric field profile being determined by the doping concentration. The maximum voltage that can be supported by the drift region is determined by the maximum electric field (E_m) reaching the critical electric field (E_c) for breakdown for the semiconductor material. The critical electric field for breakdown and the doping concentration then determine the maximum depletion width (W_{PP}).

The specific resistance (resistance per unit area) of the ideal drift region is given by:

$$R_{on.sp} = \left(\frac{W_D}{q \mu_n N_D} \right) \qquad [4.1]$$

Since this resistance was initially considered to be the lowest value achievable with silicon devices, it has historically been referred to as the *ideal specific on-resistance of the drift region*. As derived in chapter 3, the depletion width under breakdown conditions is given by:

$$W_{PP} = \frac{2BV}{E_C} \qquad [4.2]$$

where BV is the desired breakdown voltage. The doping concentration in the drift region required to obtain this breakdown voltage is given by:

$$N_D = \frac{\varepsilon_S E_C^2}{2 q BV} \qquad [4.3]$$

Combining these relationships, the specific resistance of the ideal drift region is obtained:

$$R_{on-ideal} = \frac{4BV^2}{\varepsilon_S \mu_n E_C^3} \qquad [4.4]$$

The denominator of this equation ($\varepsilon_S \mu_n E_C^3$) is commonly referred to as *Baliga's Figure of Merit (BFOM) for Power Devices*:

$$BFOM = \varepsilon_S \mu_n E_C^3 \qquad [4.5]$$

It is an indicator of the impact of the semiconductor material properties on the resistance of the drift region.

Information on impact ionization coefficients and the critical electric field for breakdown in semiconductors was not available when the theoretical analysis was first developed to relate the power device drift region resistance to the properties of the semiconductor material in 1982[1]. It was therefore assumed that the critical electric field is proportional to the energy band gap allowing its substitution for comparison of semiconductors. The *initial Baliga's Figure of Merit (BFOM) proposed for Power Devices* was:

$$BFOM(i) = \varepsilon_S \, \mu_n \, E_G^3 \qquad [4.6]$$

This equation makes it apparent that a larger figure of merit is obtained when the energy band gap of the semiconductor becomes larger. This is the basis for development of power devices from wide band gap semiconductors.

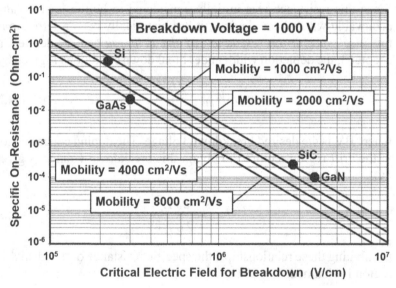

Fig. 4.2 Specific on-resistance of the *Ideal Drift Region*.

The change in the specific on-resistance for the drift region with critical electric field and mobility is shown in Fig. 4.2 for the case of a breakdown voltage of 1000 volts. The cases of mobility values of 1000, 2000, 4000, and 8000 cm²/V-s are shown. The specific on-resistance reduces rapidly with increasing critical electric field for breakdown. It is reduced when the mobility increases. In order to evaluate specific semiconductors, it is important to first determine the doping concentration for the drift region to obtain the breakdown voltage of 1000 volts. Using this doping concentration, the mobility and the critical electric field can be obtained as discussed in chapter 3.

In the case of silicon, the specific on-resistance is very high as shown by the data point in the figure. The dependence of the drift region resistance on the mobility (assumed to be for electrons here because in general they have higher mobility values than for holes) of the carriers favors semiconductors such as Gallium Arsenide whose electron mobility is 8000 cm^2/V-s. Its specific on-resistance is about 20-times smaller than for silicon as indicated by the data point in Fig. 4.2. This significant improvement was pursed at GE in the 1980s resulting in GaAs power Schottky rectifiers[2] and FETs[3,4]. These were the first power device structures demonstrated using wide band gap material.

The strong (cubic) dependence of the on-resistance on the critical electric field for breakdown favors semiconductors with even wider band gap than GaAs. This can be achieved by ternary alloys such as GaAlAs[1]. The impact is even greater with silicon carbide and gallium nitride due to their larger band gaps. The critical electric field for breakdown for these semiconductors as determined by the impact ionization coefficients for holes and electrons was discussed in the chapter 3. The location of the data points for 4H-SiC and GaN are shown in Fig. 4.2. The specific on-resistance of the drift region for SiC and GaN is 100-times lower than that for GaAs. 4H-SiC and GaN are 2000-times and 4000-times better than silicon. The design and development of SiC and GaN power devices is discussed in the rest of this book.

4.2 Ideal Specific On-Resistance for Silicon Carbide

The ideal specific on-resistance for the drift region in 4H-SiC power devices can be obtained by using Eq. [4.1] with the doping concentration and depletion layer width at breakdown derived in chapter 3. Due to the relatively high doping concentration for the drift region in 4H-SiC, it is important to include the dependence of mobility on doping level as discussed in chapter 2.

The doping concentration of the drift region can be related to the breakdown voltage by using Eq. [3.12]:

$$N_D = \frac{1.98 \times 10^{20}}{BV^{4/3}}$$
[4.7]

The depletion layer width can be related to the breakdown voltage by using Eq. [3.15]:

$$W_{PP} = 2.33 \times 10^{-7} BV^{7/6}$$
[4.8]

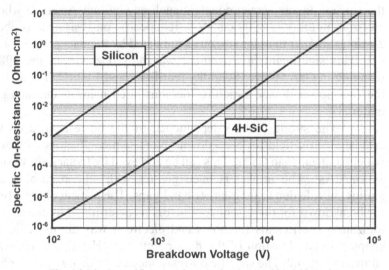

Fig. 4.3 Ideal specific on-resistance for the 4H-SiC drift region.

The ideal specific on-resistance for the drift region in 4H-SiC devices can be computed by using the above equations together with the doping dependence of mobility given by Eq. [2.22]. The results are shown in Fig. 4.3 for breakdown voltages from 100 to 100,000 volts. The case of silicon drift regions is included in this figure for comparison. A significant reduction in the specific on-resistance of drift regions is predicted by replacing silicon with 4H-SiC. The ratio of the specific on-resistance for silicon to that for 4H-SiC increases from 527 at a breakdown voltage of 100 volts to 1280 for breakdown voltages above 40,000 volts.

4.3 Ideal Specific On-Resistance for Gallium Nitride

The ideal specific on-resistance for the drift region in GaN power devices can also be obtained by using Eq. [4.1] with the doping concentration and depletion layer width at breakdown derived in chapter 3. Due to the relatively high doping concentration for the drift region in GaN, it is important to include the dependence of mobility on doping level as discussed in chapter 2.

The doping concentration of the drift region can be related to the breakdown voltage by using Eq. [3.13]:

$$N_D = \frac{3.98 \times 10^{20}}{BV^{4/3}} \qquad [4.9]$$

The depletion layer width can be related to the breakdown voltage by using Eq. [3.16]:

$$W_{PP} = 1.70 \times 10^{-7} \, BV^{7/6} \qquad [4.10]$$

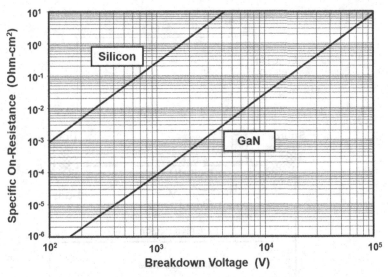

Fig. 4.4 Ideal specific on-resistance for the GaN drift region.

The ideal specific on-resistance for the drift region in GaN devices can be computed by using the above equations together with the doping dependence of mobility given by Eq. [2.23]. The results are shown in Fig. 4.4 for breakdown voltages from 100 to 100,000 volts. The case of silicon drift regions is included in this figure for comparison. A significant reduction in the specific on-resistance of drift regions is predicted by replacing silicon with GaN. The ratio of the specific on-resistance for silicon to that for GaN increases from

2230 at a breakdown voltage of 100 volts to 3110 for breakdown voltages above 5000 volts.

4.4 Ideal Specific On-Resistance for Gallium Nitride HEMT

The heterojunction high electron mobility field effect transistor (HEMT) structure made by the growth of an aluminum-gallium nitride layer on gallium nitride is an attractive concept for creating transistors with low specific on-resistances. In this approach, the drift region contains a layer of electrons at the interface between the AlGaN and GaN layers with a low sheet resistance as described in chapter 2. These lateral transistors can consequently have very low specific on-resistances.

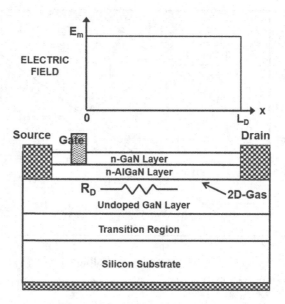

Fig. 4.5 Lateral GaN HEMT structure with ideal electric field distribution.

The ability to grow sufficiently high quality GaN layers on silicon substrates by using transition layers to account for the lattice mismatch has allowed formation of the lateral GaN-on-Si HEMT structure shown in Fig. 4.5. This approach is being pursued by many companies and will be discussed in depth in a later chapter. In this section, the lowest possible resistance that can be achieved for this concept is analyzed.

A simple analysis of the ultimate performance for the lateral HEMT structure can be performed by assuming an idealized uniform electric field distribution along the x-axis between the edge of the gate and the drain. The structure undergoes breakdown when the maximum electric field becomes equal to the critical electric field ($E_{C,L}$) for breakdown in a structure with uniform electric field along the drift region. The length of the drift region is given by:

$$L_D(HEMT) = \frac{BV}{E_{C,L}} \quad \text{[4.11]}$$

In order to obtain the ideal specific on-resistance for a lateral HEMT structure, only the resistance of the drift region will be taken into account while neglecting the space occupied by the source, gate and drain contacts. The on-resistance of the drift region is then given by:

$$R_{ON}(HEMT) = \frac{L_D}{q\,\mu\,Q_S\,Z} \quad \text{[4.12]}$$

where μ is the free carrier mobility in the 2D-gas, Q_S is the sheet carrier density, and Z is the width of the structure orthogonal to the cross-section. The specific on-resistance for the lateral HEMT structure is obtained by multiplying the on-resistance by the area ($L_D.Z$):

$$R_{ON,SP}(HEMT) = \frac{L_D^2}{q\,\mu\,Q_S} \quad \text{[4.13]}$$

Using Eq. [4.11] yields:

$$R_{ON,SP}(HEMT) = \frac{BV^2}{q\,\mu\,Q_S\,E_{C,L}^2} \quad \text{[4.14]}$$

The denominator of this equation serves as a figure-of-merit for lateral HEMT structures:

$$BFOM(HEMT) = q\,\mu\,Q_S\,E_{C,L}^2 \quad \text{[4.15]}$$

The critical electric field for breakdown for the lateral HEMT structure is a function of the breakdown voltage of the structure (as is also the case for the vertical power FET structures). The critical electric field for breakdown for the lateral HEMT structure can be derived by performing the ionization integral with a uniform electric field in the drift region. Equating the ionization integral to unity yields:

$$E_{C,L} = \left(\frac{6.667 \times 10^{41}}{BV} \right)^{1/6}$$ [4.16]

The solution indicates that the critical electric field for breakdown of a lateral HEMT structure will decrease with increasing breakdown voltage. This is due to the longer impact ionization path along the drift region with increasing breakdown voltage. Combining Eq. [4.16] with Eq. [4.11]:

$$L_D(HEMT) = 1.07 \times 10^{-7} \, BV^{7/6}$$ [4.17]

Substituting this expression into Eq. [4.13]:

$$R_{ON,SP}(HEMT) = \frac{7.154 \times 10^4 \, BV^{7/3}}{\mu \, Q_S}$$ [4.18]

The specific on-resistance for the HEMT structure can also be expressed using the sheet resistance of the two-dimensional gas layer:

$$R_{ON,SP}(HEMT) = 1.145 \times 10^{-14} \, \rho_{S,2D} \, BV^{7/3}$$ [4.19]

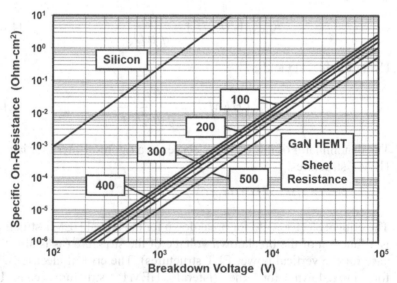

Fig. 4.6 Ideal specific on-resistance for the GaN HEMT drift region.

The ideal specific on-resistance for the drift region in GaN HEMT devices can be computed by using the above equation. The results are shown in Fig. 4.6 for breakdown voltages from 100 to 100,000 volts. Sheet resistances for the 2D-gas layer are used as parametric variables in the figure. For the typical sheet carrier density of 1.5×10^{13} cm^{-2} with an electron mobility of 1389 cm^2/Vs reported in the 2D-gas for the AlGaN/GaN HEMT structures[5], the sheet resistance is 300 ohms/sq. The case of silicon drift regions is included in this figure for comparison. A significant reduction in the specific on-resistance of drift regions is predicted by replacing silicon with GaN HEMT structures. The ratio of the specific on-resistance for silicon to that for GaN HEMT with a 2D-gas sheet resistance of 300 ohms/sq increases from 5390 at a breakdown voltage of 100 volts to 16,500 for breakdown voltage of 100,000 volts.

4.5 Ideal Specific On-Resistance for Charge Coupled Si Devices

The recent introduction of the charge-coupling concept has enabled reducing the specific resistance of the drift region for silicon devices to values well below that of the 'ideal specific on-resistance' given by Eq. [4.4]. The analysis of the specific on-resistance for the charge coupled devices is provide here. These silicon devices are the new bench mark of performance that must be exceeded by the wide band gap semiconductor based power devices.

In the case of traditional silicon devices designed with one-dimensional potential distribution, the specific on-resistance of the devices becomes limited by the boundary set by the ideal specific on-resistance. An innovative approach to break this barrier was proposed based on using two-dimensional charge coupling. The first device concept was based up on using an electrode embedded inside an oxide coated trench etched into the drift region[6,7] to create the CC-MOSFET and GD-MOSFET structures. The second concept is based up on charge coupling between adjacent vertical pillars of N and P-type drift regions to create the COOLMOS structure[8].

4.5.1 Silicon GD-MOSFET Structure

The CC-MOSFET and GD-MOSFETs are discussed in detail in a previous book[9]. The GD-MOSFET structure is illustrated in Fig. 4.7

with its doping profile. It was found that the electric field along the y-direction is not constant if a uniform doping concentration is used for the drift region[9]. A constant electric field along the y-direction can be achieved by using a linearly graded doping profile[6]. Only this case is analyzed here because it allows scaling the breakdown voltages to large values.

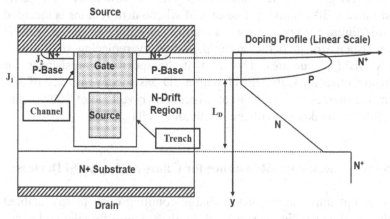

Fig. 4.7 GD-MOSFET structure and doping profile.

Without the application of a gate bias, a high voltage can be supported in the GD-MOSFET structure when a positive bias is applied to the drain. In this case, junction J_1 formed between the P-base region and the N-drift region becomes reverse biased. Simultaneously, the drain voltage is applied across the vertical MOS structure formed between the electrode in the deep trenches and the N-drift region. The MOS structure operates in the deep-depletion mode due to the presence of the reverse bias across junction J_1 between the P-base region and the N-drift junction. Consequently, depletion regions are formed across the horizontal junction J_1 and the vertical trench sidewall. This two-dimensional depletion alters the electric field distribution from the triangular shape observed in conventional parallel-plane junctions to a uniform distribution due to the linear doping profile. This allows supporting a required blocking voltage over a shorter distance. In addition, the doping concentration in the drift region can be made much greater that predicted by the one-dimensional theory. This allows very substantial reduction of the specific on-resistance to well below the ideal specific on-resistance at any desired breakdown voltage.

Drain current flow in the GD-MOSFET structure is induced by the application of a positive bias to the gate electrode. This produces an inversion layer at the surface of the P-base region along the trench sidewalls. The threshold voltage for the power GD-MOSFET structure can be controlled by adjusting the dose for the boron ion-implantation for the P-base region. The inversion layer channel provides a path for transport of electrons from the source to the drain when a positive drain voltage is applied. After transport from the source region through the channel, the electrons enter the N-drift region and are then transported through the mesa region to the N^+ substrate. The resistance of the drift region is very low in the power GD-MOSFET structure due to the high doping concentration in the mesa region. The channel resistance in the power GD-MOSFET structure is also very small due to the small cell pitch or high channel density. The input capacitance and gate charge for power MOSFET structures must be reduced to enhance their switching performance. These device parameters can be reduced by using a source connected electrode in the trenches adjacent to the drift region as illustrated in Fig. 4.7[10].

When the drain bias reaches the breakdown voltage, the maximum electric field in the semiconductor along the y-direction becomes equal to the critical electric field (E_{CU}) for breakdown of the semiconductor with uniform electric field profile. The critical electric field at which breakdown occurs can be derived for this field distribution by using the criterion that the ionization integral becomes unity at breakdown:

$$\int_0^{L_D} \alpha \, dx = 1 \qquad\qquad [4.20]$$

Using Baliga's formula for the impact ionization coefficient for silicon (see Eq. [3.8]), an expression for the critical electric field for the case of a uniform (or constant) electric field is obtained:

$$E_{CU} = 8.36 \times 10^4 \, L_D^{-1/7} \qquad\qquad [4.21]$$

where L_D is the length of the drift region as shown in Fig. 4.7. The breakdown voltage in this case is then given by:

$$BV = E_{CU} \, L_D = 8.36 \times 10^4 \, L_D^{6/7} \qquad\qquad [4.22]$$

by using Eq. [4.21].

The breakdown voltage for charge-coupled devices with a uniform electric field increases non-linearly with increasing length of the drift region as shown in Fig. 4.8 by the solid line. Using the analytical model, the drift region lengths required to achieve breakdown voltages of 50, 100, 200, 500, and 1000 volts are predicted to be 1.74, 3.9, 8.8, 25.5, and 57.2 microns, respectively. Note that the drift region length for the power GD-MOSFET structure is the length of the source electrode within the trenches as indicated in Fig. 4.7 and not the trench depth because the two-dimensional charge coupling occurs only between the source electrode and the drift region.

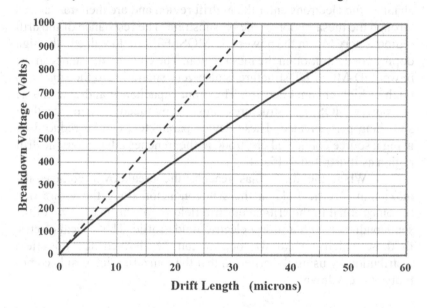

Fig. 4.8 Breakdown voltages for GD-MOSFET devices.

To compute the specific on-resistance for the power GD-MOSFET structure, it is necessary to derive a relationship for the doping profile in the drift region which will result in a uniform electric field profile in the y-direction for the two-dimensional structure[11]. Since the doping profile is optimized to achieve a uniform electric field along the y-direction, the voltage increases linearly with distance along the y-direction:

$$V(y) = E_Y\, y \qquad\qquad [4.23]$$

In the power GD-MOSFET structure, breakdown occurs by impact ionization at the middle of the mesa region (and not near the trench surface). At the on-set of breakdown, the electric field in the y-direction at the middle of the mesa region becomes equal to the critical electric field (E_{CU}) for breakdown in the uniform electric field case.

The electric field is uniform in the oxide and varies linearly along the x-direction in the semiconductor because the doping concentration is constant along the x-direction[8]. The voltage at any depth in the drift region is then given by:

$$V(y) = V_{OX} + V_S = E_X \left[\frac{\varepsilon_{Si}}{\varepsilon_{OX}} t_{TOX} + \frac{W_M}{4} \right]$$ [4.24]

where E_X is the electric field in the x-direction at the interface between the oxide and the semiconductor as shown in the figure. Applying Gausses law at any depth in the drift region:

$$\varepsilon_{Si} E_X = q N_D(y) \frac{W_M}{2}$$ [4.25]

The optimum doping profile to achieve a uniform electric field in the y-direction is obtained by combining the above relationships:

$$N_D(y) = \frac{2\varepsilon_{Si} E_{CU}}{q W_M \left[\frac{\varepsilon_{Si}}{\varepsilon_{OX}} t_{T,OX} + \frac{W_M}{4} \right]} y$$ [4.26]

It can be seen that the predicted optimum doping profile has a linear distribution along the y-direction.

The gradient (or slope) of the optimum doping profile is given by:

$$G = \frac{2\varepsilon_{Si} E_{CU}}{q W_M \left[\frac{\varepsilon_{Si}}{\varepsilon_{OX}} t_{TOX} + \frac{W_M}{4} \right]}$$ [4.27]

The optimum doping gradient is a function of the trench oxide thickness and the mesa width. In addition, the critical electric field is a function of the breakdown voltage. It is prudent to keep the maximum electric field in the trench oxide at less than 2×10^6 V/cm in order to avoid reliability problems. Based up on this criterion, the trench oxide thickness must be scaled with the desired blocking voltage capability. Using this electric field in the trench oxide, the trench oxide thicknesses for breakdown voltages of 50, 100, 200, 500, and 1000 volts are found to be 2500, 5000, 7500, 10000 and 20000 angstroms, respectively. With these values, the optimum doping gradients predicted by the analytical model for breakdown voltages of 50, 100, 200, 500, and 1000 volts are found to be 8.4, 4.04, 1.87, 0.66 and 0.30 × 10^{20} cm^{-4}, respectively, for a mesa width of 0.5 microns.

A larger optimum doping gradient is predicted for power GD-MOSFET structures with lower breakdown voltages. This is favorable for reducing the specific on-resistance of the drift region. Even in the case of power GD-MOSFET structures with larger breakdown voltages, the doping concentration is high in the vicinity of the N$^+$ substrate despite the smaller doping gradient because the drift region length is longer. As a typical example, the doping concentration for the power GD-MOSFET structure with breakdown voltage of 50 volts increases from 1 × 10^{16} cm^{-3} near the P-base/N-drift junction to 2 × 10^{17} cm^{-3} near the bottom of the trench. Such high doping concentrations are also required for the higher voltage power GD-MOSFET structures. These high doping levels result in very low specific on-resistance for the power GD-MOSFET structures.

The ideal specific on-resistance for the GD-MOSFET structure will be defined as the resistance contributed by only the drift region. In the mesa portion of the structure, the current density is uniform with a current density enhanced by the ratio of the cell pitch to the mesa width. The drift region resistance contribution from the mesa region can be computed by considering a small segment (dy) of the drift region at a depth y from the bottom of the gate electrode. The specific resistance of the drift region for the mesa portion is given by:

$$R_{on-ideal}(GDM) = \left(\frac{W_{Cell}}{W_M}\right)\int_0^{L_D} \rho_D(y)\,dy \qquad [4.28]$$

where the resistivity ρ_D is a function of the position in the drift region due to the graded doping profile. The resistivity of the drift region is given by:

$$\rho_D(y) = \frac{1}{q\mu_n(y)N_D(y)} \qquad [4.29]$$

The dependence of the electron mobility on the doping concentration must be taken into account when analyzing the drift region resistance in the power GD-MOSFET structure because the doping concentration exceeds 1 × 10^{16} cm^{-3}. The electron mobility is given by[12]:

$$\mu_n = \frac{5.1 x 10^{18} + 92\, N_D^{0.91}}{3.75 x 10^{15} + N_D^{0.91}} \qquad \text{[4.30]}$$

In order to simplify the analysis, the following approximation for the dependence of the electron mobility on doping concentration is adequate for the doping levels ranging between 1×10^{16} cm^{-3} and 3×10^{17} cm^{-3}:

$$\mu_n = \frac{5.1 x 10^{18}}{3.75 x 10^{15} + N_D^{0.91}} \qquad \text{[4.31]}$$

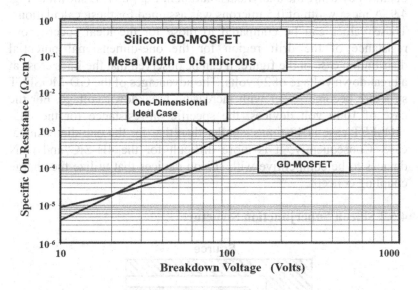

Fig. 4.9 Ideal specific on-resistance for the power GD-MOSFET structure.

Using Eq. [4.29] and Eq. [4.31] in Eq. [4.28]:

$$R_{on-ideal}(GDM) = \left(\frac{W_{Cell}}{W_M}\right) \int_0^{L_D} \left[\frac{4.6 x 10^{15}}{N_D(y)} + \frac{1.225}{N_D(y)^{0.09}}\right] dy \qquad \text{[4.32]}$$

For a linearly graded doping profile with a gradient G:

$$N_D(y) = N_0 + G y \qquad \text{[4.33]}$$

where N_0 is the initial doping concentration in the mesa region at the depth of the gate electrode. Integrating Eq. [4.32] with Eq. [4.33] yields:

$$R_{on-ideal}(GDM) = \left(\frac{W_{Cell}}{W_M}\right)\left[\begin{array}{l}\dfrac{4.6 \times 10^{15}}{G}\ln\left(\dfrac{N_0 + GL_D}{N_0}\right) + \\[3mm] \dfrac{1.346}{G}\left[(N_0 + GL_D)^{0.91} - N_0^{0.91}\right]\end{array}\right]$$

[4.34]

The specific on-resistance for the GD-MOSFET drift region obtained by using the analytical solution in Eq. [4.34] is shown in Fig. 4.9. A mesa width of 0.5 microns was assumed for these calculations. These values can be compared to those for the ideal specific on-resistance of the drift region for the one-dimensional potential distribution case in the figure. It can be observed that the lines cross at a breakdown voltage of 20 volts. The advantages of the GD-MOSFET structure become larger with increasing breakdown voltage. For the case of 1000 volt devices, the specific on-resistance for the GD-MOSFET case is 43-times smaller than that for the one-dimensional case. Consequently, it is more difficult for the 4H-SiC and GaN devices to be competitive with silicon devices that utilize the charge coupling concept.

4.5.2 Silicon Superjunction Structure

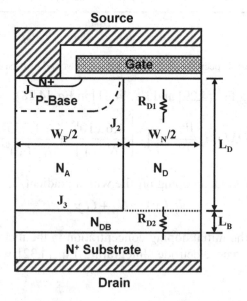

Fig. 4.10 Resistances in the SJ-MOSFET structure.

A cross-section of the super-junction power MOSFET structure is shown in Fig. 4.10. The structure was first commercialized as the COOLMOS device by Infineon[8]. The drift region of this structure includes a vertical P-type column which produces a vertically oriented junction J_2. A much larger doping concentration can be employed in the drift region of the power SJ-MOSFET structure when compared with the conventional MOSFET structures. This enables a significant reduction of the specific on-resistance for devices capable of supporting high voltages.

The doping concentration in the drift region is determined by two-dimensional charge coupling for the power SJ-MOSFET structure. In order to achieve good charge coupling, the N-type and P-type drift regions must be completely depleted when the drain bias approaches the breakdown voltage. The electric field then becomes uniform along the y-direction in the N-type and P-type drift regions. The super-junction concept has been found to be particularly effective for devices with breakdown voltages between 500 and 1000 volts.

Based up on the depletion of the N-type and P-type drift regions when the electric field at junction J_2 becomes equal to the critical electric field (E_{CU}) for breakdown for the uniform electric field distribution case:

$$Q_{Optimum} = qN_D \frac{W_N}{2} = \varepsilon_S E_{CU} = qN_A \frac{W_P}{2} \qquad [4.35]$$

where W_N and W_P are the widths of the N-type and P-type drift regions, respectively; N_D and N_A are the doping concentration of the N-type and P-type drift regions, respectively. This provides the criterion for choosing the dopant dose (product of doping concentration and thickness) in the N-drift region to achieve the desired two-dimensional charge coupling:

$$N_D W_N = N_A W_P = \frac{2\varepsilon_S E_{CU}}{q} \qquad [4.36]$$

Using Baliga's formula for the impact ionization coefficient for silicon, an expression for the critical electric field for the case of a uniform (or constant) electric field can be derived (see Eq. [4.21]). The breakdown voltage for the two dimensional charge coupled device is then given by Eq. [4.22]. The breakdown voltage for devices with a uniform electric field is plotted in Fig. 4.8. A drift region length of 25 microns is required to achieve a breakdown voltage of 500 volts. According to the analytical model, the drift region length must be increased to 57 microns to obtain a breakdown voltage of 1000 volts.

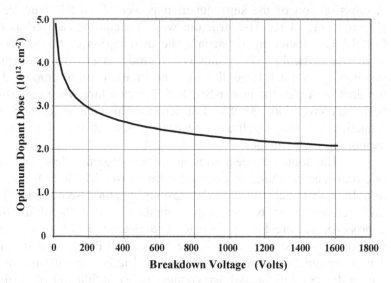

Fig. 4.11 Optimum dose for super-junction devices.

Combining Eq. [4.21], Eq. [4.22] and Eq. [4.36], yields:

$$N_D W_N = 1.106 \times 10^6 \frac{\varepsilon_S}{q} BV^{-1/6} \qquad [4.37]$$

The optimum dose predicted by this expression is shown in Fig. 4.11 as a function of the breakdown voltage. It can be observed that the optimum dose decreases gradually with increasing breakdown voltage. The optimum dose for the N-type and P-type drift regions predicted by the analytical model is 2.54×10^{12} and 2.27×10^{12} cm^{-2} for breakdown voltages of 500 and 1000 volts, respectively.

    The lowest possible specific on-resistance that can be achieved by using the super-junction concept can be obtained by neglecting the contributions from the channel, accumulation, and JFET regions, i.e. by analysis of only the drift region resistance. The drift region resistance can be analyzed with the two components R_{D1} for the N-type drift region and R_{D2} for the N-buffer layer as indicated in Fig. 4.10. For high voltage power SJ-MOSFET structures, the contribution from the N-buffer layer is much smaller than that from the N-drift region. The ideal specific on-resistance for the super-junction devices is then given by:

$$R_{on-ideal}(SJ) = R_{D1.sp} = \rho_{ND}L_D\left(\frac{W_{Cell}}{W_N}\right) = \frac{L_D}{q\mu_N N_D}\left(\frac{W_N + W_P}{W_N}\right) \qquad [4.38]$$

where ρ_{ND} is the resistivity of the N-type drift region. The length of the drift region can be related to the breakdown voltage of the super-junction device:

$$L_D = \frac{BV}{E_{CU}} \qquad [4.39]$$

because the electric field along the y-direction has a constant value equal to the critical electric for breakdown with uniform electric field in the case of super-junction devices. Using Eq. [4.36], the optimum doping concentration for super-junction devices is given by:

$$N_D = \frac{2\varepsilon_s E_{CU}}{qW_N} \qquad [4.40]$$

Combining the above relationships:

$$R_{on-ideal}(SJ) = \frac{BV}{\varepsilon_s \mu_N E_{CU}^2}\left(\frac{W_N + W_P}{2}\right) \qquad [4.41]$$

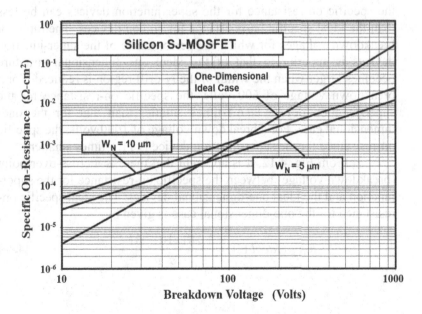

Fig. 4.12 Ideal specific on-resistance for super-junction devices.

The critical electric field for breakdown with uniform electric field is related to the breakdown voltage by Eq. [4.22]:

$$R_{on\text{-}ideal}(SJ) = \frac{1.635 \times 10^{-12} BV^{4/3}(W_N + W_P)}{\varepsilon_S \mu_N}$$ [4.42]

It is typical to use the same width for the P-type and N-type drift regions in super-junction devices. In this case, the ideal specific on-resistance for super-junction devices can be computed using:

$$R_{on\text{-}ideal}(SJ) = \frac{3.27 \times 10^{-12} BV^{4/3} W_N}{\varepsilon_S \mu_N}$$ [4.43]

The ideal specific on-resistance for super-junction devices computed by using Eq. [4.43] is provided in Fig. 4.12 for the case of two widths for the N-type drift region under the assumption that the width of the P-type drift region has the same value. Since the optimum doping concentration is relatively high for these devices, the dependence of the mobility on doping concentration was included during this analysis. The ideal specific on-resistance for the drift region in one-dimensional devices as computed by using Baliga's power law for the impact ionization coefficients is also shown in Fig. 4.12 for comparison purposes. It can be observed from this figure that the specific on-resistance for the super-junction devices can be less than the ideal specific on-resistance for silicon devices. The range of breakdown voltages for which the performance of the super-junction devices is superior to that of the ideal one-dimensional structure becomes larger when the width of the N-drift region is reduced. For a breakdown voltage of 600 volts, the specific on-resistance for the super-junction device is 6.1-times smaller than that for the one-dimensional case. For a breakdown voltage of 1000 volts, the specific on-resistance for the super-junction device is 11.1-times smaller.

From Fig. 4.12, it is apparent that there is a cross-over point (breakdown voltage) between the specific on-resistance for the super-junction and the ideal one-dimensional devices. The ideal specific on-resistance for the one-dimensional case is given by:

$$R_{ON,sp}(Ideal\ 1D) = \frac{1.181 \times 10^{-17} BV^{5/2}}{\varepsilon_S \mu_N}$$ [4.44]

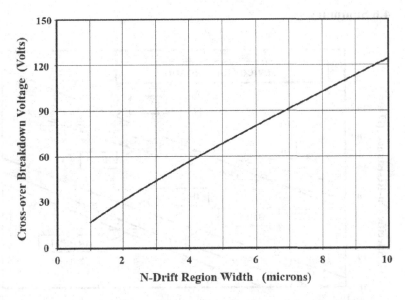

Fig. 4.13 Cross-over breakdown voltage for the super-junction devices.

An analytical expression for cross-over point can be derived by equating the specific on-resistance for the one-dimensional case to the ideal specific on-resistance for the super-junction devices if the dependence of the mobility on doping concentration is neglected. The cross-over breakdown voltage is then given by:

$$BV\,(Cross - Over) = 4.62 \times 10^4 W_N^{6/7} \qquad \text{[4.45]}$$

The cross-over breakdown voltage computed using this equation is provided in Fig. 4.13. The cross-over breakdown voltage predicted by the above equation is in good agreement with the cross-over points in Fig. 4.12 indicating the assumption of a constant mobility is reasonable. It can be observed that the cross-over breakdown voltage increases with increasing width of the N-drift region. It is therefore advantageous to utilize small widths for the drift region to maximize the performance of the super-junction devices. However, the presence of a depletion region across the vertical junction (J_3) can create an increase in the on-resistance for super-junction devices[13]. This has an adverse impact on the specific on-resistance.

4.6 Summary

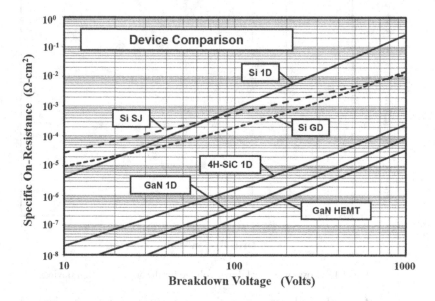

Fig. 4.14 Comparison of ideal specific on-resistance for various devices.

The ideal specific on-resistance for the drift region in power devices has been defined for one-dimensional potential distribution and related to the semiconductor material properties. This relationship indicates that semiconductors with wider band gaps, or more precisely semiconductors with larger critical electric fields for breakdown, will have substantially smaller drift region resistances. The improved performance that is achievable by replacing silicon with silicon carbide and gallium nitride has been quantified. The drift region resistance can be substantially reduced for silicon devices by taking advantage of two-dimensional charge coupling. The specific on-resistance of the GD-MOSFET and SJ-MOSFET that utilize this principle has been provided in this chapter for comparison with the 4H-SiC and GaN devices. In addition, the ideal specific on-resistance for the GaN HEMT structure has been defined and quantified as a function of the breakdown voltage.

The ideal specific on-resistance for the various devices can be compared in Fig. 4.14 for breakdown voltages ranging from 10 to 1000 volts. For devices with breakdown voltages below 100 volts, the silicon devices have excellent performance. Although the ideal specific on-resistance for the 4H-SiC and GaN devices is much superior to that for the silicon structures at breakdown voltages below

100 volts, it is very difficult to exceed the performance for the silicon devices when the resistances of other components in the device structures are taken into account. At higher breakdown voltages of 500 to 1000 volts, the ideal specific on-resistance for the 4H-SiC and GaN devices is more than two orders of magnitude better than for the traditional silicon devices. However, this margin is reduced by a factor of 10 due to the development of the silicon devices that utilize charge coupling concept.

References

[1] B. J. Baliga, "Semiconductors for High Voltage Vertical Channel Field Effect Transistors", J. Applied Physics, Vol. 53, pp. 1759-1764, 1982.

[2] B. J. Baliga et al., "Gallium Arsenide Schottky Power Rectifiers", IEEE Transactions on Electron Devices, Vol. ED-32, pp. 1130-1134, 1985.

[3] P.M. Campbell et al., "150 Volt Vertical Channel GaAs FET", IEEE International Electron Devices Meeting, Abstract 10.4, pp. 258-260, 1982.

[4] P.M. Campbell et al., "Trapezoidal Groove Schottky Gate Vertical Channel GaAs FET", IEEE International Electron Devices Meeting, Abstract 7.3, pp. 186-189, 1984.

[5] H-S. Lee et al., "3000-V 4.3 mW.cm2 InAlN/GaN MOSHHEMTs with AlGaN Back Barrier", IEEE Electron Device Letters, Vol. 33, pp. 982-984, 2012.

[6] B.J. Baliga, "Vertical Field Effect Transistors having Improved Breakdown Voltage Capability and Low On-state Resistance", U.S. Patent # 5,637,898, Issued June 10, 1997.

[7] B.J. Baliga, "Trends in Power Discrete Devices", IEEE International Symposium on Power Semiconductor Devices and ICs, Abstract P-2, pp. 5-10, 1997.

[8] L. Lorenz et al., "COOLMOS – A New Milestone in High Voltage Power MOS", IEEE International Symposium on Power Semiconductor Devices and ICs, Abstract P-1, pp. 3-10, 1999.

[9] B.J. Baliga, "Advanced Power MOSFET Concepts", Chapter 5, Springer-Science, New York, 2010.

[10] B.J. Baliga, "Power Semiconductor Devices having Improved High Frequency Switching and Breakdown Characteristics", U.S. Patent # 5,998,833, Issued December 7, 1999.

[11] S. Mahalingam and B.J. Baliga, "The Graded Doped Trench MOS Barrier Schottky Rectifier: a Low Forward Drop High Voltage Rectifier", Solid State Electronics, Vol. 43, pp. 1-9, 1999.

[12] B.J. Baliga, "Fundamentals of Power Semiconductor Devices", Springer-Science, 2008.

[13] D. Disney and G. Dolny, "JFET Depletion in Super-Junction Devices", IEEE International Symposium on Power Semiconductor Devices and ICs", pp. 157-160, 2008.

Chapter 5

Schottky Rectifiers

The main advantage of wide band gap semiconductors for power device applications is their very low resistance for the drift region even when designed to support large voltages. This favors the development of high voltage unipolar devices which have much superior switching speed than bipolar structures. The Schottky rectifier, formed by making a rectifying contact been a metal and the semiconductor drift region, is an attractive unipolar device for power applications. In the case of silicon, the maximum breakdown voltage of Schottky rectifiers has been limited by the increase in the resistance of the drift region[1]. Commercially available silicon devices are generally rated at breakdown voltages of less than 100 volts. Novel silicon structures that utilize the charge-coupling concept have allowed extending the breakdown voltage to the 200 volt range[2,3].

Many applications described in chapter 1 require fast switching rectifiers with low on-state voltage drop that can also support over 500 volts. The much lower resistance of the drift region for silicon carbide and gallium nitride enables development of such Schottky rectifiers with very high breakdown voltages. These devices offer fast switching speed and eliminate the large reverse recovery current observed in high voltage silicon P-i-N rectifiers. This reduces switching losses not only in the rectifier but also in the IGBTs used within the power circuits[4].

This chapter describes the characteristics of 4H-SiC and GaN Schottky rectifiers. Unique issues that relate to these wide band gap semiconductors must be given special consideration. The much greater electric field in these semiconductors leads to larger Schottky barrier lowering with increasing reverse voltage when compared with silicon devices. The high doping level in the drift region leads to tunneling currents at the Schottky barrier that enhances the leakage current as well. Experimental results on relevant structures are provided to define the state of the development effort on 4H-SiC and GaN Schottky rectifiers.

5.1 Schottky Rectifier Structure: Forward Conduction

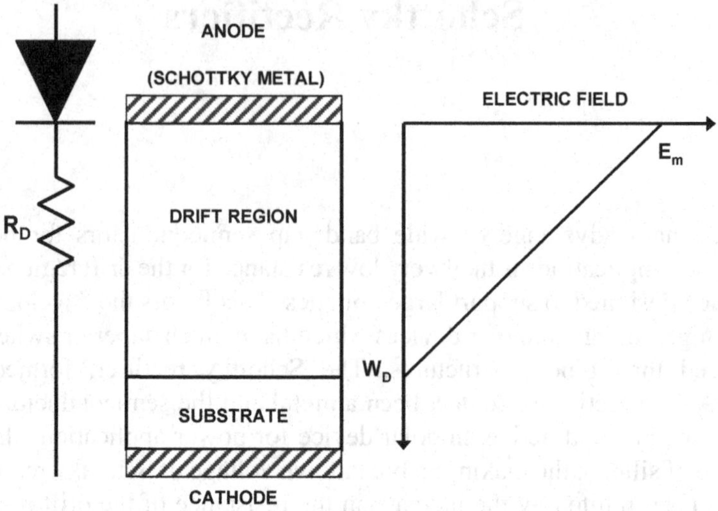

Fig. 5.1 Structure and electric field in a Schottky rectifier.

The basic one-dimensional structure of the metal-semiconductor or Schottky rectifier structure is shown in Fig. 5.1 together with electric field profile under reverse bias operation. The applied voltage is supported by the drift region with a triangular electric field distribution if the drift region doping is uniform. The maximum electric field occurs at the metal contact. The device undergoes breakdown when this field becomes equal to the critical electric field. As discussed in chapter 4, the specific on-resistance of the drift region is given by:

$$R_{on-ideal} = \frac{4BV^2}{\varepsilon_S \mu_n E_C^3} \tag{5.1}$$

The specific on-resistance of the drift region (R_{drift}) for 4H-SiC is approximately 500 to 1280 times smaller than for silicon devices for the same breakdown voltage as shown earlier in Fig. 4.3.

The on-state voltage drop for the Schottky rectifier at a forward current density J_F, including the substrate contribution (R_{subs}), is given by:

$$V_F = \frac{kT}{q} \ln\left(\frac{J_F}{J_S}\right) + \left(R_{drift} + R_{subs}\right)J_F \tag{5.2}$$

where J_S is the saturation current density. In Eq. [5.4], the first term accounts for the voltage drop across the metal-semiconductor contact while the second term accounts for the voltage drop across the series resistance (shown as R_D in Fig. 5.1).

5.1.1 Saturation Current

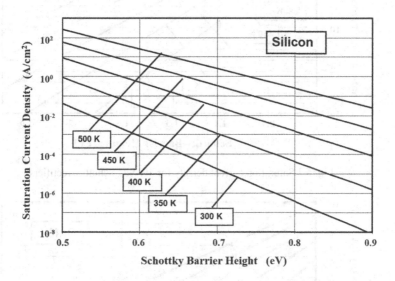

Fig. 5.2 Saturation current density for Silicon Schottky rectifiers.

The saturation current density for the Schottky barrier is related to the Schottky barrier height (ϕ_b):

$$J_S = AT^2 \exp\left(-\frac{q\phi_b}{kT}\right) \qquad [5.3]$$

where A is Richardon's constant. For silicon, the Richardson's constant is $110 \ A°K^{-2}cm^{-2}$ while that for 4H-SiC[5] and GaN[6] have been calculated to be 146 and $24 \ A°K^{-2}cm^{-2}$. Values in this range have been reported using measurements on Schottky contacts[7,8].

The saturation current density for silicon Schottky rectifiers is shown in Fig. 5.2 as a function of the barrier height for various values of the temperature. For a typical barrier height of 0.7 eV for the silicon devices, the saturation current density is 10^{-5} A/cm^2 at room temperature. It increases rapidly with reduction of the barrier height and increase in temperature.

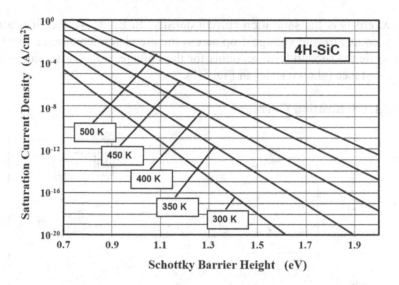

Fig. 5.3 Saturation current density for 4H-SiC Schottky rectifiers.

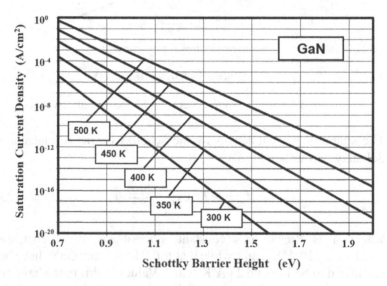

Fig. 5.4 Saturation current density for GaN Schottky rectifiers.

The saturation current density for 4H-SiC Schottky rectifiers is shown in Fig. 5.3 as a function of the barrier height for various values of the temperature. For a typical barrier height of 1.3 eV for the 4H-SiC devices, the saturation current density is 10^{-15} A/cm^2 at room temperature. It increases rapidly with reduction of the barrier height and increase in temperature. The much smaller saturation current density for 4H-SiC when compared with silicon produces reduced leakage current in devices and allows their operation at higher temperatures.

The saturation current density for GaN Schottky rectifiers is shown in Fig. 5.4 as a function of the barrier height for various values of the temperature. For a barrier height of 1.3 eV for the GaN devices, the saturation current density is 10^{-15} A/cm^2 at room temperature. It increases rapidly with reduction of the barrier height and increase in temperature. This value is close to that for 4H-SiC. However, actual Schottky contacts on GaN have low barrier heights of 0.5 to 0.7 eV making their leakage currents similar to silicon devices.

5.1.2 On-State Characteristics

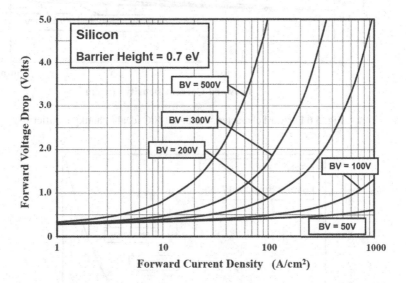

Fig. 5.5 Forward Characteristics of ideal silicon Schottky rectifiers.

The calculated forward conduction characteristics for ideal silicon Schottky rectifiers are shown in Fig. 5.5 for various breakdown voltages. The resistance contribution from the substrate was excluded from these characteristics. A Schottky barrier height of 0.7 eV was chosen because this is a typical value used in actual devices. It can be seen that the series resistance of the drift region does not adversely impact the on-state voltage drop for the device with breakdown voltage of 50 volts at a nominal on-state current density of

200 A/cm². However, the drift region resistance becomes significant when the breakdown voltage exceeds 100 volts. This has limited the application of silicon Schottky rectifiers to systems, such as switch-mode power supply circuits, operating at voltages below 100 V.

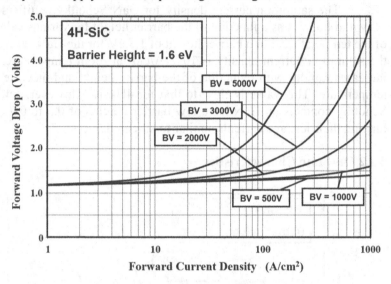

Fig. 5.6 Forward characteristics of 4H-SiC ideal Schottky rectifiers.

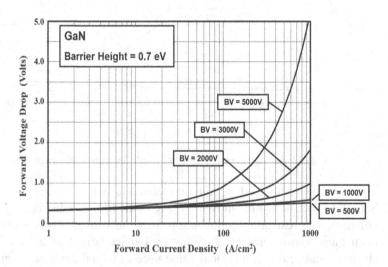

Fig. 5.7 Forward characteristics of GaN ideal Schottky rectifiers.

The significantly smaller resistance of the drift region enables scaling of the breakdown voltage of 4H-SiC and GaN Schottky rectifiers to much larger voltages typical of medium and high power electronic systems, such as those used for motor control. The forward characteristics of high voltage 4H-SiC Schottky rectifiers are shown in Fig. 5.6 for the case of a Schottky barrier height of 1.6 eV. This value was chosen because it is representative of nickel Schottky contacts commonly used for products. The N^+ substrate resistance was not included to obtain these ideal characteristics. It can be seen that the drift region resistance does not produce a significant increase in on-state voltage drop until the breakdown voltage exceeds 3000 volts. From these results, it can be concluded that 4H-SiC Schottky rectifiers are excellent companion diodes for medium and high power electronic systems that utilize Insulated Gate Bipolar Transistors (IGBTs). Their fast switching speed and absence of reverse recovery current can reduce power losses and improve the efficiency in motor control applications[4].

The forward characteristics of high voltage GaN Schottky rectifiers are shown in Fig. 5.7 for the case of a Schottky barrier height of 0.7 eV. This value was chosen because it is representative of Schottky contacts commonly used for HEMT products. The N^+ substrate resistance was not included to obtain these ideal characteristics. It can be seen that the drift region resistance does not produce a significant increase in on-state voltage drop until the breakdown voltage exceeds 5000 volts.

All practical Schottky rectifiers are fabricated by the growth of a thin epitaxial layer on a highly doped N^+ substrate that is sufficiently thick for handling in process equipment during device fabrication. It is important to include the resistance associated with the thick, highly doped N^+ substrate because this is comparable to that for the drift region in many instances. The specific resistance of the N^+ substrate can be determined by taking the product of its resistivity and thickness. For silicon, N^+ substrates with resistivity of 1 mΩ-cm are available. If the thickness of the substrate is 200 microns, the specific resistance contributed by the N^+ substrate is 2×10^{-5} Ω-cm^2. The impact of adding this substrate resistance on the forward characteristics for silicon Schottky rectifiers is shown in Fig. 5.8 for the cases of breakdown voltages of 50 and 100 volts. It can be concluded that the impact is very small in this case.

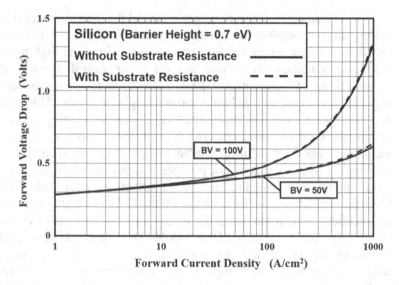

Fig. 5.8 Forward characteristics of silicon Schottky rectifiers including the substrate resistance.

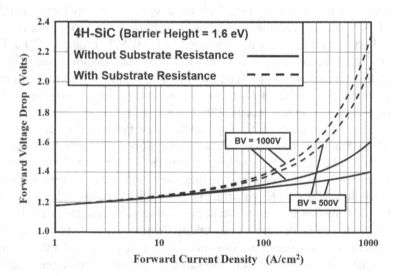

Fig. 5.9 Forward characteristics of 4H-SiC Schottky rectifiers including the substrate resistance.

For silicon carbide, the available resistivity of the N^+ substrates is substantially larger. For the available substrates with a typical resistivity of 0.02 Ω-cm and thickness of 350 microns, the substrate contribution is 7×10^{-4} Ω-cm^2. The impact of adding this substrate resistance on the forward characteristics for 4H-SiC Schottky rectifiers is shown in Fig. 5.9 for the cases of breakdown voltages of 500 and 1000 volts. It can be concluded that the impact is significant in this case with an increase in on-state voltage drop of about 0.3 volts.

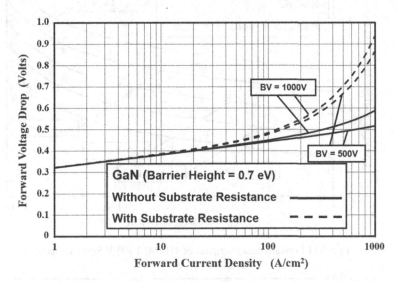

Fig. 5.10 Forward characteristics of GaN Schottky rectifiers including the substrate resistance.

In the case of GaN, the available substrates have a typical resistivity of 0.01 Ω-cm and thickness of 350 microns resulting in a substrate contribution of 3.5×10^{-4} Ω-cm^2. The impact of adding this substrate resistance on the forward characteristics for GaN Schottky rectifiers is shown in Fig. 5.10 for the cases of breakdown voltages of 500 and 1000 volts. It can be concluded that the impact is significant in this case with an increase in on-state voltage drop of about 0.1 volts.

The effect of changing the barrier height on the on-state characteristics of the Schottky rectifier can be understood by substitution of Eq. [5.3] into Eq. [5.2]:

$$V_F = \phi_B + \frac{kT}{q}\ln\left(\frac{J_F}{AT^2}\right) + \left(R_{drift} + R_{subs}\right).J_F \qquad [5.4]$$

It then becomes obvious that the on-state characteristics will shift to higher values in proportion to the increase in the barrier height. This is illustrated in Figs. 5.11 and 5.12 for the cases of 4H-SiC and GaN

Schottky rectifiers with breakdown voltages of 1000 volts. The substrate resistance was not included in these plots.

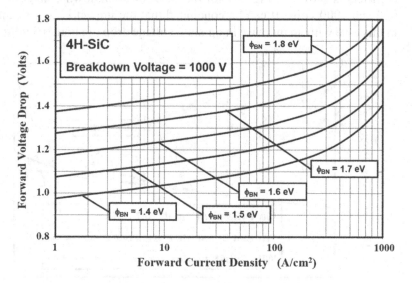

Fig. 5.11 Forward characteristics of 4H-SiC 1000-V Schottky rectifiers.

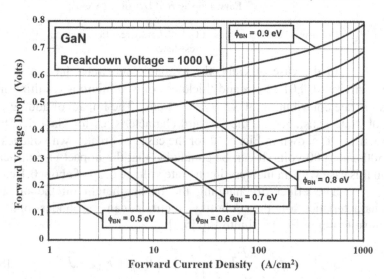

Fig. 5.12 Forward characteristics of GaN 1000-V Schottky rectifiers.

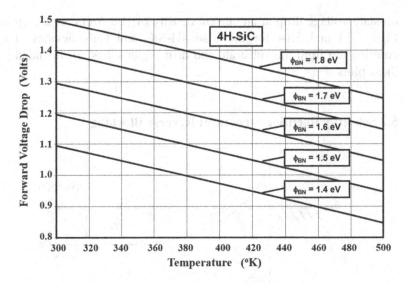

Fig. 5.13 Temperature dependence of on-state voltage drop for 4H-SiC Schottky Rectifiers.

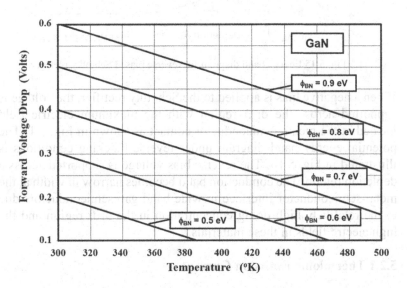

Fig. 5.14 Temperature dependence of on-state voltage drop for GaN Schottky Rectifiers.

The on-state voltage drop for the Schottky rectifiers reduces with increasing temperature. This is in accordance with Eq. [5.4] because the second term has a negative value. This reduction in on-state voltage drop at an current density of 100 A/cm^2 is shown in Figs. 5.13 and 5.14 for the case 4H-SiC and GaN devices. The contribution from the substrate and drift regions was not included in these plots.

5.2 Schottky Rectifier Structure: Reverse Blocking

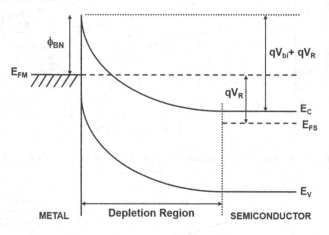

Fig. 5.15 Energy band diagram for reverse biased Schottky rectifier.

When a negative bias is applied to the Schottky rectifier, the voltage is supported across the drift region with the maximum electric field located at the metal-semiconductor contact as shown in Fig. 5.1. The potential energy band diagram under reverse blocking conditions is illustrated in Fig. 5.15. The reverse bias voltage is supported across a depletion region. The conduction band becomes narrow in width at the metal-semiconductor interface in wide band gap semiconductors due to the relatively high doing concentration in the drift region and the high electrc fields in these materials.

5.2.1 Thermionic Emission Current

According to thermionic emission theory, the leakage current is independent of the reverse bias voltage as determined by the saturation current given in Eq. [5.3][1]. The saturation current for 4H-SiC calculated by using this equation is shown in Fig. 5.3 for various temperatures. In comparison with silicon, it is possible to obtain a much larger barrier height in 4H-SiC metal-semiconductor contacts due to its larger band gap. It is possible to achieve very low saturation current density by using a Schottky barrier height of about 1.5 eV. However, it is preferable to use a low barrier height to obtain a low on-state voltage drop as shown in Fig. 5.11. For 4H-SiC Schottky rectifiers with blocking voltage above 500 volts, a good compromise occurs for a barrier height of about 1.5 eV. This can be achieved by using Nickel as the Schottky contact metal as discussed in section 2.15.4. A lower barrier height of about 1.1 eV can be obtained by using Titanium.

The saturation current density for GaN is shown in Fig. 5.4. The Schottky barrier height for metals on GaN have been reported to have relatively small values despite its large energy band gap as discussed in section 2.15.6. The Schottky barrier height for Nickel and Tantalum Nitride on GaN is only about 0.7 eV. This results in a high saturation current density (and leakage current) for GaN Schottky rectifiers.

5.2.2 Schottky Barrier Lowering

In silicon Schottky rectifiers, it is well established that the leakage current is exacerbated by the Schottky barrier lowering phenomenon[1]. The barrier lowering is determined by the electric field at the metal-semiconductor interface:

$$\Delta\phi_B = \sqrt{\frac{qE_m}{4\pi\varepsilon_S}} \qquad [5.5]$$

where E_m is the maximum electric field located at the metal-semiconductor interface. For a one-dimensional structure, the maximum electric field is related to the applied reverse bias voltage (V_R) by:

$$E_m = \sqrt{\frac{2qN_D}{\varepsilon_S}(V_R + V_{bi})} \qquad [5.6]$$

In addition, it is necessary to include the effect of pre-avalanche multiplication on the leakage current[9]. The multiplication coefficient (M) can be determined from the maximum electric field at the metal-semiconductor contact:

$$M = \left\{1 - 1.52\left[1 - \exp\left(-\frac{4.33\times10^{-24}E_m^{4.93}W_D}{6}\right)\right]\right\}^{-1} \qquad [5.7]$$

where W_D is the depletion layer width. Commercially available silicon devices exhibit an order of magnitude increase in leakage current from low reverse bias voltages to the rated voltage (about 80 percent of the breakdown voltage) due to these phenomena.

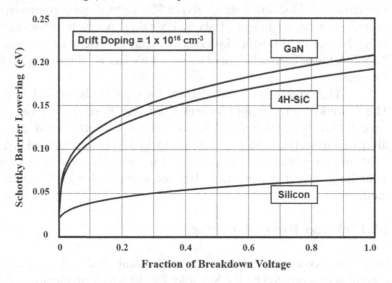

Fig. 5.16 Schottky barrier lowering for 4H-SiC and GaN Schottky rectifiers.

As discussed in chapter 4, the low specific on-resistance of the drift region in 4H-SiC and GaN devices is associated with the much larger electric fields in the material before the on-set of impact ionization. Consequently, the Schottky barrier lowering in 4H-SiC and GaN Schottky rectifiers is significantly larger than in silicon devices. This is illustrated in Fig. 5.16 for the case of a drift region doping level of 1×10^{16} cm^{-3}. It is worth pointing out the 4H-SiC and GaN devices have much larger breakdown voltages than the silicon device requiring normalization of the reverse voltage to the breakdown voltage. It can be seen that the barrier lowering is three times greater in 4H-SiC and GaN at the breakdown voltage.

5.2.3 Tunneling Current

The leakage current for 4H-SiC and GaN Schottky rectifiers increases much more rapidly with reverse bias than for silicon devices due to the enhanced Schottky barrier lowering. However, even this effect is insufficient to account the extremely rapid increase in leakage current with reverse bias voltage observed in actual devices. In order to explain this more rapid increase in leakage current, it is necessary to include the field emission (or tunneling) component of the leakage current[10].

The thermionic field emission model for the tunneling current leads to a barrier lowering effect proportional to the square of the electric field at the metal-semiconductor interface. When combined with the thermionic emission model, the leakage current density can be written as:

$$J_S = AT^2 \exp\left(-\frac{q\phi_b}{kT}\right).\exp\left(\frac{q\Delta\phi_b}{kT}\right).\exp\left(C_T E_m^2\right) \qquad [5.8]$$

where C_T is a tunneling coefficient. A tunneling coefficient of 8×10^{-13} cm^2/V^2 for 4H-SiC was found to yield an increase in leakage current by six orders of magnitude consistent with the behavior observed in the literature[1]. In the case of GaN, a tunneling coefficient of 7.1×10^{-12} cm^2/V^2 was found to yield a match with experimental results[11].

5.2.4 Leakage Current

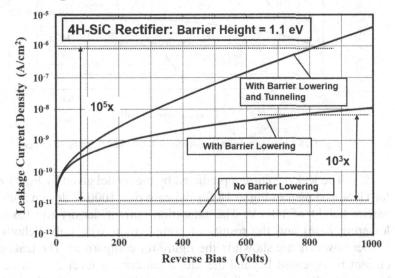

Fig. 5.17 Leakage current density for a 1-kV 4H-SiC Schottky rectifier.

The enhanced Schottky barrier lowering in silicon carbide devices leads to a more rapid increase in leakage current with increasing reverse bias as shown in Fig. 5.17 for the case of a 1000-V device with a barrier height of 1.1 eV. The saturation current density (no barrier lowering case) and the results of computation with only Schottky barrier lowering are shown in the figure for comparison. The leakage current is increased beyond the saturation current level even at zero bias due to the presence of electric field at the contact from the built-in potential. The leakage current is predicted by the model given by Eq. [5.8] to increase by five orders of magnitude when the reverse voltage approaches the breakdown voltage. The Schottky barrier lowering effect produces an increase by three-orders of magnitude as shown in the figure. Both the Schottky barrier lowering effect and the tunneling effect can be mitigated by using the JBS rectifier structure discussed in the next chapter.

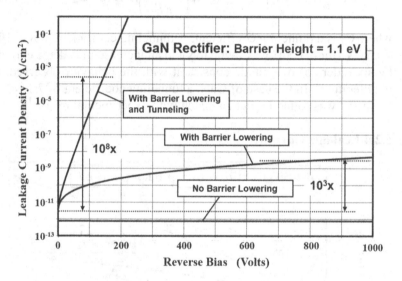

Fig. 5.18 Leakage current density for a 1-kV GaN Schottky rectifier.

The leakage current predicted by the model given by Eq. [5.8] for GaN is shown in Fig. 5.18 for the case of a 1000-V device with a barrier height of 1.1 eV. The saturation current density (no barrier lowering case) and the results of computation with only Schottky barrier lowering are shown in the figure for comparison. The leakage current is increased beyond the saturation current level even at zero bias due to the presence of electric field at the contact from the built-in potential. As in the case of 4H-SiC, the Schottky barrier lowering effect produces an increase by three-orders of magnitude as shown in the figure. The leakage current is predicted by the model given by Eq. [5.8] to increase by eight orders of magnitude when the reverse voltage is just 15% of the breakdown voltage. This indicates a major problem with developing Schottky rectifiers from GaN material unless the electric field at the Schottky contact can be reduced as discussed in the next chapter.

5.3 4H-SiC Vertical Diode Experimental Results

The first high voltage Schottky rectifiers with low on-state voltage drop and excellent switching behavior were reported by using 6H-SiC material[12] by using Platinum as the Schottky contact metal. These devices had a breakdown voltage of 400-V without any edge termination. The breakdown voltage was later improved to 1000-V by using an argon ion plant at the periphery of the diodes[13]. These devices had an on-state voltage drop of 1.1 volts at an on-state current density of 100 A/cm^2. It was found that the leakage current increased by about 5 orders of magnitude with increase in reverse bias voltage. Similar results were later demonstrated for Schottky rectifiers fabricated at using 4H-SiC[14] by using Titanium as the Schottky contact metal. These diodes had an on-state voltage drop of less than 1.1 volts at an on-state current density of 100 A/cm^2. It was found that the leakage current for these Schottky diodes also increased by about 5 orders of magnitude with increase in reverse bias voltage. Similar results were subsequently reported by other groups[15,16,17,18].

Schottky rectifiers with higher breakdown voltages have been developed over the years due to the low specific on-resistance of the drift region in 4H-SiC. Devices with breakdown voltages of 1.5 – 2.5 kV were reported from 4H-SiC in 1999[19] with on-state voltage drop of 2.4 volts at an on-state current density of 100 A/cm^2. Schottky rectifiers with breakdown voltage of 4-kV were reported[20] using 50 micron thick epitaxial layers with a high on-state voltage drop of over 6 volts an on-state current density of 100 A/cm^2. 4H-SiC based Schottky rectifiers with breakdown voltage of 4.9-kV were reported[21] in 2000 using 50 micron thick epitaxial layers with a doping concentration of 7×10^{14} cm^{-3}. These diodes used a 100 micron wide boron-implanted edge termination. They also had a high on-state voltage drop of 2.4 volts at an on-state current density of 25 A/cm^2. Rectifiers were fabricated using Molybdenum Schottky contacts on 4H-SiC with a breakdown voltage of 4.15-kV[22]. These diodes used a

3-zone JTE edge termination to achieve the breakdown voltage within 90% of the ideal value. The on-state voltage drop for these diodes was good at 1.89 volts at an on-state current density of 100 A/cm^2. All of these Schottky diodes exhibited a increase in leakage current by 5 orders of magnitude with increasing reverse bias voltage.

5.4 GaN Vertical Diode Experimental Results

Early work on GaN Schottky rectifiers was performed with material grown on insulating substrates[23,24,25,26,27] such as sapphire and aluminum oxide. These devices have a quasi-vertical design where the current flows vertically from the Schottky contact to a N+ layer and is then redirected to the surface to an ohmic contact surrounding the Schottky contact. This type of design is not suitable for power devices because of substantial parasitic series resistances resulting in high on-state voltage drop. The leakage current for these diodes increases by 4 orders of magnitude with increasing reverse bias as expected from the model in Eq. [5.8] but with a C_T value similar to 4H-SiC. More recently, high voltage Schottky rectifiers were demonstrated on GaN grown on silicon substrates[28]. They were also quasi-vertical devices that cannot be scaled for power electronic applications.

　　　　The availability of free standing 250 micron thick GaN substrates of sufficient quality allowed the fabrication of 450-V vertical Schottky diode structures[29] shown in Fig. 5.1. The back-side ohmic contact was created using Ti/Al annealed at 750 °C for 20 seconds. The Pt/Au Schottky contacts were found to have an on-state voltage drop of about 3 volts. GaN Schottky rectifiers with breakdown voltage of 600 volts were fabricated[30,31] on 250 micron thick GaN substrates with doping level of 7×10^{15} cm^{-3}. An on-state voltage drop of 1.2 volts was measured at an on-state current density of 100 A/cm^2. However, this is due to a low series resistance because of current spreading in the substrate for the small diameter diodes. In practice, it is necessary to make the diodes on thin epitaxial layers grown on highly doped substrates as discussed in section 5.1.2.

　　　　The first vertical Schottky rectifiers fabricated on epitaxial layers (with a doping concentration of 1×10^{14} cm^{-3}) grown on N$^+$ GaN substrates were reported in 2011 with an ideal edge termination produced by using argon ion-implantation[32]. The breakdown voltage was improved from 300 volts to 1600 volts by using this edge termination with an argon dose of 2×10^{16} cm^{-2}. The breakdown voltage was limited by the punch of the 2 micron thick epitaxial layer. It was subsequently demonstrated that the argon implanted zone must have a width of 50 microns to achieve nearly ideal breakdown voltage[33].

High performance vertical GaN Schottky rectifiers were reported in 2013 by using Ni/Au as the Schottky contact metal[34]. Breakdown voltage of 600 to 1100 volts was achieved on 5 micron thick epitaxial layers grown on 330 micron thick GaN substrates. An on-state voltage drop of 1.46 volt was observed at an on-state current density of 500 A/cm^2. The leakage current for these diodes increases by 4 orders of magnitude with increasing reverse bias from 600 volts to 1100 volts. This is consistent with the model in Eq. [5.8] but with a C_T value reported for GaN[11]. The reverse recovery performance of these GaN Schottky diodes was found to be superior to that for silicon and 4H-SiC diodes. A reduction of reverse recovery charge to 20% of that for the silicon diode and 60% of that for the 4H-SiC diode was reported[35].

5.5 GaN Lateral Diode Experimental Results

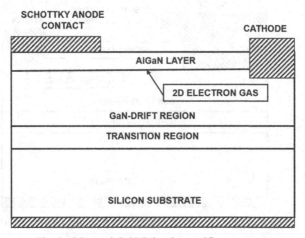

Fig. 5.19 Lateral GaN Schottky rectifier structure.

As previously discussed in chapter 1, the GaN HEMT structure has very promising characteristics due to the low sheet resistance of the drift region created by a 2-D electron gas. This highly conductive drift region can also be applied towards making GaN lateral Schottky rectifiers, whose structure is shown in Fig. 5.19. Lateral GaN Schottky diodes on the AlGaN/GaN hetrojunction were reported with breakdown voltage of 1000 volts with the structure shown in   Fig. 5.19[36]. The ohmic contact and Schottky contact were formed using Ti/Al/Ni/Au and Pt/Au/Ti, respectively. The devices had poor on-state voltage drop.

Devices with blocking voltages of 350 volts were achieved by using CHF$_3$ plasma treatment to introduce Flourine ions under the Schottky anode contact[37]. The ohmic contact and Schottky contact were formed using Ti/Al/Ni/Au and Ni/Au as discussed in chapter 2. No information on the on-state characteristics was provided. Similar devices were reported with current limiting capability by adding a Schottky contact on the cathode side[38]. These devices had a low on-state voltage drop of about 1 volt at an on-state current density of 100 A/cm^2.

Lateral GaN Schottky diodes with the AlGaN/GaN hetero-structure with breakdown voltages of 3500 volts[39] were achieved in 2011. They had the basic device structure shown in Fig. 5.19 with ohmic contact and Schottky contacts formed using Ti/Al/ Ni/Au and Ni/Au, respectively. The on-state voltage drop for these diodes was only 3 volts at an on-state current density of 100 A/cm^2.

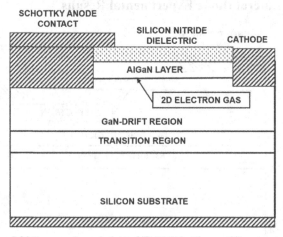

Fig. 5.20 Recessed anode lateral GaN Schottky rectifier structure.

Lateral GaN Schottky rectifiers with the anode contact recessed through the AlGaN layer, as shown in Fig. 5.20, were reported with high breakdown voltages. Devices with breakdown voltages of 1000 volts were achieved with an anode-cathode spacing of 8 microns[40]. The devices had an average electric field of 1.25×10^6 V/cm along the surface. 600-V devices with anode-cathode spacing of

4 microns had a series specific on-resistance of 1.5×10^{-4} Ω-cm^2. This value is only one order of magnitude higher than the ideal specific on-resistance discussed in chapter 4 (see Fig. 4.14).

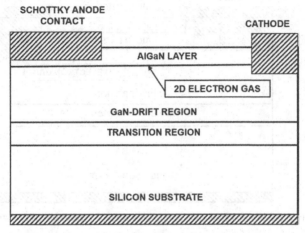

Fig. 5.21 Alternate recessed anode lateral GaN Schottky rectifier structure.

An alternate lateral GaN Schottky rectifier with anode recessed partially through the AlGaN layer, shown in Fig. 5.21, was used to fabricate devices with breakdown voltages up to 1440 volts[41]. The Ni/Au anode contact is made after reducing the thickness of the AlGaN layer to 50 angstroms to allow depletion of the 2-D gas by the Schottky barrier. A low knee voltage of 0.6 volts is obtained with this structure. The leakage current increases by only one-order of magnitude with this structure indicating that the Schottky contact is being shielded from high electric fields in the GaN drift region. The breakdown voltage of the devices increases by about 100 volts per micron of drift region length. This corresponds to an average lateral electric field of 1×10^6 V/cm which is smaller than the critical electric field for GaN (given by Eq. [4.16]) because of high local electric fields at the edges of the anode and cathode contacts.

A lateral GaN Schottky diode structure, shown in Fig. 5.22, with the edges of the anode contact passivated using silicon nitride has been reported with excellent on-state and reverse blocking characteristics[42]. These diodes were fabricated using TaN as the Schottky contact. The Si$_3$N$_4$ passivation layer at the contact edges had a thickness of 15 nm. This reduced the reverse leakage current by 4 orders of magnitude when compared with the conventional structure. The on-state voltage drop for these diodes was found to be 1.1 volts.

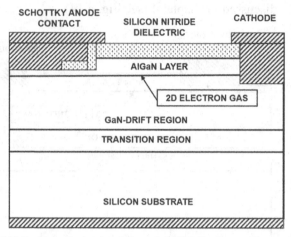

Fig. 5.22 Passivated edge anode lateral GaN Schottky rectifier structure.

5.6 Summary

The physics of operation of the 4H-SiC and GaN Schottky rectifiers is similar to that of silicon devices in the forward conduction direction. The low specific on-resistance for the drift region for these materials allows development of devices with very high breakdown voltages compared with silicon devices. However, the reverse leakage current in 4H-SiC and GaN Schottky rectifiers is significantly enhanced by the larger Schottky barrier lowering effect and thermionic field emission (or tunneling) current at high reverse bias voltages. This is due to the much larger electric field in 4H-SiC and GaN drift regions when compared with silicon. Fortunately, the availability of larger Schottky barrier heights for metal contacts on 4H-SiC, at the expense of an increased on-state voltage drop, allows reducing the absolute values for the leakage currents to a level where the reverse bias power dissipation is below the forward bias power dissipation enabling stable operation in circuits.

A better approach is to shield the Schottky contact from the high electric field developed in the 4H-SiC and GaN drift region as discussed in the next chapter. This is achieved by using the JBS structure for 4H-SiC and the lateral HEMT-based structure for GaN.

References

[1] B.J. Baliga, "Fundamentals of Power Semiconductor Devices", Springer-Science, 2008.

[2] B.J. Baliga, "Schottky Barrier Rectifiers and Methods of Forming the Same", U.S. Patent 5,612,567, March 18, 1997.

[3] B.J. Baliga, "The Future of Power Semiconductor Technology", Proceedings of the IEEE, Vol. 89, pp. 822-832, 2001.

[4] B.J. Baliga, "Power Semiconductor Devices for Variable Frequency Drives", Proceedings of the IEEE, Vol. 82, pp. 1112-1122, 1994.

[5] A. Itoh, T. Kimoto and H. Matsunami, "High Performance of High Voltage 4H-SiC Schottky Barrier Diodes", IEEE Electron Device Letters, Vol. 16, pp. 280-282, 1995.

[6] J. Piprek, "Nitride Semiconductor Devices", Wiley, New York, 2007.

[7] Y. Lv *et al.*, "Extraction of AlGaN/GaN Heterostructure Schottky Diode Barrier Heights from Forward Current-Voltage Characteristics", Journal of Applied physics, Vol. 109, pp. 074512-1 − 074512-6, 2011.

[8] A. Kumar *et al.*, "Temperature Dependence of Electrical Characteristics of Pt/GaN Schottky Diode Fabricated by UHV e-beam Evaporation", Nanoscale Research Letters, Vol. 8, pp. 8.1-8.8, 2013.

[9] S. L. Tu and B. J. Baliga, "On the Reverse Blocking Characteris-tics of Schottky Power Diodes", IEEE Transactions on Electron Devices, Vol. 39, pp. 2813-2814, 1992.

[10] T. Hatakeyama and T. Shinohe, "Reverse Characteristics of a 4H-SiC Schottky Barrier Diode", Silicon Carbide and Related Materials – 2001, Material Science Forum, Vol. 389-393, pp. 1169-1172, 2002.

[11] M. Ozbek and B.J. Baliga, "Tunneling Coefficient for GaN Schottky Barrier Diodes", Solid State Electronics, Vol. 62, pp. 1-4, 2011.

[12] M. Bhatnagar, P.K. McLarty and B.J. Baliga, "Silicon Carbide High Voltage (400V) Schottky Barrier Diodes", IEEE Electron Device Letters, Vol. EDL-13, pp. 501-503, 1992.

[13] D. Alok, B.J. Baliga and P.K. McLarty, "A Simple Edge Termination for Silicon Carbide with Nearly Ideal Breakdown Voltage", IEEE Electron Device Letters, Vol. EDL-15, pp. 394-395, 1994.

[14] R. Raghunathan, D. Alok and B.J. Baliga, "High Voltage 4H-SiC Schottky Barrier Diodes", IEEE Electron Device Letters, Vol. EDL-16, pp. 226-227, 1995.

[15] A. Itoh, T. Kimoto and H. Matsunami, "Efficient Power Schottky Rectifiers of 4H-SiC", IEEE International Symposium on Power Semiconductor Devices and ICs, pp. 101-106, 1995.

[16] A. Itoh, T. Kimoto and H. Matsunami, "Excellent Revesre Blocking Characteristics of High-Voltage 4H-SiC Schottky Rectifiers with Boron-Implanted Edge Termination", IEEE Electron Device Letters, Vol. 17, pp. 139-141, 1996.

[17] K.J. Schoen *et al.*, "Design Considerations and Experimental Analysis of High Voltage SiC Schottky Barrier Rectifiers", IEEE Transactions on Electron Devices, Vol. 45, pp. 1595-1604, 1998.

[18] V. Saxena *et al.*, "High-Voltage Ni and Pt SiC Schottky Diodes utilizing Metal Field Plate Terminations", IEEE Transactions on Electron Devices, Vol. 46, pp. 456-464, 1999.

[19] R.K. Chilukuri and B.J. Baliga, "High Voltage Ni/4H-SiC Schottky Rectifiers", IEEE International Symposium on Power Semiconductor Devices and ICs, pp. 161-164, 1999.

[20] H.M. McGlothlin *et al.*, "4 kV Silicon Carbide Schottky Diodes for High Frequency Switching Applications", IEEE Device Research Conference, pp. 42-43, 1999.

[21] R. Singh *et al.*, "SiC Power Schottky and PiN Diodes", IEEE Transactions on Electron Devices, Vol. 49, pp. 665-672, 2002.

[22] T. Nakamura *et al.*, "A 4.15 kV 9.07 mΩ-cm^2 4H-SiC Schottky Barrier Diode using Mo Contact Annealed at High Temperature", IEEE Electron Device Letters, Vol. 26, pp. 99-101, 2005.

[23] Z.Z. Bandic *et al.*, "High Voltage (450-V) GaN Schottky Rectifiers", Applied Physics Letters, Vol. 74, pp. 1266-1268, 1999.

[24] G.T. Dang *et al.*, "High Voltage GaN Schottky Rectifiers", IEEE Transactions on Electron Devices, Vol. 47, pp. 692-696, 2000.

[25] B.S. Shelton *et al.*, "Simulation of the Electrical Characteristics of High-Voltage Mesa and Planar GaN Schottky and p-i-n Rectifiers", IEEE Transactions on Electron Devices, Vol. 48, pp. 1498-1502, 2001.

[26] T.G. Zhu *et al.*, "GaN and AlGaN High-Voltage Rectfiers grown by MOCVD", Journal of Electronic Materials, Vol. 31, pp. 406-410, 2002.

[27] A-P. Zhang *et al.*, "GaN and AlGaN High Voltage Power Rectifiers", Chapter 2 in 'Wide Energy Bandgap Electronic Devices', Ed. F. Ren and J.C. Zolper, World Scientific Press, Singapore, 2003.

[28] L. Voss *et al.*, "Electrical Performance of GaN Schottky Rectifiers", Journal of the Electrochemical Society, Vol. 153, pp. G681-G684, 2006.

[29] J.W. Johnson *et al.*, "Schottky Rectifiers Fabricated on Free-Standing GaN Substrates", Solid-State Electronics, Vol. 45, pp. 405-410, 2001.

[30] Y. Zhou *et al.*, "High Breakdown Voltage Schottky Rectifier Fabricated on Bulk n-GaN Substrate", Solid-State Electronics, Vol. 50, pp. 1744-1747, 2006.

[31] Y. Zhou *et al.*, "Temperature-Dependent Electrical Characteristics of Bulk n-GaN Schottky Rectifier", Journal of Applied Physics, Vol. 101, pp. 024506-1-024506-4, 2007.

[32] A.M. Ozbeck and B.J. Baliga, "Planar Nearly Ideal Edge-Termination Technique for GaN Devices", IEEE Electron Device Letters, Vol. 32, pp. 300-302, 2011.

[33] A.M. Ozbeck and B.J. Baliga, "Finite-Zone Argon Implant Edge-Termination for High-Voltage GaN Schottky Rectifiers", IEEE Electron Device Letters, Vol. 32, pp. 1361-1363, 2011.

[34] K. Sumiyoshi *et al.*, "Low On-Resistance High Breakdown Voltage GaN Diodes on Low Dislocation Density GaN Substrates", SEI Technical Review, Vol. 77, pp. 113-117, 2013.

[35] S. Yoshimoto *et al.*, "Fast Recovery Performance of Vertical Schottky Barrier Diodes on Low Dislocation Density Freestanding GaN Substrates", SEI Technical Review, Vol. 80, pp. 35-39, 2015.

[36] A. Kamada *et al.*, "High-Voltage AlGaN/GaN Schottky Barrier Diodes on Si Substrates with Low-Temperature GaN Cap layer for Edge Termination", IEEE International Symposium on Power Semiconductor Devices and ICs, pp. 225-228, 2008.

[37] S-W Peng *et al.*, "A High Breakdown Voltage and Low Switching Loss GaN Schottky Diode using CHF3 Plasma Treatment", IEEE International Conference of Electron Devices and Solid-State Circuits, pp. 1-3, 2010.

[38] K.J. Chen and C. Zhou, "GaN Smart Discrete Power Devices", IEEE International Conference on Solid-State and IC Technology, pp. 1303-1306, 2010.

[39] G-Y Lee, H-H Liu and J-I Chyi, "High-Performance AlGaN/GaN Schottky Diodes with an AlGaN/AlN Buffer Layer", IEEE Electron Device Letters, Vol. 32, pp. 1519-1521, 2011.

[40] E. Bahat-Treidel *et al.*, "Fast-Switching GaN-Based Lateral Power Schottky Barrier Diodes with Low On-set Voltage and Strong Reverse Blocking", IEEE Electron Device Letters, Vol. 33, pp. 357-359, 2012.

[41] J-G. Lee *et al.*, "Low Turn-on Voltage AlGaN/GaN-on-Si Rectifier with gated Ohmic Anode", IEEE Electron Device Letters, Vol. 34, pp. 214-216, 2013.

[42] S. Lenci *et al.*, "Au-Free AlGaN/GaN Power Diode on 8-in Si Substrate with gated Edge Termination", IEEE Electron Device Letters, Vol. 34, pp. 1015-1037, 2013.

Chapter 6

Shielded Schottky Rectifiers

In the case of silicon Schottky rectifiers, it has been traditional to trade-off the on-state (or conduction) power loss against the reverse blocking power loss by optimizing the Schottky barrier height. As the Schottky barrier height is reduced, the on-state voltage drop decreases producing a smaller conduction power loss. At the same time, the smaller barrier height produces an increase in the leakage current leading to larger reverse blocking power loss. It has been demonstrated that the power loss can be minimized by reducing the Schottky barrier height at the expense of a reduced maximum operating temperature[1].

This optimization process is exacerbated by the rapid increase in the leakage current with increasing reverse bias voltage due to the Schottky barrier lowering phenomenon. The first method proposed to ameliorate the barrier lowering effect in vertical silicon Schottky rectifiers utilized shielding by incorporation of a P-N junction[2,3]. Since the basic concept was to create a potential barrier to shield the Schottky contact against high electric fields generated in the semiconductor by using closely spaced P+ regions around the contact, this structure was named the *'Junction-Barrier controlled Schottky (JBS) rectifier'*. In the JBS rectifier, the on-state current was designed to flow in the un-depleted gaps between the P+ regions when the diode is forward biased to preserve unipolar operation. In addition to reducing the leakage current, the presence of the P+ regions was also shown to enhance the ruggedness of the diodes. Detailed optimization of the silicon JBS rectifier characteristics was achieved with sub-micron dimensions between the P+ regions[4].

Subsequently, a novel silicon structure that utilizes the charge-coupling concept was proposed to reduce the resistance in the drift region[5]. Since these structures utilized an electrode embedded within an oxide coated trench region that surrounds the metal-semiconductor contact, this structure was named the *'Trench MOS*

Barrier controlled Schottky (TMBS) rectifier'. The performance of this structure was further enhanced by using a graded doping profile to create a device named the *'Graded-Doped Trench MOS Barrier controlled Schottky (GD-TMBS) rectifier'*[6]. This has allowed extending the breakdown voltage of silicon Schottky rectifiers to the 200 volt range[7,8].

Yet another method proposed to create a potential barrier under the Schottky contact was the utilization of a second metal-semiconductor contact with large barrier height surrounding the main Schottky contact with a low barrier height[9]. Since a stronger potential barrier could be created by locating the high barrier metal with a trench, this structure was named the *'Trench Schottky Barrier controlled Schottky (TSBS) rectifier'*.

In the previous chapter it was demonstrated that the increase in leakage current with reverse bias voltage is much stronger for the Schottky rectifiers made from 4H-SiC. This is due to a larger Schottky barrier lowering effect associated with the larger electric field in silicon carbide drift regions and the onset of field emission (or tunneling) current. It is therefore obvious that the methods proposed to suppress the electric field at the Schottky contact in silicon diodes will have even greater utility in silicon carbide rectifiers.

This chapter discusses the application of shielding techniques to reduce the electric field at the metal-semiconductor contact in silicon carbide Schottky rectifiers. It is demonstrated that the JBS and TSBS concepts can be extended to silicon carbide to achieve significant improvement in performance. However, the TMBS and GD-TMBS approaches are not appropriate for silicon carbide because of the high electric field generated in the oxide leading to its rupture. Experimental results on relevant structures are provided to define the state of the development effort on shielded silicon carbide Schottky rectifiers.

6.1 Junction Barrier Schottky (JBS) Rectifier: 4H-SiC

The Junction Barrier controlled Schottky (JBS) rectifier structure is illustrated in Fig. 6.1. It consists of a P^+ region placed around the Schottky contact to generate a potential barrier under the metal-semiconductor contact in the reverse blocking mode. The space between the P^+ regions is chosen so that there is an un-depleted region below the Schottky contact to enable unipolar conduction through the structure in the on-state. The voltage drop across the diode is not sufficient to forward bias the P-N junction during normal on-state operation. Low voltage silicon devices operate with on-state voltage drops around 0.4 volts which is well below the 0.7 volts needed for inducing strong injection across the P-N junction. The margin is even larger for 4H-SiC due to its much larger band gap. Typical silicon carbide Schottky rectifiers have on-state voltage drops of less than 1.5 volts due to low specific resistance of the drift region. This is well below the 3 volts required to induce injection from the P-N junction. Consequently, the JBS concept is well suited for development of silicon carbide structures with very high breakdown voltages.

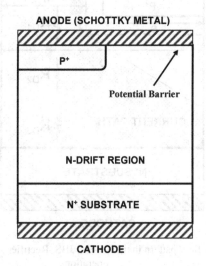

Fig. 6.1 Silicon carbide Junction Barrier controlled Schottky (JBS) rectifier structure.

The P$^+$ region is usually formed by ion-implantation of P-type dopants using a mask with appropriate spacing to leave room for the Schottky contact. For processing convenience, the same metal layer is used to make an ohmic contact to the highly doped P$^+$ region as well as to make the Schottky barrier contact to the N- drift region. The space between the P$^+$ regions must be optimized to obtain the best compromise between the on-state voltage drop, which increases as the space is reduced, and the leakage current, which decreases as the space is reduced.

The shape of the P-N junction for silicon carbide JBS rectifiers is different than that in silicon devices because no diffusion of the p-type dopant occurs during the annealing of the ion implants.

The junction takes a rectangular profile as opposed to a cylindrical shape in silicon devices. This shape is preferable for creating a potential barrier below the Schottky contact to suppress the electric field at the contact.

6.1.1 On-State Characteristics

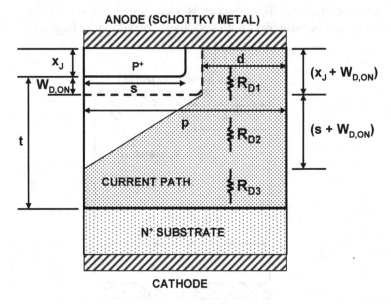

Fig. 6.2 Current flow pattern in the 4H-SiC JBS Rectifier structure during on-state operation.

The current flow path in the 4H-SiC JBS rectifier can be modelled using the shaded area shown in Fig. 6.2. In this model, it is assumed that the current through the Schottky contact flows only within the un-depleted portion (with dimension 'd') of the drift region at the top surface. Consequently, the current density at the Schottky contact (J_{FS}) is related to the cell (or cathode) current density (J_{FC}) by:

$$J_{FS} = \left(\frac{p}{d}\right)J_{FC}$$ [6.1]

where p is the cell pitch. The dimension 'd' is determined by the cell pitch (p), the size of the ion-implant window (s), and the depletion width ($W_{D,ON}$):

$$d = p - s - W_{D,ON}$$ [6.2]

In deriving this equation, it has been assumed that there is no straggle in the ion-implant. Depending up on the lithography used for device fabrication to minimize the size (dimension 's') of the P^+ region, the current density at the Schottky contact can be enhanced by a factor of two or more. This must be taken into account when computing the voltage drop across the Schottky contact:

$$V_{FS} = \phi_B + \frac{kT}{q} \ln\left(\frac{J_{FS}}{AT^2}\right)$$ [6.3]

After flowing across the Schottky contact, the current flows through the un-depleted portion of the drift region. In the model, it is assumed that the current flows through a region with a uniform width 'd' until it reaches the bottom of the depletion region and then spreads to the entire cell pitch (p) at a 45 degree spreading angle. The current paths overlap at a distance $(s + W_{D,ON})$ from the bottom of the depletion region. The current then flows through a uniform cross-sectional area.

The net resistance to current flow can be calculated by adding the resistance of the three segments. The resistance of the first segment of uniform width 'd' is given by:

$$R_{D1} = \frac{\rho_D \cdot (x_J + W_{D,ON})}{d.Z}$$ [6.4]

The resistance of the second segment is given by:

$$R_{D2} = \frac{\rho_D}{Z} \ln\left(\frac{p}{d}\right)$$ [6.5]

The resistance of the third segment with a uniform cross-section of width p is given by:

$$R_{D3} = \frac{\rho_D \cdot (t - s - 2W_{D,ON})}{p.Z}$$ [6.6]

The specific resistance for the drift region can be calculated by multiplying the total cell drift region resistance ($R_{D1} + R_{D2} + R_{D3}$) by the cell-area (p.Z):

$$R_{sp,drift} = \frac{\rho_D \cdot p \cdot (x_J + W_{D,ON})}{d} + \rho_D \cdot p \cdot \ln\left(\frac{p}{d}\right) + \rho_D \cdot (t - s - 2W_{D,ON})$$

[6.7]

In addition, it is important to include the resistance associated with the thick, highly doped N^+ substrate because this is substantially larger than for silicon devices. The specific resistance of the N^+ substrate can be determined by taking the product of its resistivity and thickness. For 4H-SiC, the lowest available resistivity for N^+ substrates is 20 mΩ-cm. If the thickness of the substrate is 350 microns, the specific resistance contributed by the N^+ substrate is 7×10^{-4} Ω-cm^2.

The on-state voltage drop for the JBS rectifier at a forward cell current density J_{FC}, including the substrate contribution, is then given by:

$$V_F = \phi_B + \frac{kT}{q}\ln\left(\frac{J_{FS}}{AT^2}\right) + \left(R_{sp,drift} + R_{sp,subs}\right)J_{FC}$$

[6.8]

The depletion layer width for silicon carbide is also substantially larger than for silicon due to the larger built-in potential. However, these structures operate at a relatively larger on-state voltage drop of over 1 volt. Consequently, when computing the on-state voltage drop using the above equation, it is satisfactory to make the approximation that the depletion layer width can be computed by subtracting an on-state voltage drop of 1.5 volts from a built-in potential (V_{bi}) of about 3.2 volts for the P-N junction:

$$W_{D,ON} = \sqrt{\frac{2\varepsilon_S (V_{bi} - 1.5)}{qN_D}}$$

[6.9]

Due to the abrupt junctions formed in silicon carbide by the ion implant process, it is appropriate to assume that the entire depletion occurs on the lightly doped N-side of the junction. For a 4H-SiC JBS rectifier fabricated using a doping concentration of 2×10^{16} cm^{-3}, corresponding to a breakdown voltage of about 1000 V, the zero bias depletion width is about 0.4 microns. Based upon this, it can be concluded that it is important to take the depletion width into account for silicon carbide JBS rectifiers.

The forward characteristics of 1-kV 4H-SiC JBS rectifiers, calculated using the above analytical model with a Schottky barrier height of 1.6 eV, are shown in Fig. 6.3 with the cell pitch (p) as a parameter. The width of the P⁺ region (2s) was kept at 1 micron for this analysis. A junction depth of 1 micron was used with a drift region thickness of 10 microns below the junction to support the 1000 volts. The doping concentration of the drift region was 2×10^{16} cm^{-3}. In comparison with the Schottky rectifier characteristics (shown by the dashed line in the figure), the increase in on-state voltage drop at a forward current density of 200 A/cm^2 is small (less than 0.1 volts) as long as the cell pitch is more than 1.5 microns. This cell pitch is sufficient to obtain substantial reduction of the electric field at the metal-semiconductor contact as shown in the next section.

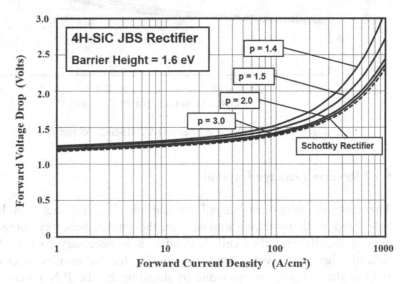

Fig. 6.3 Forward characteristics of 1-kV 4H-SiC JBS rectifiers.

The forward characteristics of 1-kV 4H-SiC JBS rectifiers, calculated for the same structure using the above analytical model with a Schottky barrier height of 1.1 eV, are shown in Fig. 6.4 with the cell pitch (p) as a parameter. The reduced barrier height allows reduction of the on-state voltage drop. In comparison with the Schottky rectifier characteristics (shown by the dashed line in the figure), the increase in on-state voltage drop at a forward current density of 200 A/cm^2 is small (less than 0.1 volts) as long as the cell pitch is more than 1.5 microns. The on-state voltage drop is reduced to 1.0 with the smaller barrier height compared with 1.5 V for the previous structure.

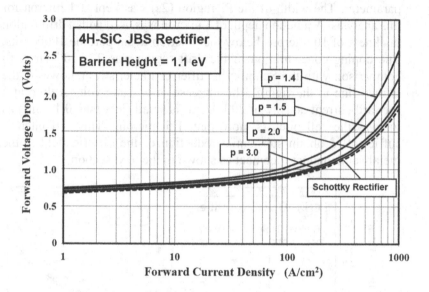

Fig. 6.4 Forward characteristics of 1-kV 4H-SiC JBS rectifiers.

6.1.2 Reverse Leakage Current

The leakage current in the silicon carbide JBS rectifier can be calculated by firstly taking into account the smaller Schottky contact area in the JBS rectifier cell. Secondly, it is necessary to include Schottky barrier lowering while accounting for the smaller electric field at the Schottky contact due to shielding by the P-N junction. Third, the thermionic field emission current must be included while accounting for the smaller electric field at the Schottky contact due to shielding by the P-N junction.

After making these adjustments, the leakage current for the silicon carbide JBS rectifier can be calculated by using:

$$J_L = \left(\frac{p-s}{p}\right) A T^2 \exp\left(-\frac{q\phi_b}{kT}\right).\exp\left(\frac{q\Delta\phi_{bJBS}}{kT}\right).\exp\left(C_T E_{JBS}^2\right)$$

[6.10]

where C_T is a tunneling coefficient (8×10^{-13} cm^2/V^2 for 4H-SiC). In contrast to the Schottky rectifier, the barrier lowering for the JBS rectifier is determined by the reduced electric field E_{JBS} at the contact:

$$\Delta\phi_{bJBS} = \sqrt{\frac{qE_{JBS}}{4\pi\varepsilon_S}} \qquad\qquad [6.11]$$

The electric field at the Schottky contact in the JBS rectifier varies with distance away from the P-N junction. The highest electric field is observed at the middle of the Schottky contact with a progressively smaller value closer to the P-N junction. When developing an analytical model with a worst case scenario, it is prudent to use the electric field at the middle of the contact to compute the leakage current.

Until the depletion regions from the adjacent P-N junctions produce a potential barrier under the Schottky contact, the electric field at the metal-semiconductor interface in the middle of the contact increases with the applied reverse bias voltage as in the case of the Schottky rectifier. A potential barrier is established by the P-N junctions after depletion of the drift region below the Schottky contact at the pinch-off voltage. The pinch-off voltage (V_P) can be obtained from the device cell parameters:

$$V_P = \frac{qN_D}{2\varepsilon_S}(p-s)^2 - V_{bi} \qquad\qquad [6.12]$$

It is worth pointing out that the built-in potential for 4H-SiC is much larger than for silicon. Although the potential barrier begins to form after the reverse bias exceeds the pinch-off voltage, the electric field continues to rise at the Schottky contact due to encroachment of the potential to the Schottky contact. This problem is less acute for the silicon carbide structure than in the silicon JBS rectifier because of the rectangular shape of the P-N junction resulting from the very low diffusion coefficients for dopants in 4H-SiC. In order to analyze the impact of this on the reverse leakage current, the electric field E_{JBS} can be related to the reverse bias voltage by:

$$E_{JBS} = \sqrt{\frac{2qN_D}{\varepsilon_S}(\alpha V_R + V_{bi})} \qquad\qquad [6.13]$$

where α is a coefficient used to account for the build-up in the electric field after pinch-off.

Due to the two-dimensional nature of the P-N junction in the JBS rectifier structure, it is difficult to derive an analytical expression for alpha. However, the value for alpha for 4H-SiC has been related to the aspect ratio of the structure[10] by using numerical simulations:

$$\alpha\left(4H-SiC\right)=e^{-5.75*(AR)} \qquad [6.14]$$

with the aspect ratio defined as:

$$(AR)=\frac{x_J}{2(p-s)} \qquad [6.15]$$

For a junction depth of 1 micron, the aspect ratio is 0.5 for a pitch of 2.0 microns if s is 1 micron. This results in an alpha value of only 0.0564 which implies a greatly reduced electric field according to Eq. [6.13].

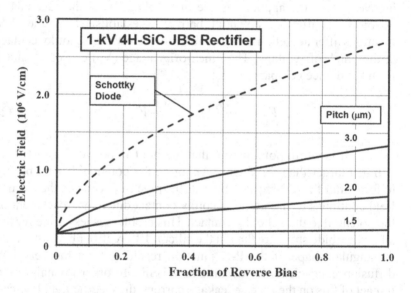

Fig. 6.5 Electric field at the Schottky contact for 1 kV 4H-SiC JBS rectifiers with various cell pitch.

As an example, consider the case of the 1-kV silicon carbide JBS rectifier discussed in the previous section with a junction depth of 1 micron and a P$^+$ region with dimension 's' of 1 micron. The reduction of the electric field at the Schottky contact can be predicted by calculating values for alpha as given by Eq. [6.14]. The alpha values for a pitch of 1.5, 2.0, and 3.0 are 0.00318, 0.0564, and 0.2375, respectively. The electric field at the middle of the Schottky contact in the 4h-SiC JBS rectifier structure obtained by using these values for alpha are plotted in Fig. 6.5. An alpha of unity corresponds to the Schottky rectifier structure with no shielding. It can be observed that substantial reduction of the electric field at the Schottky contact is obtained as the pitch is reduced. Even for a pitch of 2 microns, which results in an on-state characteristic close to that of the Schottky diode as shown in Fig. 6.4, the electric field at the Schottky contact is reduced to only 6.6×10^5 V/cm at 100% of the breakdown voltage compared with 2.7×10^6 V/cm for the Schottky diode.

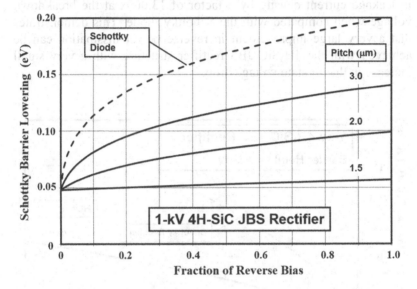

Fig. 6.6 Schottky barrier lowering in 1 kV 4H-SiC JBS rectifiers with various cell pitch.

The reduction of the Schottky barrier lowering due to the reduction of the electric field at the Schottky contact in the 4H-SiC JBS structure is shown in Fig. 6.6 for the case of the 1-kV device. Without the shielding by the P-N junction, a barrier lowering of 0.20 eV occurs in the Schottky rectifier at close to the breakdown voltage. The barrier lowering is reduced to 0.10 eV with an cell pitch of 2 microns in the 4H-SiC JBS rectifier structure.

As discussed in the previous chapter, the large Schottky barrier lowering for 4H-SiC, in conjunction with the thermionic field emission current and tunneling current, results in an increase in leakage current by five-orders of magnitude when the voltage increases to the breakdown voltage. The leakage current is greatly reduced by the shielding in the 4H-SiC JBS rectifier structure as shown in Fig. 6.7. For the 4H-SiC JBS rectifier structure with pitch of 2.0

microns and an implant window (s) of 1 microns, the Schottky contact area is reduced to 50 percent of the cell area. This results in a proportionate reduction of leakage current at low reverse bias voltages. More importantly, the suppression of the electric field at the Schottky contact, by the presence of the P-N junction, greatly reduces the rate of increase in leakage current with increasing reverse bias. The reverse leakage increases by only one order of magnitude when the reverse bias is increased to the breakdown voltage. The net effect is a reduction in leakage current density by a factor of 13,000x at the breakdown voltage when compared with the Schottky diode. This demonstrates that a very large improvement in reverse power dissipation can be achieved with the 4H-SiC JBS rectifier structure with a very small increase in the on-state voltage drop.

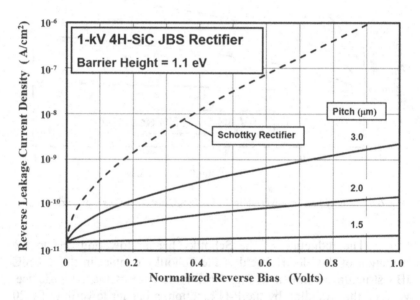

Fig. 6.7 Reverse leakage current for 1 kV 4H-SiC JBS rectifiers with cell pitch.

6.1.3 Experimental Results

The first high voltage silicon carbide JBS rectifiers were reported in 1997 after the evolution of adequate technology for ion-implantation and activation of P-type dopants[11,12]. These structures were sometimes erroneously labeled as Merged PN Schottky (MPS) rectifiers[11]. The MPS concept was originally proposed for improving the performance of high voltage silicon rectifiers[13]. In the MPS concept, the P-N junction is used to inject minority carriers into the drift region and modulate (reduce) the resistance in series with the Schottky contact. When compared to PiN rectifiers, the MPS structure has been found to contain a smaller stored charge leading to an improved trade-off between on-state voltage drop and reverse recovery characteristics for silicon rectifiers. As already pointed out, the very low resistance in the drift region of high voltage silicon carbide rectifiers does not warrant the modulation of its conductivity with the injection of minority carriers that degrade the switching performance. It is therefore appropriate to use the JBS nomenclature for the high voltage silicon carbide structures rather than the MPS moniker. However, these JBS rectifiers can operate in the MPS mode at surge current levels. This is beneficial for reducing the power losses and creating a more rugged operation.

In the first JBS rectifiers[12] with breakdown voltage of 700 volts achieved by using 9 micron thick epitaxial layers with doping level of 3×10^{15} cm^{-3}, the addition of the P$^+$ regions to the Schottky contact was found to result in an improvement in the breakdown voltage and leakage current. In spite of a reduction of the electric field at the titanium Schottky contact consistent with that shown in Fig. 6.5, the reduction of the leakage current was only by two orders of magnitude. However, the increase in the on-state voltage drop was small, similar to that shown in Fig. 6.3. Subsequently, the breakdown voltage of the JBS rectifier was extended to 2.8 kV[14] by using 27 micron thick epitaxial layers with doping concentration of 3×10^{15} cm^{-3}. The on-state voltage drop for these JBS diodes was 2.0 volts versus 1.8 volts observed for the Schottky diodes. In this study, the leakage current for the Schottky diode was reported to increase with reverse voltage by a factor of 10^6 while that for the JBS diode increased by only a factor of 10^3.

Subsequent work[15,16] on JBS rectifiers confirmed the benefits of using the junction barrier concept for reducing the leakage current. In addition, it was found that the surge current capability of the diodes was enhanced when compared with Schottky diodes. In Schottky rectifiers, the on-state voltage drop becomes extremely large (~30 volts) under surge current levels (current density above 1000 A/cm^2) leading to the observation of destructive failures. In contrast, the injection of minority carriers at applied voltages above 3.5 volts in the JBS rectifier (MPS mode) enables reduction of the on-state voltage drop under surge current levels preventing destructive failures. Further, unlike Schottky rectifiers, the JBS rectifiers were found to exhibit a positive temperature coefficient for the on-state voltage drop at a current density of 100 A/cm^2. Thus, the increased resistance created by the P$^+$ regions in the JBS rectifiers had the benefit of compensating for the reduction in voltage drop across the Schottky contact with increasing temperature.

Excellent 4H-SiC JBS rectifiers with breakdown voltage of 1400 volts were reported in 2002 by using linear and honeycombed P$^+$ grid structures[17]. Ion implantation of aluminum was used to form the P$^+$ regions and the Schottky contact was made using Ti/Pt/Au. The on-state voltage drop for the 4H-SiC JBS rectifiers was about 1.2 volts at an on-state current density of 100 A/cm^2. The reverse recovery current of these diodes was found to be much smaller than for silicon P-i-N rectifiers resulting in improved efficiency for DC-DC converters.

The breakdown voltage of the JBS rectifier was extended[18] to 4.3 kV using a 30 micron thick drift layer with doping concentration of 2×10^{15} cm^{-3}. The leakage current for these diodes was reported to increase by 3 orders of magnitude with reverse bias when compared with 6 orders of magnitude for the Schottky diode. For this reduced doping concentration, an optimum spacing of 9 microns between the P$^+$ regions was observed to provide the best characteristics.

JBS rectifiers manufactured from 4H-SiC were reported in 2006 with much better ruggedness than previous Schottky rectifier products[19]. The operation of the JBS diodes in the MPS mode was shown to reduce power losses and improve the ability to operate at surge current levels.

3.5 kV JBS rectifiers fabricated using epitaxial layers with 1.5×10^{15} cm^{-3} doping and thickness of 31 microns were reported in 2007[20]. These devices could operate at up to 300 °C with increase in leakage current by only 10x up to the full blocking voltage. An increase in the reverse recovery charge by a factor of 10x was observed from 30 °C to 200 °C.

The blocking voltage capability of 4H-SiC JBS rectifiers was extended to 10-kV by using epitaxial layers with 6×10^{14} cm^{-3} doping and thickness of 120 microns[21]. These diodes were operated an on-state current density of 20 A/cm^2 to obtain an on-state voltage drop of 4 volts. The P$^+$ grid was fabricated using aluminum ion implants and Nickel was used as the Schottky contact. The leakage current for the diodes increased by three-orders of magnitude with increasing reverse bias voltage indicating shielding of the Schottky contact.

When the spacing between the P$^+$ grid regions in the 4H-SiC JBS rectifiers is reduced, it becomes important to include the impact of lateral straggle of the aluminum ion-implants. Monte-Carlo simulations of the lateral extension of aluminum ions indicates a spread of about 0.2 microns from the mask edges[22]. This phenomenon is important when optimizing the 4H-SiC JBS rectifier structure.

In conclusion, the JBS concept has been found to be very valuable for reducing the leakage current in 4H-SiC Schottky rectifiers with a modest increase in the on-state voltage drop. All the products available from manufacturers rely up on the JBS structure to achieve low leakage current and ruggedness.

6.2 Trench MOS Barrier Schottky (TMBS) Rectifier: 4H-SiC

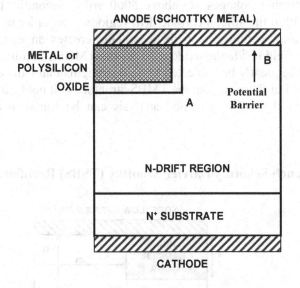

Fig. 6.8 Trench MOS barrier controlled Schottky (TMBS) rectifier structure.

The Trench MOS barrier Schottky Rectifier (TMBS) structure, shown in Fig. 6.8, was originally proposed for silicon Schottky rectifiers[5,23] to achieve charge coupling between the electrode in the trench and the dopant charge in the drift region. The charge coupling allows for very high doping concentration in the drift region while maintaining a breakdown voltage above the theoretically predicted values for one-dimensional parallel plane junctions[10]. The high drift region doping reduces   the   series   resistance   of   the   drift   region resulting in low on-state voltage drop for Schottky rectifiers with higher breakdown voltages. This concept, particularly when enhanced with a linearly graded doping profile[6,7], has been shown to produce excellent silicon Schottky rectifiers with breakdown voltages of 50 to 100 volts[24]. In addition, the electric field at the Schottky contact can be reduced by the formation of a potential barrier due to extension of the depletion region from the sidewalls of the trench.

Recently, there have been attempts to apply the TMBS concept to silicon carbide Schottky rectifiers[25,26]. This approach is misdirected because of two problems. Firstly, the doping concentration in the drift region for high voltage silicon carbide structures is very high when compared with silicon as shown earlier. It is therefore unnecessary to utilize charge coupling to enhance the doping unless the operating voltages are above 5000 volts. Secondly, the electric field within the silicon carbide drift regions is an order of magnitude larger than within silicon devices. This creates an equally higher electric field inside the oxide used for the TMBS structure. The oxide can consequently be subjected to field strengths that induce failure by rupture. For these reason, the TMBS structure will not be discussed in detail in this book. Detailed analysis can be found in a previous book[27].

6.3 Trench Schottky Barrier Schottky (TSBS) Rectifier: 4H-SiC

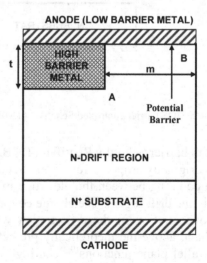

Fig. 6.9 Trench Schottky barrier controlled Schottky (TSBS) rectifier structure.

In the section 6.1.2, it has been demonstrated that the shielding of the metal-semiconductor contact to reduce the electric field suppresses leakage current at high voltages. The potential barrier that must be created for the shielding of the contact can also be produced by using another Schottky contact placed adjacent to the main Schottky contact. The second Schottky contact must have a larger barrier height than the main Schottky contact so that the leakage current from this contact is small despite the high electric field at its interface with the semiconductor. In addition, it is preferable to locate the high barrier height metal contact within a trench surrounding the main low barrier metal Schottky contact as illustrated in Fig. 6.9. The trench architecture enhances the potential barrier under the main Schottky contact in comparison with a planar structure.

All of the above features were first proposed for improvement of Schottky barrier rectifiers in a patent[9] issued in 1993. This patent describes devices with a high barrier metal located in trenches etched around the main low barrier metal Schottky contact with various geometries, such a stripes or cellular configurations. A simple method for fabrication of the device was also described. It consists of depositing the low barrier metal and patterning it to expose areas for etching the trenches. The trenches are then formed using the metal as the masking layer. The high barrier metal is then deposited into the trenches and over the low barrier metal to complete the cell structure. The TSBS concept was not pursued for silicon devices due to the success of the JBS and TMBS concepts. However, it was particularly suitable for silicon carbide devices because it eliminated the need for formation of P-N junctions which require very high temperature anneals that can degrade the semiconductor surface. Due to the large band gap of silicon carbide, it is possible to produce high barrier Schottky contacts by using metals such as Nickel in conjunction with low barrier main Schottky contacts such as Titanium. Detailed analysis of the 4H-SiC TSBS is available in a previously book[27]. Once ion implantation and annealing of p-type dopants such as aluminum and boron was optimized, the JBS concept was successfully implemented making further development of the TSBS concept unnecessary.

The TSBS concept for improving the performance of silicon carbide Schottky rectifiers was first simultaneously explored at PSRC[28,29] and Purdue University[30]. Titanium was used as the main Schottky contact metal due to its relatively low barrier height (1.1 eV) on 4H-SiC while Nickel was used in the trenches as the high barrier height (1.7 eV) metal. Using this approach with an epitaxial layer with doping concentration of 3×10^{15} cm^{-3} and thickness of 13 microns, a breakdown voltage of 400 volts was observed[31]. This was far below the capability of the material and compares poorly with 1720 volts observed for the planar nickel Schottky rectifier. The analysis discussed in the previous section indicates that this breakdown problem is not endemic to the TSBS structure but may be related to the chip design and layout. However, a reduction of the leakage current by a factor of 75x was observed confirming the ability of the TSBS concept in mitigating the barrier lowering and tunneling currents from the main Schottky contact. In addition, the on-state characteristics were found to be similar to those for a planar Titanium Schottky contact as expected from the description of the simulated structures in the previous section.

More recently[32], a reduction of leakage current by a factor of 30x has also been reported by using the planar TSBS structure (i.e. zero trench depth) with little impact on the forward characteristics. The smaller improvement in the leakage current when compared with the trench structure is consistent with less effective potential barrier observed in the simulation results described in the previous section. The breakdown voltages of these diodes ware far below the capability of the underlying drift region due to absence of adequate edge termination.

6.4 Trench Schottky Barrier Schottky (TSBS) Rectifier: GaN

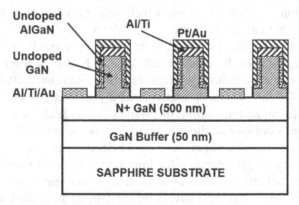

Fig. 6.10 GaN TSBS rectifier trench structure.

The TSBS concept has also been applied to make GaN Schottky rectifiers with reduced leakage currents[33]. The authors refer to this idea as the 'field effect Schottky barrier' but the principle of operation is the same as the TSBS concept. Two implementations of the GaN device were demonstrated. A vertical trench based structure shown in Fig. 6.10 and a lateral planar structure shown in Fig. 6.11. Al/Ti was used to form the low barrier height Schottky contact and Pt/Au was used to form the high barrier height contact. The ohmic contact was created using Al/Ti/Au. On-state characteristics with a knee at only 0.2 volts were observed for devices that could block up to 400 volts.

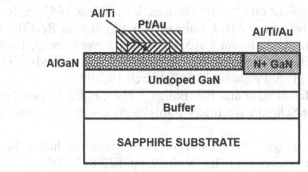

Fig. 6.11 GaN TSBS rectifier planar structure.

6.5 Summary

The JBS rectifier concept was first proposed for improving the trade-off between on-state voltage drop and leakage current for silicon Schottky rectifiers. This idea has been widely adopted for 4H-SiC Schottky rectifiers because the larger electric field in the drift region produces much greater Schottky barrier lowering and induces tunneling current as well. It is demonstrated in this chapter that a significant improvement in the leakage current of silicon carbide Schottky rectifiers can be achieved by shielding the Schottky contact against high electric fields generated in the semiconductor. The JBS structures are found to be very effective in reducing leakage current without significant degradation of the forward characteristics.

References

[1] B.J. Baliga, "Fundamentals of Power Semiconductor Devices", Springer-Science, 2008.

[2] B.J. Baliga, "The Pinch Rectifier: A Low Forward Drop High Speed Power Diode", IEEE Electron Device Letters, Vol. 5, pp. 194-196, 1984.

[3] B.J. Baliga, "Pinch Rectifier", U. S. Patent 4,641,174, Issued February 3, 1987.

[4] M. Mehrotra and B. J. Baliga, "Very Low Forward Drop JBS Rectifiers Fabricated using Sub-micron Technology", IEEE Transactions on Electron Devices, Vol. 41, pp. 1655-1660, 1994.

[5] M. Mehrotra and B. J. Baliga, "Schottky Barrier Rectifier with MOS Trench", U. S. Patent 5,365,102, Issued November 15, 1994.

[6] B.J. Baliga, "Schottky Barrier Rectifiers and Methods of Forming the Same", U.S. Patent 5,612,567, March 18, 1997.

[7] S. Mahalingam and B.J. Baliga, "The Graded Doped Trench MOS Barrier Schottky Rectifiers", Solid State Electronics, Vol. 43, pp. 1-9, 1999.

[8] B.J. Baliga, "The Future of Power Semiconductor Technology", Proceedings of the IEEE, Vol. 89, pp. 822-832, 2001.

[9] L. Tu and B. J. Baliga, "Schottky Barrier Rectifier including Schottky Barrier regions of Differing Barrier Heights", U. S. Patent 5,262,668, November 16, 1993.

[10] B.J. Baliga, "Advanced Power Rectifier Concepts", Springer-Science, 2009.

[11] R. Held, N. Kaminski and E. Niemann, "SiC Merged p-n/Schottky Rectifiers for High Voltage Applications", Silicon Carbide and Related Materials – 1997, Material Science Forum, Vol. 264-268, pp. 1057-1060, 1998.

[12] F. Dahlquist *et al.*, "Junction Barrier Schottky Diodes in 4H-SiC and 6H-SiC", Silicon Carbide and Related Materials – 1997, Material Science Forum, Vol. 264-268, pp. 1061-1064, 1998.

[13] B. J. Baliga, "Analysis of a High Voltage Merged PiN/Schottky (MPS) Rectifier", IEEE Electron Device Letters, Vol. 8, pp. 407-409, 1987.

[14] F. Dahlquist *et al.*, "A 2.8kV JBS Diode with Low Leakage", Silicon Carbide and Related Materials – 1999, Material Science Forum, Vol. 338-342, pp. 1179-1182, 2000.

[15] D. Peters *et al.*, "Comparison of 4H-SiC pn, Pinch, and Schottky Diodes for the 3kV Range", Silicon Carbide and Related Materials – 2001, Material Science Forum, Vol. 389-393, pp. 1125-1128, 2002.

[16] F. Dahlquist, H. Lendenmann and M. Ostling, "A JBS Diode with controlled Forward Temperature Coefficient and Surge Current Capability", Silicon Carbide and Related Materials – 2001, Material Science Forum, Vol. 389-393, pp. 1129-1132, 2002.

[17] R. Singh *et al.*, "High Power 4H-SiC JBS Rectifiers", IEEE Transactions on Electron Devices, Vol. 49, pp. 2054-2063, 2002.

[18] J. Wu *et al.*, "4,308V, 20.9 mO-cm2 4H-SiC MPS Diodes Based on a 30 micron Drift Layer", Silicon Carbide and Related Materials – 2003, Material Science Forum, Vol. 457-460, pp. 1109-1112, 2004.

[19] R. Rupp *et al.*, "2nd Generation" SiC Schottky Diodes", IEEE International Symposium on Power Semiconductor Devices and ICs, pp. 1-4, 2006.

[20] P. Brosselard *et al.*, "High Temperature behavior of 3.5 kV 4H-SiC JBS Diodes", IEEE International Symposium on Power Semiconductor Devices and ICs, pp. 285-288, 2007.

[21] B.A. Hull, "Performance and Stability of Large Area 4H-SiC 10-kV Junction Barrier Schottky Rectifiers", IEEE Transactions on Electron Devices, Vol. 55, pp. 1864-1870, 2008.

[22] K. Mochizuki *et al.*, "Influence of Lateral Spreading of Implanted Aluminum Ions and Implantation-Induced Defects on Forward Current-Voltage Characteristics of 4H-SiC Junction Barrier Schottky Diodes", IEEE Transactions on Electron Devices, Vol. 56, pp. 992-997, 2009.

[23] M. Mehrotra and B. J. Baliga, "The Trench MOS Barrier Schottky Rectifier", IEEE International Electron Devices Meeting, Abstract 28.2.1, pp. 675-678, 1993.

[24] S. Mahalingam and B. J. Baliga, "A Low Forward Drop High Voltage Trench MOS Barrier Schottky Rectifier with Linearly Graded Doping Profile", IEEE International Symposium on Power Semiconductor Devices and ICs, Paper 10.1, pp. 187-190, 1998.

[25] V. Khemka, V. Ananthan and T. P. Chow, "A 4H-SiC Trench MOS Barrier Schottky (TMBS) Rectifier", IEEE International Symposium on Power Semiconductor Devices and ICs, pp. 165-168, 1999.

[26] Q. Zhang, M. Madangarli and T. S. Sudarshan, "SiC Planar MOS-Schottky Diode", Solid State Electronics, Vol. 45, pp. 1085-1089, 2001.

[27] B.J. Baliga, "Silicon Carbide Power Devices", World Scientific Press, Singapore, 2005.

[28] M. Praveen, S. Mahalingam and B. J. Baliga, "Silicon Carbide Dual Metal Schottky Rectifiers", PSRC Technical Working Group Meeting, Report TW-97-002-C, 1997

[29] B.J. Baliga, "High Voltage Silicon Carbide Devices", Material Research Society Symposium Proceedings, Vol. 512, pp. 77-88, 1998.

[30] K. J. Schoen *et al.*, "A Dual Metal Trench Schottky Pinch-Rectifier in 4H-SiC", IEEE Electron Device Letters, Vol. 19, pp. 97-99, 1998.

[31] F. Roccaforte *et al.*, "Silicon Carbide Pinch Rectifiers using a Dual-Metal Ti-NiSi Schottky Barrier", IEEE Transactions on Electron Devices, Vol. 50, pp. 1741-1747, 2003.

[32] F. Roccaforte *et al.*, "Silicon Carbide Pinch Rectifiers using a Dual-Metal Ti-NiSi Schottky Barrier", IEEE Transactions on Electron Devices, Vol. 50, pp. 1741-1747, 2003.

[33] S. Yishida *et al.*, "A New GaN based Field Effect Schottky Barrier Diode with a Very Low On-State Voltage Operation", IEEE International Symposium on Power Semiconductor Devices and ICs, Paper P-41, pp. 323-326, 2004.

Chapter 7

P-i-N Rectifiers

Power device applications, such as motor control, require rectifiers with blocking voltages ranging from 300 volts to 5000 volts. Silicon P-i-N rectifiers are widely used for these high voltage applications. In a P-i-N rectifier, the reverse blocking voltage is supported across a depletion region formed with a P-N junction structure. The voltage is primarily supported within the n-type drift region with the properties of the p-type region optimized for good on-state current flow. Any given reverse blocking voltage can be supported across a thinner drift region by utilizing the punch-through design[1]. Since it is beneficial to use a low doping concentration for the n-type drift region in this design, it is referred to as an i-region (implying that the drift region is intrinsic in nature). The silicon P-i-N rectifiers that are designed to support large voltages rely upon the high level injection of minority carriers into the drift region. This phenomenon greatly reduces the resistance of the thick, very lightly doped drift region necessary to support high voltages in silicon. Consequently, the on-state current flow is not constrained by the low doping concentration in the drift region. A reduction of the thickness of the drift region, by utilizing the punch-through design, is beneficial for decreasing the on-state voltage drop.

In the case of silicon carbide rectifiers, the drift region doping level is relatively large and its thickness is much smaller than for silicon devices to achieve very high breakdown voltages as discussed in chapter 3. This enables the design of 4H-SiC based Schottky rectifiers with reverse blocking capability of at least 5000 volts with low on-state voltage drop. Based upon the inherent fast switching capability of Schottky rectifiers, it is anticipated that 4H-SiC JBS rectifiers will displace silicon P-i-N rectifiers for applications with reverse blocking capability of up to 5000 volts[2]. The 4H-SiC P-i-N rectifiers are of interest for applications that require blocking voltage of 10-kV and higher. Their performance is discussed in this chapter.

7.1 P-i-N Rectifier Structure

The reverse blocking voltage capability for the P-i-N rectifier is achieved using a punch-through drift region design[1]. In this case, the drift region has a low doping concentration to make the electric field as uniform as possible across the drift region. The thickness of the drift region is chosen to achieve the desired reverse breakdown voltage. The punch-through design was discussed in section 3.2.

The on-state current flow in the P-i-N rectifier is governed by three current transport mechanisms: (a) at very low current levels, the current transport is dominated by the recombination process within the space charge layer of the P-N junction — referred to as the *recombination current*; (b) at low current levels, the current transport is dominated by the diffusion of minority carriers injected into the drift region — referred to as the *diffusion current*; and (c) at high current levels, the current transport is dictated by the presence of a high concentration of both electrons and holes in the drift region — referred to as *high-level injection current*. These current transport phenomena are discussed in detail in the textbook[1]. At the on-state operating current levels, current flow in the P-i-N rectifier is governed by the third process with injection of mobile carriers with concentrations far greater than the background doping concentration of the drift region. Only this process is considered here.

7.2 Reverse Blocking

The reverse blocking capability for the P-i-N rectifier is governed by Eq. [3.20]. As discussed in chapter 3, the thickness of the drift region in the P-i-N rectifier can reduced by using this approach. The lowest doping concentration in the drift (or i-region) is determined by technological considerations.

7.2.1 Silicon 10-kV Device

The example of a 10-kV design will be analyzed in this chapter for comparison of 4H-SiC and silicon devices. The lowest doping concentration feasible for silicon devices is 1×10^{12} cm^{-3}. Using this value, the drift region thickness is found to be 908 microns to achieve a reverse breakdown voltage of 10-kV.

7.2.2 4H-SiC 10-kV Device

In the case of 4H-SiC, the lowest doping concentration that is feasible is 1×10^{14} cm^{-3}. Using this value, the drift region thickness is found to be 75 microns to achieve a reverse breakdown voltage of 10-kV. The large reduction in drift region thickness produces a reduction in the stored charge within the device. This in turn improves the reverse recovery as shown in this chapter.

7.3 On-State Stored Charge

P-i-N rectifiers operate at on-state current density in excess of 10 A/cm^2 and typically at 100 A/cm^2. At these levels, the drift region operates under high level injection conditions. The physics of operation under high level injection conditions is discussed in detail in the textbook[1].

7.3.1 High Level Injection Physics

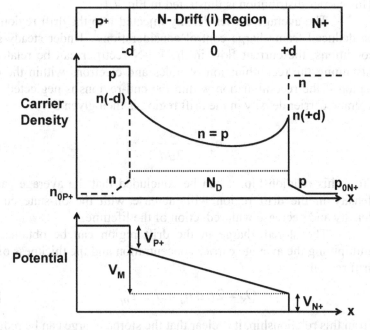

Fig. 7.1 Carrier and potential distribution under high-level injection conditions for a P-i-N rectifier.

The N-drift region in the P-i-N rectifier must be lightly doped in order to support high voltages in the reverse blocking mode. When the forward bias applied to the rectifier increases, the injected minority carrier concentration also increases in the drift region until it ultimately exceeds the background doping concentration (N_D) in the drift region. This is defined as *high level injection*. The carrier distribution in the drift region under high-level injection conditions is given by[1]:

$$n(x) = p(x) = \frac{\tau_{HL} J_{ON}}{2qL_a} \left[\frac{\cosh(x/L_a)}{\sinh(d/L_a)} - \frac{\sinh(x/L_a)}{B\cosh(d/L_a)} \right] \qquad [7.1]$$

where J_{ON} is the on-state current density, τ_{HL} is the high-level lifetime in the drift region, q is the charge for the electron, and 2d is the width of the drift region, and L_a is the ambipolar diffusion length, The parameter B is given by:

$$B = \frac{\left(\mu_n + \mu_p\right)}{\left(\mu_n - \mu_p\right)} \qquad [7.2]$$

This carrier distribution is illustrated in Fig. 7.1.

The average carrier density injected into the drift region can be deduced from charge control considerations[1]. Under steady-state conditions, the current flow in the P-i-N rectifier can be related to sustaining the recombination of holes and electrons within the drift region if the recombination within the end-regions is neglected. The average carrier density in the drift region is then given by:

$$n_a = \frac{J_T \tau_{HL}}{2qd} \qquad [7.3]$$

From this relationship, it can be concluded that the average carrier density in the drift region will increase with the on-state current density and decrease with reduction of the lifetime.

The stored charge in the drift region can be obtained by multiplying the average carrier concentration and the thickness of the drift region:

$$Q_S = 2qd\, n_a = J_{ON}\, \tau_{HL} \qquad [7.4]$$

From this relationship, it is clear that the stored charge can be reduced by decreasing the lifetime.

7.3.2 High Level Injected Carriers: Silicon

The injected carrier profile for the 10-kV silicon P-i-N rectifier can be calculated by using Eq. [7.2]. The value for B for silicon is 2.14. The results obtained by using various high-level lifetime values are shown in Fig. 7.2 for the case of an on-state current density of 100 A/cm². In all cases, the injected concentration of holes and electrons is far greater than the doping concentration (1×10^{12} cm^{-3}) in the drift region. The lifetime values were chosen to maintain a reasonable on-state voltage drop as discussed later. The carrier distribution has a minimum close to the center of the drift region.

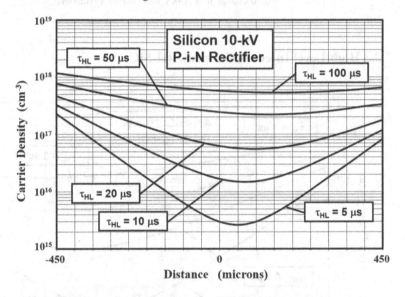

Fig. 7.2 Carrier distribution under high-level injection conditions for a 10-kV silicon P-i-N Rectifier with various high-level lifetime values.

The stored charge in the drift region for the 10-kV silicon P-i-N rectifier can be computed using Eq. [7.4]. It has the values shown in Table 7.1. The stored charge in the silicon 10-kV Pi-N rectifier is large due to the large high level-lifetime values required to maintain a reasonable on-state voltage drop. This results in very large reverse recovery time and slow switching speed for the silicon 10-kV P-i-N rectifier as discussed later.

High Level Lifetime (microseconds)	Stored Charge (Coulombs/cm^2)
100	1e-2
50	5e-3
20	2e-3
10	1e-3
5	5e-4

Table 7.1 Stored charge in a 10-kV silicon P-i-N rectifier.

7.3.3 High Level Injected Carriers: 4H-SiC

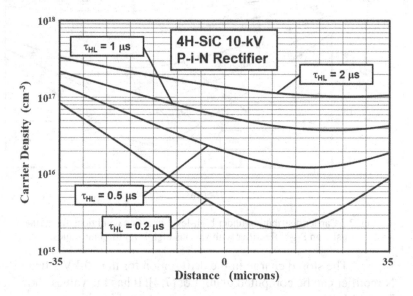

Fig. 7.3 Carrier distribution under high-level injection conditions for a 10-kV 4H-SiC P-i-N Rectifier with various high-level lifetime values.

The injected carrier profile for the 10-kV 4H-SiC P-i-N rectifier can also be calculated by using Eq. [7.2]. The value for B for 4H-SiC is 1.235. The results obtained by using various high-level lifetime values are shown in Fig. 7.3 for the case of an on-state current density of 100 A/cm^2. In all cases, the injected concentration of holes and electrons is far greater than the doping concentration (1×10^{14} cm^{-3}) in the drift region. The lifetime values were chosen to maintain a reasonable on-state voltage drop as discussed later.

High Level Lifetime (microseconds)	Stored Charge (Coulombs/cm^2)
2	2e-4
1	1e-4
0.5	5e-5
0.2	2e-5

Table 7.2 Stored charge in a 10-kV 4H-SiC P-i-N rectifier.

The stored charge in the drift region for the 10-kV 4H-SiC P-i-N rectifier can be computed using Eq. [7.4]. It has the values shown in Table 7.2. The stored charge in the 4H-SiC 10-kV Pi-N rectifier is much smaller than that in the 10-kV silicon P-i-N rectifier due to the smaller high level-lifetime values required to maintain a reasonable on-state voltage drop. This results in very small reverse recovery time and fast switching speed for the 4H-SiC10-kV P-i-N rectifier as discussed later.

7.4 On-State Voltage Drop

The on-state voltage drop for a P-i-N rectifier can be derived from the carrier distribution profile[1]. It is given by:

$$V_{ON} = \frac{2kT}{q} \ln\left[\frac{J_T d}{2qD_a n_i F(d/L_a)} \right]$$ [7.5]

In this expression, the function F(d/L$_a$) is a strongly dependent on the high-level lifetime. It can be calculated using the expressions in the textbook and has a maximum value when d/L$_a$ becomes equal to unity. Consequently, the lowest on-state voltage drop occurs when the high-level lifetime is chosen so that the ambipolar diffusion length becomes equal to half the width of the drift region.

7.4.1 Silicon 10-kV Device

The calculated on-state voltage drop for a silicon 10-kV P-i-N rectifier with a drift region width of 908 microns is shown in Fig. 7.4 at an on-state current density of 100 A/cm^2. As expected, the on-state voltage drop exhibits a minimum at a (d/L$_a$) ratio of unity and increases rapidly when the (d/L$_a$) ratio exceeds 3. The cases of high level-lifetime of 5, 10, 20, 50, and 100 microseconds are indicated on the figure to relate the on-state voltage drop and the stored charge.

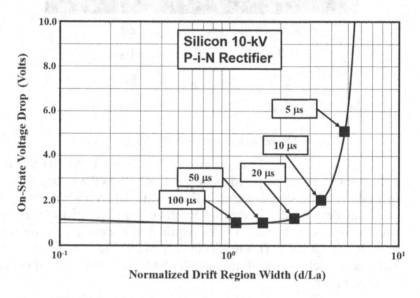

Fig. 7.4 On-state voltage drop for a 10-kV silicon P-i-N rectifier.

7.4.2 4H-SiC 10-kV Device

The calculated on-state voltage drop for a 4H-SiC 10-kV P-i-N rectifier with a drift region width of 75 microns is shown in Fig. 7.5 at an on-state current density of 100 A/cm^2. As expected, the on-state voltage drop exhibits a minimum at a (d/L$_a$) ratio of unity and increases rapidly when the (d/L$_a$) ratio exceeds 3. However, the on-state voltage drop is 3 times larger than for the silicon device at small vales for the d/L$_a$ ratio. The cases of high level-lifetime of 0.2, 0.5, 1, and 2 microseconds are indicated on the figure to relate the on-state voltage drop and the stored charge. The on-state voltage drop remains low even for these very small high-level lifetime values due to the small thickness of the drift region in the 4H-SiC device. Even when the lifetime is reduced to 0.2 microseconds, the on-state voltage drop for the 4H-SiC P-i-N rectifier is only 3.9 volts.

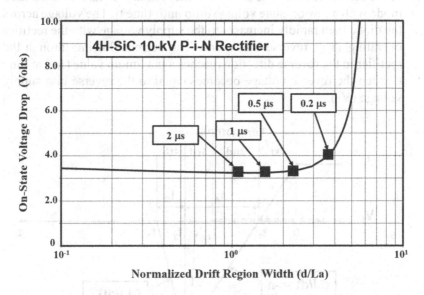

Fig. 7.5 On-state voltage drop for a 10-kV 4H-SiC P-i-N rectifier.

7.5 Reverse Recovery

Power rectifiers control the direction of current flow in circuits used in various power conditioning applications. They operate for part of the time in the on-state when the bias applied to the anode is positive and for the rest of the time in the blocking state when the bias applied to the anode is negative. During each operating cycle, the diode must be rapidly switched between these states to minimize power losses. Much greater power losses are incurred when the diode switches from the on-state to the reverse blocking state than when it is turned on. The stored charge within the drift region of the power rectifier produced by the on-state current flow must be removed before it is able to support high voltages. This produces a large reverse current for a short time duration. This phenomenon is referred to as *reverse recovery*.

In power electronic circuits, it is common-place to use power rectifiers with an *inductive load*. In this case, the current reduces at a constant ramp rate ('a') as illustrated in Fig. 7.6 until the diode is able to support voltage. Consequently, a large *peak reverse recovery current* (J_{PR}) occurs due to the stored charge followed by the reduction of the current to zero. The power rectifier remains in its forward biased mode with a low on-state voltage drop until time t_1. The voltage across the diode then rapidly increases to the supply voltage with the rectifier operating in its reverse bias mode. The current flowing through the rectifier in the reverse direction reaches a maximum value (J_{PR}) at time t_2 when the reverse voltage becomes equal to the reverse bias supply voltage (V_S).

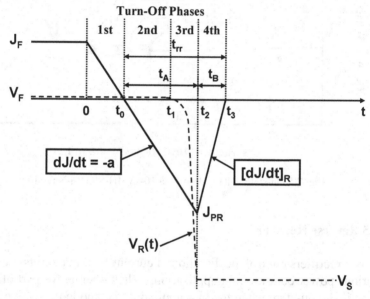

Fig. 7.6 Anode current and voltage waveforms for the P-i-N Rectifier during the reverse recovery process.

The simultaneous presence of a high current and voltage produces large instantaneous power dissipation in the power rectifier. The peak reverse recovery current also flows through the power switch that is controlling the switching event. This increases the power losses in the transistor[1]. In the case of typical motor control PWM circuits that utilize IGBTs as power switches, a large reverse recovery current can trigger latch-up failure that can destroy both the transistor and the rectifier. It is therefore desirable to reduce the magnitude of the peak reverse recovery current and the time duration of the recovery transient. This time duration is referred to as the *reverse recovery time* (t_{rr}).

An analytical model for the reverse recovery process for the turn-off of a P-i-N rectifier under a constant rate of change of the current (*current ramp-rate*) was created by assuming that the concentration of the free carriers in the drift region can be linearized[1]. The time t_0 at which the current crosses zero is given by:

$$t_0 = \frac{J_F}{a} \qquad [7.6]$$

The second phase of the turn-off process occurs from the time t_0 at which the current crosses zero up to the time t_1 when the P^+/N junction can begin to support voltage. This time is given by[1]:

$$t_1 = \frac{J_F}{a} \sqrt{\frac{L_a}{(Kd - L_a)} + 1} \qquad [7.7]$$

where

$$K = \left[\frac{\cosh(-d/L_a)}{\sinh(d/L_a)} - \frac{\sinh(-d/L_a)}{B\cosh(d/L_a)} \right] \qquad [7.8]$$

During the third phase of the turn-off transient, the P-i-N rectifier begins to support an increasing voltage. This requires the formation of a space-charge region $W_{SC}(t)$ at the P^+/N junction that expands with time. The expansion of the space-charge region is achieved by further extraction of the stored charge in the drift region resulting in the reverse current continuing to increase after time t_1. The growth of the reverse bias voltage across the P-i-N rectifier can be analytically modeled under the assumption that the sweep out of the stored charge is occurring at an approximately constant current. The evolution of the space charge region is given by[1]:

$$W_{SC}(t) = \frac{a}{2qn_a}\left(t^2 - t_1^2\right) - \frac{J_{ON}}{qn_a}\left(t - t_1\right) \qquad [7.9]$$

The voltage supported across this space-charge region can be obtained by solving Poisson's equation:

$$V_R(t) = \frac{q\left[N_D + p(t)\right]}{2\varepsilon_S} W_{SC}(t)^2 \qquad [7.10]$$

This expression, in conjunction with Eq. [7.9] for the expansion of the space-charge width, indicates a rapid rise in the voltage supported by the P-i-N rectifier after time t_1. The end of the third phase occurs when the reverse bias across the P-i-N rectifier becomes equal to the supply voltage (V_S). Using this value in Eq. [7.10] together with   Eq. [7.9], the time t_2 (and hence J_{PR}) can be obtained.

During the fourth phase of the turn-off process, the reverse current rapidly reduces at approximately a constant rate as illustrated in Fig. 7.6 while the voltage supported by the P-i-N rectifier remains constant at the supply voltage. The stored charge within the drift region after the end of the third phase is removed during this time. The equations for obtaining time t_B are provided in the textbook[1].

7.5.1 Silicon 10-kV Device

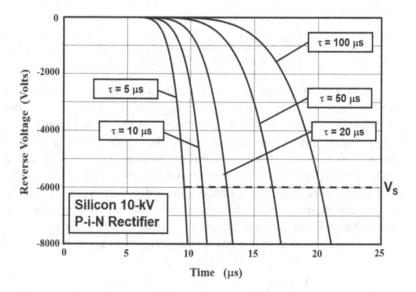

Fig. 7.7 Analytically calculated voltage waveforms for a 10 kV P-i-N rectifier during the reverse recovery process for various lifetime values.

The reverse recovery waveforms for the 10-kV silicon P-i-N rectifier are discussed here for the case of a ramp rate of 2×10^7 A/cm²-s. The voltage waveforms calculated using the analytical solutions are shown in Fig. 7.7 for the case of a high level lifetimes of 5, 10, 20, 50, and

100 microseconds. The voltages rises more rapidly when the lifetime is reduced.

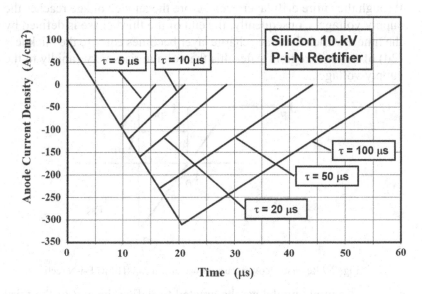

Fig. 7.8 Analytically calculated current waveforms for a 10 kV silicon P-i-N rectifier during the reverse recovery process using various lifetime values.

The peak reverse recovery current occurs at the end of the third phase. The reverse recovery current waveforms obtained using the analytical models are shown in Fig. 7.8 for the various values of lifetime. The peak reverse recovery current densities predicted by the analytical model are 90, 120, 160, 230, and 310 A/cm^2 for lifetime values of 5, 10, 20, 50, and 100 microseconds, respectively. The reverse recovery time (t_{RR}) increases from 11 to 16 to 23 to 39 to 55 microseconds as the high level lifetime is increased from 5 to 10 to 20 to 50 to 100 microseconds. These very long reverse recovery times limit the maximum operating frequency for the 10-kV silicon rectifier as discussed later.

7.5.2 4H-SiC 10-kV Device

The reverse recovery waveforms for the 10-kV 4H-SiC P-i-N rectifier are discussed here. The same ramp rate of 2×10^7 A/cm^2-s was used again for comparison with the silicon 10 kV P-i-N rectifier. The 4H-SiC P-i-N rectifier does not behave the same way as the silicon diode because of the much thinner drift layer and reduced stored charge. The first two phases for turn-off are the same as for the silicon device. However, during the third phase, the space charge region expands through the entire drift layer even before the anode voltage reaches the supply voltage. Consequently, the end of the third phase is defined by the time at which the space charge layer becomes equal to the thickness (2d) of the drift region. After this the voltage rises very rapidly to the supply voltage.

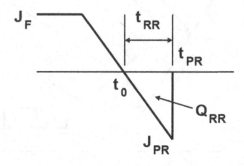

Fig. 7.9 Reverse recovery current waveform for 4H-SiC P-i-N rectifier.

A simple model can be created to define the end of the third phase for the 4H-SiC P-i-N rectifier by assuming that all the stored charge in the drift region is removed during the reverse recovery time (t_{RR}). The charge removed during reverse recovery is given by:

$$Q_{RR} = \frac{1}{2} J_{PR} t_{PR} \qquad\qquad [7.11]$$

The peak reverse recovery current is also related to the ramp rate by:

$$J_{PR} = a t_{PR} \qquad\qquad [7.12]$$

Combining these relationships with Eq. [7.4] for the stored charge in the drift region yields:

$$t_{RR} = \sqrt{\frac{2\tau_{HL} J_{ON}}{a}} \qquad\qquad [7.13]$$

and

$$J_{PR} = \sqrt{2 a \tau_{HL} J_{ON}} \qquad\qquad [7.14]$$

These expressions allow obtaining the peak reverse recovery current and reverse recovery times for the 4H-SiC P-i-N rectifier.

Based up on the above model, the voltage waveforms calculated for the 4H-SiC 10-kV rectifier, with a drift region doping concentration of 10^{14} cm^{-3}, using the analytical solutions with a ramp rate of 2×10^7 A/cm^2-s are shown in Fig. 7.10 for the case of a high level lifetimes of 5, 10, 20, 50, and 100 microseconds. The voltages rises at an earlier time when the lifetime is reduced. For all the cases, the abrupt rise in voltage occurs when it reaches a value of less than 500 volts well below the supply voltage of 6 kV.

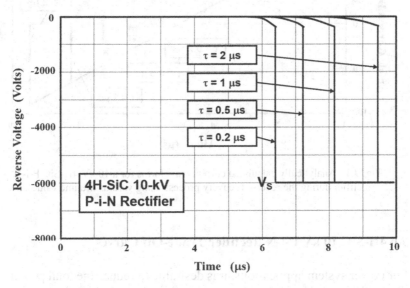

Fig. 7.10 Analytical voltage waveforms for 10 kV 4H-SiC P-i-N rectifiers.

The peak reverse recovery current occurs at the end of the third phase. The reverse recovery current waveforms obtained using the analytical model for the 4H-SiC P-i-N rectifier are shown in Fig. 7.11 for the various values of lifetime. The peak reverse recovery current densities predicted by the analytical model are 28.3, 44.7, 63.2, and 89.4 A/cm^2 for lifetime values of 0.2, 0.5, 1, and 2 microseconds, respectively. The reverse recovery time (t_{RR}) increases from 1.4 to 2.2 to 3.2 to 4.5 microseconds as the high level lifetime is increased from 0.2 to 0.5 to 1 to 2 microseconds. These very short reverse recovery times allow the 4H-SiC 10-kV P-i-N rectifier to operate at high frequencies as discussed later. However, the abrupt (or snappy) reverse recovery of the current creates large voltages across stray inductances in power circuits and produces oscillations in the current.

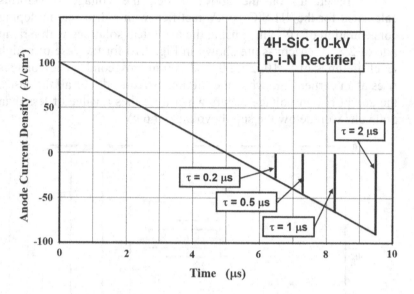

Fig. 7.11 Analytically calculated current waveforms for a 10 kV 4H-SiC P-i-N rectifier during the reverse recovery process using various lifetime values.

7.6 4H-SiC 10 kV P-i-N Rectifier Trade-Off Curves

For power system applications, it is desirable to reduce the total power dissipation produced in the rectifiers to maximize the power conversion efficiency. This also reduces the heat generated within the power devices maintaining a lower junction temperature which is desirable to prevent thermal runaway and reliability problems. In the previous sections, it was demonstrated that the peak reverse recovery current and the turn-off time can be reduced by reducing the minority carrier lifetime in the drift region of the P-i-N rectifier structure. This enables reduction of the power losses during the switching transient. However, the on-state voltage drop in a P-i-N rectifier increases when the minority carrier lifetime is reduced, which produces an increase in the power dissipation during on-state current flow. To minimize the power dissipation, it is common-place to perform a trade-off between on-state and switching power losses for power P-i-N rectifiers by developing trade-off curves.

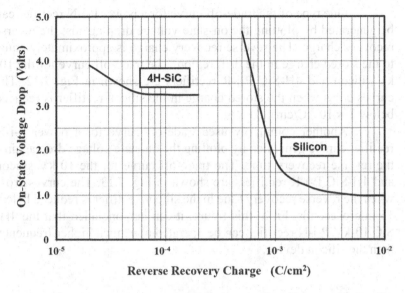

Fig. 7.12 Trade-off curves for the 10-kV P-i-N rectifiers using reverse recovery charge.

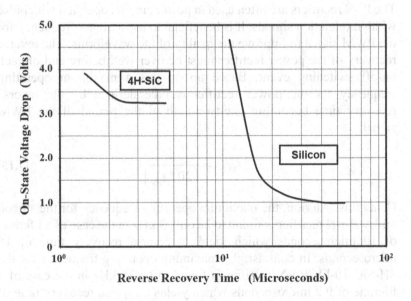

Fig. 7.13 Trade-off curves for the 10-kV P-i-N rectifiers using reverse recovery time.

One type of the trade-off curve for a power P-i-N rectifier can be generated by plotting the on-state voltage drop against the reverse recovery charge. The reverse recovery charge is approximately equal to the stored charge in the drift region. The trade-off curve for the 10-kV silicon and 4H-SiC P-i-R rectifiers are shown in Fig. 7.12. The curves cross when the stored charge in the silicon rectifier is reduced below 7×10^{-4} C/cm^2.

Another commonly used trade-off curve for a power P-i-N rectifier can be generated by plotting the on-state voltage drop against the reverse recovery time. The trade-off curve for the 10-kV silicon and 4H-SiC P-i-R rectifiers are shown in Fig. 7.13. The curves cross when the reverse recovery time in the silicon rectifier is reduced below 15 microseconds. From this figure, it can be concluded that the 4H-SiC 10-kV P-i-N rectifier can be operated at a much higher frequency than the silicon device.

7.7 4H-SiC 10 kV PiN Rectifier Maximum Operating Frequency

The Pi-N rectifiers are often used in power circuits operated with pulse width modulation signals. It is beneficial to use a higher frequency for the PWM signal to improve the quality of the waveforms. The reverse recovery of the power rectifier must be over well before end of each on-off switching event. Based up on this a maximum operating frequency for the power rectifier can be defined by the reverse recovery time becoming one-twentieth of the period of the PWM signal:

$$f_{MAX} = \frac{1}{[20 * t_{RR}]} \qquad [7.15]$$

Using this criterion, the maximum operating frequency for the silicon 10-kV P-i-N rectifier is found to be only 3 kHz in the case of a lifetime of 10 microseconds which yields a reverse recovery time of 16 microseconds. In contrast, the maximum operating frequency for the 4H-SiC 10-kV P-i-N rectifier is found to be 35 kHz in the case of a lifetime of 0.2 microseconds which yields a reverse recovery time of 1.4 microseconds.

7.8 4H-SiC P-i-N Rectifier Experimental Results

Many groups have worked on the development of high voltage P-i-N rectifiers over the years. In early studies, the quality of the P^+ region was poor due to problems with achieving a high doping concentration[3]. The best results were obtained by epitaxial growth of the anode followed by etching steps to create the edge termination[4]. More recently, the anode has been formed by ion implantation of Al or B or both dopants. The forward characteristic of implanted junction diodes was found to be sensitive to the junction depth and activation process[5]. The on-state voltage drop of a 4H-SiC diode fabricated using 100 micron thick drift region with doping concentration of $1-3 \times 10^{14}$ cm^{-3} has been reported[3] at 7.1 volts for a current density of 100 A/cm^2. These diodes had a breakdown voltage of 8.6 kV. This is significantly worse than that obtained by the numerical simulations despite the reported lifetime in the diodes being about 2 microseconds. Some of this can be attributed to the high contact resistance to the anode.

The on-state voltage drop of diodes formed by co-implantation of aluminum, carbon, and boron to create the anode region[6] has been reported to be about 4.7 volts at room temperature. These diodes, fabricated using a 40 micron thick drift region with doping concentration of 1×10^{15} cm^{-3}, had a breakdown voltage of about 4.5 kV. The on-state voltage drop was found to reduce with increasing temperature at a current density of 100 A/cm^2. This was attributed to a reduction in the contact resistance to the anode region. The lifetime extracted from the reverse recovery measurements was about 21 nanoseconds. These results confirm the conclusion that fast switching PiN rectifiers can be developed from 4H-SiC.

By making improvements to the epitaxial growth process, 4H-SiC power rectifiers with blocking voltage of up to 20 k V were reported in 2004[7]. This required continuous growth of a 100 micron thick drift layer with doping concentration of 2×10^{14} cm^{-3} followed by the anode layer of 2 microns in thickness and doping concentration of 8×10^{18} cm^{-3}. The 10-kV 4H-SiC P-i-N rectifier had an on-state voltage drop of only 3.9 volts at an on-state current density of 100 A/cm^2. The reverse recovery time for these diodes was 0.2 microseconds with a peak reverse recovery current equal to the on-state current density. The diode yield was greatly improved by reduction of basal plane dislocations.

The use of carbon implantation and thermal oxidation to improve the lifetime in the drift region of P-i-N rectifiers was reported in 2012[8]. A 120 micron thick epitaxial layer with doping concentration of 7×10^{13} cm^{-3} was grown for a targeted breakdown voltage of 18.5 kV. An on-state voltage drop of only 4 volts at an on-state current density of 100 A/cm^2 was achieved. The reverse recovery time was reported to be only 0.1 microseconds with the peak reverse current equal to the on-state current level. However, these tests were performed at a low supply voltage of 200 volts.

The reverse recovery behavior of 4H-SiC 6.5 kV, 1 kA P-i-N rectifiers has been compared with those of silicon diodes for medium voltage converter applications[9,10]. The 6 kV 4H-SiC diodes had an on-state voltage drop of 3.42 volts at 100 A/cm^2. The 4H-SiC diode module was constructed using 3.5 mm × 3.5 mm diode chips with an anode area of 7.1 mm^2. The on-state characteristics of the silicon and 4H-SiC P-i-N rectifiers crossed at a current density of 40 A/cm^2 and on-voltage of 3.5 volts. The 4H-SiC devices had superior on-state voltage drop above this current density. The reverse recovery current and time for the 4H-SiC rectifiers was demonstrated to be far better than the silicon diodes. However, it was found that the 4H-SiC rectifiers exhibit a snappy reverse recovery which produces severe ringing. This is consistent with the model in section 7.5.2.

A record high 21.7 kV breakdown voltage was reported for a 4H-SiC diode in 2012[11]. A space-modulated JTE edge termination was employed to achieve a breakdown voltage within 81% of the ideal breakdown voltage of the 186 micron thick epitaxial layer with doping concentration of 2.3×10^{14} cm^{-3}. The on-state voltage drop for the diodes was 9.34 volts at an on-state current density of 50 A/cm^2.

7.9 4H-SiC P-i-N Rectifier Forward Bias Instability

An unusual increase in on-state voltage drop after forward current flow through the 4H-SiC power P-N junction diodes for long periods of time, called the *bipolar degradation phenomenon*, has been reported[12]. The mechanism for this phenomenon[13] is: (a) electron-hole recombination occurring in the i-region of the diode provides energy to the 4H-SiC lattice allowing the basal plane dislocations (BPDs) to create stacking faults; (b) the stacking fault traps electrons becoming negatively charged; (c) the injection of electrons from the cathode is inhibited by the negative charge in the vicinity of the stacking faults; (d) the electron-hole injected concentration near the stacking faults is greatly reduced suppressing current flow through these regions; (e) the loss of current carrying area increases the current density over the rest of the device area; and (f) the increased current density results in the observed rise in on-state voltage drop. The creation and propagation of the stacking fault from the bottom of the epitaxial layer near the cathode junction towards the P-N junction at the top of the device has been observed using light emission images[12]. The lifetime is not reduced by this phenomenon nor is the leakage current degraded[14].

7.10 GaN P-i-N Rectifier Experimental Results

GaN P-i-N diodes have been recently fabricated by the growth of epitaxial layers on GaN substrates[15]. Devices with breakdown voltages ranging from 600 to 3700 volts were reported. A breakdown voltage of 1700 volts was obtained using epitaxial layers with doping concentration of 1×10^{16} cm^{-3} and thickness of 18 microns. This is 60% of the breakdown voltage predicted by using the impact ionization coefficients (see chapter 3). The diodes exhibited the expected knee voltage of 3.0 volts and an on-state voltage drop of 3.2 volts at a current density of 100 A/cm^2. Devices with breakdown voltage of 3.7 kV were obtained using epitaxial layers with doping concentration of 5 to 6×10^{15} cm^{-3} and thickness of 40 microns. This is close to the breakdown voltage predicted by using the impact ionization coefficients (see chapter 3). The diodes exhibited the expected knee voltage of 3.0 volts and an on-state voltage drop of 3.3 volts at a current density of 100 A/cm^2. The low on-state voltage drop was observed despite a very low lifetime of 2 ns measured in the drift region. The reverse leakage current for the diodes increased by only 2 orders of magnitude with increasing reverse bias.

7.11 Summary

The physics of operation of the silicon carbide PiN rectifier has been shown to be similar to that used to describe silicon devices. High level injection in the drift region enables modulating its conductivity to reduce the on-state voltage drop. Due to the much smaller width of the drift region required to obtain a given breakdown voltage when compared with silicon devices, a much smaller lifetime can be used in silicon carbide devices. This is favorable for reducing switching losses. However, the large band gap of silicon carbide results in a large on-state voltage drop of 3 to 4 volts. Consequently, 4H-SiC PiN rectifiers are superior to 4H-SiC Schottky rectifiers only when the blocking voltage exceeds 10 kV.

References

[1] B.J. Baliga, "Fundamentals of Power Semiconductor Devices", Springer Scientific, New York, 2008.

[2] B.J. Baliga, "Advanced Power Rectifier Concepts", Springer-Science, New York, 2009.

[3] R. Singh, "Silicon Carbide Bipolar Power Devices – Potentials and Limits", MRS Symposium Proceedings, Vol. 640, pp. H4.2.1-H4.2.12, 2001.

[4] Y. Sugawara, K. Asano, R. Singh and J.W. Palmour", "6.2 kV 4H-SiC pin Diode with Low Forward Voltage Drop", Silicon Carbide and Related Materials – 1999, Material Science Forum, Vol. 338-342, pp. 1371-1374, 2000.

[5] R.K. Chilukuri, P. Ananthanarayanan, V. Nagapudi and B.J. Baliga, "High Voltage P-N Junction Diodes in Silicon Carbide using Field Plate Edge Termination", MRS Symposium Proceedings, Vol. 572, pp. 81-86, 1999.

[6] J.B. Fedison *et al.*, "Al/C/B Co-implanted High Voltage 4H-SiC PiN Junction Rectifiers", Silicon Carbide and Related Materials – 1999, Material Science Forum, Vol. 338-342, pp. 1367-1370, 2000.

[7] M.K. Das *et al.*, "High Power, Drift-Free 4H-SiC PIN Diodes", International Journal of High Speed Electronics and Systems, Vol. 14, pp. 860-864, 2004.

[8] K. Nakayama *et al.*, "Characteristics of a 4H-SiC PiN Diode with Carbon Implantation/Thermal Oxidation", IEEE Transactions on Electron Devices, Vol. 59, pp. 895-901, 2012.

[9] F. Filsecker, R. Alvarez and S. Bernet, "Comparison of 6.5 kV Silicon and SiC Diodes", IEEE Energy Conversion Congress and Exhibition, pp. 2261-2267, 2012.

[10] F. Filsecker, R. Alvarez and S. Bernet, "Characterization of a New 6.5 kV 1000 A SiC Diode for Medium Voltage Converters", IEEE Energy Conversion Congress and Exhibition, pp. 2253-2260, 2012.

[11] H. Niwa *et al.*, "Breakdown Characteristics of 12-20 kV class 4H-SiC PiN Diodes with Improved Junction Termination Structure",

IEEE International Symposium on Power Semiconductor Devices and ICs, pp. 381-384, 2012.

[12] R. Singh, "Reliability and Performance Limitations in SiC Power Devices", Microelectronics Reliability, Vol. 46, pp. 713-730, 2006.

[13] M.K. Das *et al.*, "Ultra High Power 10 kV, 50 A SiC PiN Diodes", IEEE International Symposium on Power Semiconductor Devices and ICs, pp. 159-162, 2005.

[14] P. Brosselard *et al.*, "The Effect of the Temperature on the Bipolar Degradation of 3.3 kV 4H-SiC PiN Diodes", IEEE International Symposium on Power Semiconductor Devices and ICs, pp. 237-240, 2008.

[15] I.C. Kizilyalli *et al.*, "Vertical Power p-n Diodes based on Bulk GaN", IEEE Transactions on Electron Devices, Vol. 62, pp. 414-422, 2015.

Chapter 8

MPS Rectifiers

In the previous chapter, it was shown that the reverse recovery transient in P-i-N rectifier produces large power dissipation in the diodes and the switching transistors. The reverse recovery transient was shown to be related to the presence of stored charge in the drift region during on-state current flow. The MPS rectifier structure was created to reduce the stored charge within silicon power rectifiers[1] in the 1980s by merging the physics of the P-i-N rectifier and the Schottky rectifier. The MPS rectifier structure is illustrated in Fig. 8.1. In this structure, the drift region is designed using the same criteria as used for P-i-N rectifiers in order to support the desired reverse blocking voltage. The device structure contains a P-N junction over a portion under the metal contact and a Schottky contact for the remaining portion. It is convenient to utilize the same metal layer for making an ohmic contact to the P^+ region and a Schottky contact to the N- drift region.

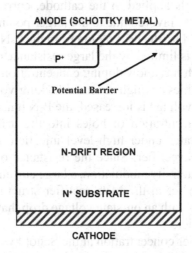

Fig. 8.1 The MPS rectifier structure.

8.1 Device Physics

As shown in Fig. 8.1, the P-i-N rectifier and Schottky rectifier are located in close proximity in the MPS rectifier structure allowing the mingling of the physics of operation of both devices. The space between the P-N junctions in the MPS rectifier structure is designed so that it is pinched off at a relatively small reverse bias voltage. After the depletion of the space between the P-N junctions, a potential barrier is formed under the Schottky metal which screens the contact from the reverse bias applied to the cathode terminal. The electric field at the Schottky contact can be greatly reduced in the MPS rectifier when compared with the normal Schottky rectifier by judicious choice of the space between the P-N junctions. This allows suppression of the Schottky barrier lowering reducing the leakage current in silicon devices well below that for Schottky rectifiers. An even greater reduction of the leakage current can be achieved for silicon carbide devices due to suppression of the thermionic field emission current. Consequently, it is possible to achieve good reverse blocking characteristics in the MPS rectifier structure. This behavior is the same as discussed for the JBS rectifier in a previous chapter.

In the JBS rectifier, on-state current flow occurs only via the Schottky contact. In contrast, the on-state current flow in the MPS rectifier occurs via the P-N junction and the Schottky contact. At low on-state bias levels applied to the cathode, current transport via the Schottky contact is favored due to the larger potential required for the injection of holes into the drift region from the P-N junction. However, this current flow is limited by the large resistance of the un-modulated drift region which has a low doping concentration and large thickness as required to achieve a high reverse blocking voltage capability. As the forward bias voltage is increased, the P-N junction begins to inject a significant concentration of holes into the drift region. The drift region then operates under high-level injection conditions as in the case of the P-i-N rectifier. Since the resistance of the drift region is reduced by conductivity modulation, a large current flow can occur via the Schottky contact in the MPS rectifier structure. This allows on-state current flow with an on-state voltage drop that is smaller than that for a P-i-N rectifier.

The carrier concentration at the Schottky contact remains low in the MPS rectifier because it does not inject a significant concentration of minority carriers into the drift region. The resulting carrier profile is superior to that observed in the P-i-N rectifier in terms of charge removal during the turn-off transient. The MPS rectifier exhibits a smaller peak reverse recovery current and charge leading to reduced switching power loss. Moreover, the trade-off between the on-state voltage drop and reverse recovery power loss can be generated by changing the relative area of the P-N junction and Schottky contact in the MPS rectifier structure. A further refinement of the trade-off curve can be achieved by using the lifetime control techniques commonly used for the P-i-N rectifier.

8.2 On-State Characteristics

The MPS rectifier operates like the JBS rectifier at low current levels because there is insufficient voltage to produce injection from the P-N junction. The P-N junction begins to inject holes into the drift region at high current densities. The hole concentration exceeds the doping concentration resulting in operation under high-level injection conditions like in a P-i-N rectifier. The resulting modulation of the conductivity of the drift layer allows efficient current flow via the Schottky contact with a lower on-state voltage drop than a P-i-N rectifier.

8.2.1 Low Forward Bias Conditions

The N-drift region in the MPS rectifier must be lightly doped in order to support a high voltage in the reverse blocking mode. At small forward bias voltages, the voltage across the P-N junction produces low-level injection of holes into the N-type drift region. There is no conductivity modulation of the drift region at these small forward bias voltages. The resistance of the drift region is therefore determined by the doping concentration. Current transport through the MPS rectifier occurs by the thermionic emission process at the Schottky contact followed by current flow through the drift region. The high resistance of the drift region in MPS rectifiers designed to support large reverse bias voltages initially limits the current transport. The MPS rectifier characteristics resemble those of a Schottky rectifier at these small forward bias voltages.

The forward conduction *i-v* characteristics of the MPS rectifier can be analytically modeled as a metal-semiconductor (Schottky) contact with the series resistance of the drift region. In the analytical model, it is important to include the impact of the larger current density at the Schottky contact when compared with the cathode current density because the P-N junction occupies a portion of the upper surface area. The current constriction in the region between the P-N junctions and current spreading from this region into the drift region enhances the series resistance. This resistance can be modeled as previous performed for the JBS rectifier. Unlike for the JPS rectifiers, the P-N junction depth is a very small fraction of the drift region thickness in the case of the MPS rectifiers because they are designed to support large voltages. The drift region resistance in the MPS rectifier is quite close to the one-dimensional resistance of the drift region. In order to keep this resistance as small as possible, it is necessary to employ a non-punch-through design for the drift region.

Due to the large thickness of the drift region in relation to the junction depth and the width of the window used for the P^+ diffusions in the MPS rectifier structure, the current path in the drift region will invariably overlap before reaching the N^+ substrate. The current flow pattern was illustrated in Fig. 6.2 with the shaded area. In the MPS rectifier structure, the thickness of the drift region (t in Fig. 6.2) is much larger than that for the JBS rectifier.

The current density at the Schottky contact (J_{FS}) is enhanced due to the presence of the P^+ region and the depletion layer at the P-N junction. This increases the voltage drop across the Schottky contact. The current through the Schottky contact flows only within the un-depleted portion (with dimension 'd') of the drift region at the top surface. Consequently, the current density at the Schottky contact (J_{FS}) is related to the cell (or cathode) current density (J_{FC}) by:

$$J_{FS} = \left(\frac{p}{d}\right) J_{FC} \qquad [8.1]$$

where p is the cell pitch. The dimension 'd' is determined by the cell pitch (p), the size of the P^+ ion-implant window (2s), and the on-state depletion width ($W_{D,ON}$):

$$d = p - s - W_{D,ON} \qquad [8.2]$$

Depending up on the lithography used for device fabrication to minimize the size (dimension 's') of the P^+ region, the current density at the Schottky contact can be considerably enhanced. This must be taken into account when computing the voltage drop across the Schottky contact given by:

$$V_{FS} = \phi_B + \frac{kT}{q} \ln\left(\frac{J_{FS}}{AT^2}\right) \qquad [8.3]$$

The specific resistance for the drift region can be calculated by using Eq. [6.7] derived for the JBS rectifier:

$$R_{sp,drift} = \frac{\rho_D \cdot p \cdot (x_J + W_{D,ON})}{d} + \rho_D \cdot p \cdot \ln\left(\frac{p}{d}\right) + \rho_D \cdot (t - s - x_J - 2W_{D,ON})$$

$$[8.4]$$

The on-state voltage drop for the MBS rectifier at a small forward bias, including the substrate contribution, is then given by:

$$V_F = \phi_B + \frac{kT}{q} \ln\left(\frac{J_{FS}}{AT^2}\right) + \left(R_{sp,drift} + R_{sp,subs}\right) J_{FC} \qquad [8.5]$$

When computing the on-state voltage drop using this equation, it is satisfactory to make the approximation that the depletion layer width can be computed by subtracting the on-state voltage drop from the built-in potential of the P-N junction.

8.2.2 High Level Injection Conditions

When the forward bias applied to the MPS rectifier increases, the injected minority carrier concentration from the P-N junction also increases in the drift region until it ultimately exceeds the background doping concentration (N_D) in the drift region resulting in *high level injection*. When the injected hole concentration in the drift region becomes much greater than the background doping concentration, charge neutrality requires that the concentrations for electrons and holes become equal:

$$n(x) = p(x) \qquad [8.6]$$

The large concentration of free carriers reduces the resistance of the drift region resulting in *conductivity modulation* of the drift region. As in the case of the P-i-N rectifier, conductivity modulation of the drift region is beneficial for allowing the transport of a high current density through lightly doped drift regions with a low on-state voltage drop.

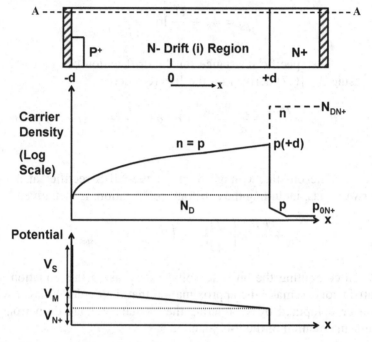

Fig. 8.2 Carrier and potential distribution under high-level injection conditions
for the MPS rectifier.

The carrier distribution within the drift region for the MPS
rectifier is different from that observed for the P-i-N rectifier because
of the presence of the Schottky contact. The carrier distribution $p(x)$
can be obtained by solving the continuity equation for holes in the N-
region[1]:

$$\frac{d^2 p}{dx^2} - \frac{p}{L_a^2} = 0 \qquad [8.7]$$

where L_a is the ambipolar diffusion length given by:

$$L_a = \sqrt{D_a \tau_{HL}} \qquad [8.8]$$

The general solution for the carrier concentration governed by
Eq. [8.7] is given by:

$$p(x) = A\cosh\left(\frac{x}{L_a}\right) + B\sinh\left(\frac{x}{L_a}\right) \qquad [8.9]$$

with the constants A and B determined by the boundary conditions for the N-drift region.

For the MPS rectifier, it is appropriate to solve for the carrier profile along the path indicated by the dashed line marked 'A-A' in Fig. 8.2 which is located through the Schottky contact. At the interface between the N-drift region and the N$^+$ cathode region (located at x = +d in Fig. 8.2), the total current flow occurs exclusively by electron transport:

$$J_{FC} = J_n(+d)$$ [8.10]

and

$$J_p(+d) = 0$$ [8.11]

Using these equations:

$$J_{FC} = 2qD_n \left(\frac{dp}{dx}\right)_{x=+d}$$ [8.12]

The second boundary condition occurs at the junction between the N-drift region and the Schottky contact (located at x = –d in Fig. 8.2). Here, the hole concentration becomes zero due to negligible injection at the Schottky contact:

$$p(-d) = 0$$ [8.13]

The above boundary conditions can be used to obtain the constants A and B in Eq. [8.13]:

$$A = -\frac{L_a J_{FC}}{2qD_n} \left[\frac{sinh(-d/L_a)}{cosh(-d/L_a)cosh(d/L_a) - sinh(-d/L_a)sinh(d/L_a)} \right]$$ [8.14]

$$B = -\frac{L_a J_{FC}}{2qD_n} \left[\frac{cosh(-d/L_a)}{cosh(-d/L_a)cosh(d/L_a) - sinh(-d/L_a)sinh(d/L_a)} \right]$$ [8.15]

Using these constants in Eq. [8.9] and simplifying the expression yields:

$$p(x) = n(x) = \frac{L_a J_{FC}}{2qD_n} \frac{sinh\left[(x+d)/L_a\right]}{cosh\left[2d/L_a\right]}$$ [8.16]

The carrier distribution described by this equation was schematically illustrated in Fig. 8.2. It has a maximum value at the interface between the drift region and the N^+ substrate with a magnitude of:

$$p(+d) = p_M = \frac{L_a J_{FC}}{2qD_n} \frac{sinh[2d/L_a]}{cosh[2d/L_a]} = \frac{L_a J_{FC}}{2qD_n} tanh\left(\frac{2d}{L_a}\right) \qquad [8.17]$$

and reduces monotonically when proceeding towards the Schottky contact in the negative x direction. The concentration becomes equal to zero at the Schottky contact as required to satisfy the boundary condition used to derive the expression.

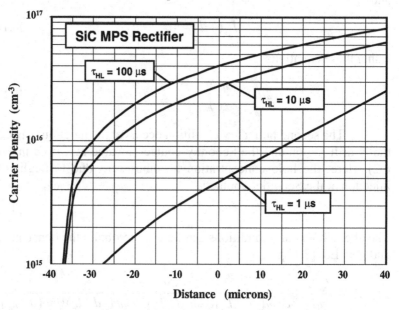

Fig. 8.3 Carrier distribution under high-level injection conditions for the 10 kV SiC MPS rectifier with various high-level lifetime values.

The carrier distributions calculated at an on-state current density of 100 A/cm^2 by using Eq. [8.9] are shown in Fig. 8.3 for the case of three values for the high-level lifetime for a 10-kV 4H-SiC MPS rectifier. These device have a drift region thickness of 80 microns and doping concentration of 5×10^{14} cm^{-3}. The largest concentrations for the electrons and holes in the drift region occur at its boundary with the N^+ end-region. The carrier concentration at this boundary decreases when the lifetime is reduced. It has a magnitude of 8.2×10^{16} cm^{-3} for a high lifetime of 100 microseconds, 6.3×10^{16} cm^{-3} for a moderate lifetime of 10 microseconds, and 2.6×10^{16} cm^{-3} for a low lifetime of 1 microsecond. When the lifetime is reduced to 1 microsecond, the analytical model predicts that the entire drift region is not conductivity modulated. The carrier profiles for the higher lifetime cases are close to linear in shape.

8.2.3 On-State Voltage Drop

At low on-state current density levels, the on-state characteristics for the silicon carbide MPS rectifier will resemble those for the Schottky rectifier with an enhanced current density at the Schottky contact. At larger current levels with high-level injection in the drift region, the on-state voltage drop for the silicon carbide MPS rectifier can be obtained by summing the voltage drops along the path marked 'A-A' in Fig. 8.2 through the Schottky contact. The total voltage drop along this path consists of the voltage drop across the Schottky contact (V_{FS}), the voltage drop across the drift region (middle region voltage V_M), and the voltage drop at the interface with the N$^+$ substrate (V_{N+}):

$$V_{ON} = V_{FS} + V_M + V_{N+} \qquad [8.18]$$

The voltage drop across the Schottky contact is given by:

$$V_{FS} = f_{BN} + \frac{kT}{q} ln\left(\frac{J_{FS}}{AT^2} \right) \qquad [8.19]$$

where the current density at the Schottky contact (J_{FS}) is related to the cell or cathode current density by Eq. [8.1]. The voltage drop in the drift (middle) region with inclusion of recombination is given by:

$$V_M = \left\{ \frac{4D_n}{(\mu_n + \mu_p)} \left[\frac{(d / L_a)}{tanh(2d / L_a)} \right] - \frac{kT}{2q} \right\} ln\left(\frac{2d}{x_J} \right) \qquad [8.20]$$

The voltage drop across the interface between the drift region and the N$^+$ substrate is given by:

$$V_{N+} = \frac{kT}{q} ln\left[\frac{P_M}{N_D} \right] = \frac{kT}{q} ln\left[\frac{J_{FC}L_a tanh(2d / L_a)}{2qD_n N_D} \right] \qquad [8.21]$$

The on-state voltage drop for the MPS rectifier can be computed by utilizing the three components discussed above:

$$V_{ON} = f_{BN} + \frac{kT}{q} ln\left(\frac{J_{FC} p}{AT^2 d}\right)$$

$$+ \left\{\frac{4D_n}{(\mu_n + \mu_p)}\left[\frac{(d/L_a)}{tanh(2d/L_a)}\right] - \frac{kT}{2q}\right\} ln\left(\frac{2d}{x_J}\right) \qquad \text{[8.22]}$$

$$+ \frac{kT}{q} ln\left[\frac{J_{FC} L_a tanh(2d/L_a)}{2qD_n N_D}\right]$$

The on-state voltage drop for the MPS rectifier is a function of the lifetime in the drift region because the middle region and N/N^+ interface voltage drops change with lifetime.

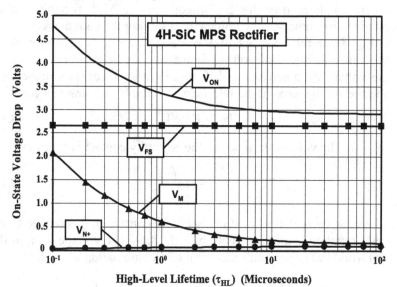

Fig. 8.4 Voltage drops in the 10 kV 4H-SiC MPS rectifier.

As an example, consider the case of a silicon carbide MPS rectifier with a drift region doping concentration of 5×10^{14} cm^{-3} and thickness of 80 microns capable of supporting 10,000 volts in the reverse blocking mode. The variation of the on-state voltage drop with high-level lifetime in the drift region for this case as predicted by the analytical model is provided in Fig. 8.4 for the case of a Schottky barrier height of 2.95 eV. It can be observed that the on-state voltage drop begins to increase when the lifetime is reduced below 10 microseconds. This is due to the relatively low diffusion length for holes in silicon carbide. The three components of the on-state voltage drop are also shown in the figure. The voltage drop at the Schottky contact is independent of the lifetime. Due to the large Schottky barrier height used in the analytical model, this contribution to the on-state voltage drop is large. The voltage drop at the N/N^+ interface increases slightly when the lifetime becomes larger than 1 microsecond but makes only a small contribution to the total on-state voltage drop. The most significant increase in the voltage drop occurs for the middle region when the lifetime is reduced below 10 microseconds. It is therefore necessary to achieve lifetime values approaching 10 microseconds to obtain a low on-state voltage drop in the 4H-SiC MPS rectifier. The on-state voltage drop of the 10 kV silicon carbide MPS rectifier is found to be just under 3 volts for a drift region lifetime of 10 microseconds. This is remarkable because it is less than the on-state voltage drop for a 4H-SiC P-i-N rectifier.

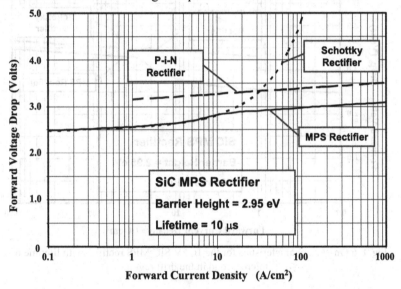

Fig. 8.5 On-state characteristics for the 10 kV SiC MPS rectifier with lifetime of 10 microseconds in the drift region.

At low current densities, the on-state voltage drop for the 4H-SiC MPS rectifier resembles that for the Schottky rectifier as observed in Fig. 8.5 for current density up to 10 A/cm^2. At larger forward current densities, the injected carrier density in the drift region exceeds the background doping concentration leading to high level injection conditions. In this mode of operation, the injected carrier concentration in the drift region increases in proportion to the current density resulting in a constant voltage drop across the drift region for the analytical model without recombination in the N^+ end region. From Eq. [8.22], an expression for the on-state current density can be derived:

$$J_{FC} = \sqrt{\frac{2qAT^2 D_n N_D}{p} \frac{(d/L_a)}{tanh(2d/L_a)}} e^{-\frac{q(f_{BN}+V_M)}{2kT}} e^{\frac{qV_{ON}}{2kT}} \qquad [8.23]$$

The current flow is observed to become proportional to ($qV_{ON}/2kT$), similar to that observed for the P-i-N rectifier under high-level injection conditions, as observed in Fig. 8.5. It is worth pointing out that the on-state voltage drop for the MPS rectifier is lower than that for the P-i-N rectifier.

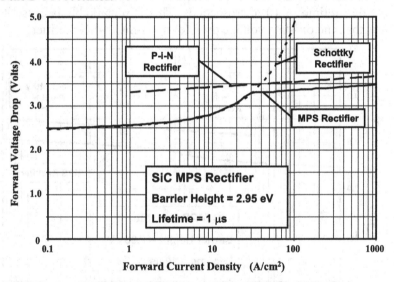

Fig. 8.6 On-state characteristics for the 10 kV SiC MPS rectifier with lifetime of 1 microsecond in the drift region.

The impact of reducing the lifetime to 1 microsecond in the drift region on the on-state characteristics for the 10 kV 4H-SiC MPS rectifier is shown in Fig. 8.6. The on-state voltage drop for the silicon carbide MPS rectifier increases by more than that for the P-i-N rectifier making its on-state voltage drop closer to that for the P-i-N rectifier. However, the stored charge in the 4H-SiC MPS rectifier is smaller making its switching performance superior to that for the P-i-N rectifier as shown later in the chapter.

The on-state voltage drop for the silicon carbide MPS rectifier is also dependent on the barrier height of the Schottky metal. According to the analytical model, a reduction in the on-state voltage drop of the silicon carbide MPS rectifier can be achieved by reducing the barrier height as illustrated in Fig. 8.7 where the characteristics of devices with two barrier heights can be compared. The analytical model predicts a decrease in on-state voltage drop that is equal to the reduction of the barrier height.

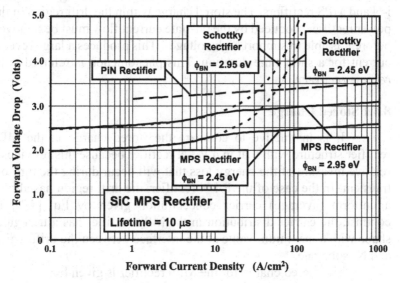

Fig. 8.7 On-state characteristics for 10 kV SiC MPS rectifiers with different Schottky barrier heights.

8.3 Reverse Blocking

The reverse blocking characteristics for the MPS rectifier can be modelled by using the same approach as described for the JBS rectifiers in chapter 6. The leakage current in the MPS rectifiers is low due to suppression of the Schottky barrier lowering and tunneling by the shielding provided to the Schottky contacts by the P-N junctions.

8.4 Switching Performance

Power rectifiers operate for part of the time in the on-state when the bias applied to the anode is positive and for the rest of the time in the blocking state when the bias applied to the anode is negative. During each operating cycle, the diode must be rapidly switched between these states to minimize power losses. Much greater power losses are incurred when the diode switches from the on-state to the reverse blocking state than when it is turned on. The presence of a large concentration of free carriers in the drift region during on-state current flow is responsible for the low on-state voltage drop of high voltage P-i-N and MPS rectifiers. The stored charge within the drift region of the power rectifier produced by the on-state current flow must be removed before it is able to support high voltages. This produces a large reverse current for a short time duration. This phenomenon is referred to as *reverse recovery*.

8.4.1 Stored Charge

It is instructive to compare the stored charge in the MPS rectifier structure with that in a P-i-N rectifier because this provides a relative measure of the energy loss that will occur during the turn-off transient. In the case of the P-i-N rectifier, the drift region has almost a uniform (average) carrier concentration given by Eq. [7.3]. In contrast, the carrier distribution in the MPS rectifier has a triangular shape with a maximum value at the interface between the drift region and N^+ substrate.

The stored charge for the MPS rectifier is given by:

$$Q_S = \frac{1}{2} q\, p_M \left(2d\right) = \frac{q\, d\, L_a\, J_{FC}}{2 D_n} tanh\left(\frac{2d}{L_a}\right) \qquad [8.24]$$

For a lifetime of 10 microseconds in the drift region, the maximum carrier concentration is found to be 6.3×10^{16} cm^{-3} at an on-state current density of 100 A/cm^2. The total stored charge in the drift region of the 4H-SiC MPS rectifier is then found to be 40 μC/cm^2. The stored charge in the 4H-SiC MPS rectifier is 2.5-times smaller than for the 4H-SiC P-i-N rectifier with a lifetime of 1 microsecond in the drift region. Moreover, the carrier concentration for the MPS rectifier is zero at the P-N junction during on-state operation. This allows the device to begin supporting a reverse voltage much faster than in the case of the P-i-N rectifier which shortens the reverse recovery process making the peak reverse recovery current of the MPS rectifier much smaller than that of the P-i-N rectifier.

8.4.2 Reverse Recovery: Highly Doped Drift Layer

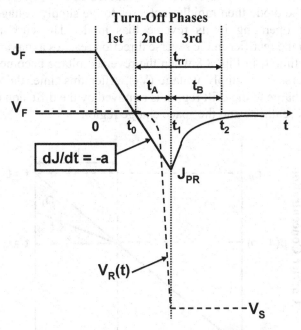

Fig. 8.8 Reverse recovery waveforms for the MPS Rectifier.

As discussed previously in chapter 6, it is common-place to use power rectifiers with an *inductive load* with the current reducing at a constant ramp rate ('a'). A large *peak reverse recovery current* (J_{PR}) occurs due to the stored charge followed by the reduction of the current to zero. In the case of the P-i-N rectifier, the device remains in its forward biased mode with a low on-state voltage drop even after the current reverses direction as illustrated in Fig. 7.6 due to the high minority carrier concentration at the junction in the initial on-state. Until the minority carrier concentration reduces to zero (at time t_1 in Fig. 7.6), the junction is unable to support a reverse blocking voltage. The voltage across the diode then rapidly increases to the supply voltage with the rectifier operating in its reverse bias mode. The current flowing through the rectifier in the reverse direction reaches a maximum value (J_{PR}) (at time t_2 in Fig. 7.6) when the rev-erse voltage becomes equal to the reverse bias supply voltage (V_S).

In the case of the MPS rectifier, the carrier concentration at the Schottky contact is zero in the on-state making the carrier concentration at the junction also close to zero. Consequently, this device is able to support a reverse blocking voltage immediately after the current reverses direction as illustrated in Fig. 8.8. The voltage across the diode then rapidly increases to the supply voltage with the rectifier operating in its reverse bias mode. The current flowing through the rectifier in the reverse direction reaches a maximum value (J_{PR}) at time t_1 in Fig. 8.8 when the reverse voltage becomes equal to the reverse bias supply voltage (V_S). After this time, the remaining stored charge in the drift region is removed by the diffusion of carriers to the N^+ cathode and the space charge region.

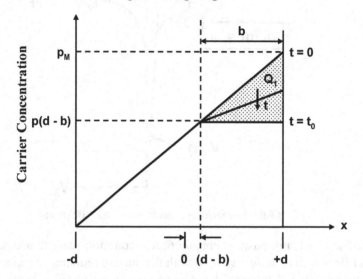

Fig. 8.9 Carrier distribution profiles in the MPS rectifier during the first phase of the reverse recovery process.

An analytical model for the reverse recovery process for the turn-off of the MPS rectifier under a constant rate of change of the current (*current ramp-rate*) can be created by assuming that the initial concentration of the free carriers in the drift region can be linearized as illustrated in Fig. 8.9. As shown in the figure, the initial carrier distribution established by the on-state current flow is linear with the concentration increasing from zero at $x = -d$ to the concentration of p_M at $x = +d$:

$$p(x) = n(x) = \left(\frac{x+d}{2d}\right)p_M \qquad [8.25]$$

The hole and electron concentrations in the drift region can be assumed to be equal in the on-state and during the turn-off transient because of charge neutrality. In the case of finite lifetime in the drift region, the maximum carrier concentration p_M is given by Eq. [8.21].

During the first phase of the turn-off process, the current flow reduces from the on-state current density to zero. Since the current flow remains in the forward direction, the Schottky contact and the P-N junction remain in forward bias during this time interval. The hole and electron concentrations remain close to zero at the Schottky contact and the P-N junction to satisfy the boundary conditions at the Schottky contact (as in the case of the on-state analysis). However, electrons are extracted from the cathode side during this time interval. The current flowing at the cathode due to the diffusion of electrons is given by:

$$J = 2qD_n \left(\frac{dn}{dx}\right)_{x=+d}$$
[8.26]

If the carrier distribution during the first phase is linearized as illustrated in Fig. 8.9, the slope of the carrier profile near the cathode is given by:

$$\left(\frac{dn}{dx}\right)_{x=+d} = \left(\frac{dp}{dx}\right)_{x=+d} = \frac{J(t)}{2qD_n} = \frac{J_{ON} - at}{2qD_n}$$
[8.27]

where 'a' is the current ramp rate. At the end of the first phase (time t_0 in Fig. 8.8), the current becomes equal to zero leading to a zero slope for the carrier profile as illustrated in Fig. 8.9. As in the case of the turn-off analysis for the P-i-N rectifier, it will be assumed that the carrier profile pivots around a fixed point located at a distance 'b' from the interface between the drift region and the N^+ substrate.

The distance 'b' in Fig. 8.9 can be obtained by relating the charge Q_1 removed during the first phase to the current flow. Note that the x-values are defined from the center of the drift region as shown in Fig. 8.9. The hole concentration at $x = (d - b)$ can be assumed to remain the same as during the on-state operation if recombination is neglected during the turn-off transient. This assumption is justified because the turn-off time is much shorter than the lifetime in the drift region. The change in the stored charge within the drift region during the first phase can be obtained from the shaded area, indicated by Q_1, in the figure:

$$Q_1 = \frac{qb}{2}\left[p_M - p(d-b)\right]$$
[8.28]

Using Eq. [8.28] for the initial carrier profile, the hole concentration at x = (d − b) is given by:

$$p(d-b) = \left(\frac{2d-b}{2d}\right) p_M \qquad [8.29]$$

Substituting this into Eq. [8.31] yields:

$$Q_1 = \frac{qp_M}{4d} b^2 \qquad [8.30]$$

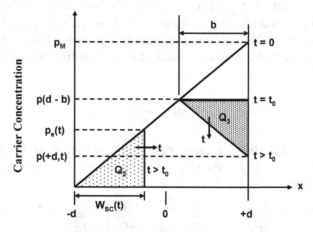

Fig. 8.10 Carrier distribution profiles in the MPS rectifier during the second phase of the reverse recovery process.

According to the charge control principle, this charge can be related the current flow during the turn-off transient from t = 0 to t = t_0:

$$Q_1 = \int_0^{t_0} J(t)\,dt = \int_0^{t_0} (J_{ON} - at)\,dt = J_{ON} t_0 - \frac{at_0^2}{2} = \frac{J_{ON}^2}{2a} \qquad [8.31]$$

because the time t_0 at which the current crosses zero is given by:

$$t_0 = \frac{J_F}{a} \qquad [8.32]$$

Combining the above relationships:

$$b = \sqrt{\frac{2d}{qap_M} J_{ON}} \qquad [8.33]$$

The distance 'b' can therefore be calculated from the on-state current density and the ramp rate 'a'.

Once the current density becomes negative at the start of the second phase of the turn-off process, the MPS rectifier immediately begins to support a reverse voltage as illustrated in Fig. 8.8 with the formation of a space-charge region $W_{SC}(t)$ at the P^+/N junction that expands with time as illustrated in Fig. 8.10. The expansion of the space-charge region is achieved by extraction of the stored charge in the drift region in the vicinity of the junction resulting in the reverse current continuing to increase after time t_0. It can be assumed that the initial hole distribution in the vicinity of the junction does not change during the second phase of the turn-off process in the conductivity modulated portion of the N-base region because the lifetime in the drift region is much greater than the switching time interval. Consequently, the concentration of holes at the edge of the space-charge region (p_e) increases during the turn-off process as the space-charge width increases:

$$p_e(t) = p_M \left[\frac{W_{SC}(t)}{2d} \right] \qquad [8.34]$$

According to the charge-control principle, the charge removed by the expansion of the space-charge layer must equal the charge removed due to collector current flow:

$$J(t) = q p_e(t) \frac{dW_{SC}(t)}{dt} = q p_M \left[\frac{W_{SC}(t)}{2d} \right] \frac{dW_{SC}(t)}{dt} \qquad [8.35]$$

by using Eq. [8.37]. The collector current density increases linearly at the ramp rate during the second phase of the turn-off process. Consequently:

$$q p_M \left[\frac{W_{SC}(t)}{2d} \right] \frac{dW_{SC}(t)}{dt} = at \qquad [8.36]$$

where 'a' is the ramp rate. Integrating this equation and applying the boundary condition of zero width for the space-charge layer at time zero provides the solution for the evolution of the space-charge region width with time:

$$W_{SC}(t) = \sqrt{\frac{2da}{q p_M}} t \qquad [8.37]$$

According to this analysis, the space-charge layer expands towards the right-hand-side at a constant rate as indicated by the horizontal time arrow in Fig. 8.10.

The collector voltage supported by the MPS rectifier structure is related to the space charge layer width by:

$$V_R(t) = \frac{q(N_D + p_{SC})W_{SC}^2(t)}{2\varepsilon_S} = \left(\frac{N_D + p_{SC}}{p_M}\right)\left(\frac{da}{\varepsilon_S}\right)t^2 \qquad [8.38]$$

The hole concentration in the space-charge layer (p_{SC}) can be related to the collector current density under the assumption that the carriers are moving at the saturated drift velocity for holes in the space-charge layer:

$$p_{SC}(t) = \frac{J_R(t)}{qv_{sat,p}} \qquad [8.39]$$

In the MPS rectifier, the hole concentration in the space-charge increases during the voltage rise-time because the current density is increasing. The analytical model for turn-off of the MPS rectifier structure predicts a quadratic increase in the collector voltage with time.

The end of the second phase of the turn-off process occurs when the collector voltages reaches the reverse bias supply voltage (V_S). This time interval (t_A in Fig. 8.8) can be obtained by making the reverse bias voltage equal to the supply voltage in Eq. [8.42]:

$$t_A = \sqrt{\frac{\varepsilon_S p_M V_S}{ad(N_D + p_{SC})}} \qquad [8.40]$$

According to the analytical model, the voltage rise-time is proportional to the square root of the reverse bias supply voltage and inversely proportional to square root of the ramp time.

The width of the space-charge layer at the end of the voltage transient can be obtained by using the collector supply voltage:

$$W_{SC}(t_A) = \sqrt{\frac{2\varepsilon_S V_{CS}}{q(N_D + p_{SC})}} \qquad [8.41]$$

The width of the space-charge layer at the end of the second phase depends upon the reverse bias supply voltage and the peak reverse recovery current (via p_{SC}).

Since the end of the second phase occurs when the reverse bias across the MPS rectifier becomes equal to the supply voltage (V_S):

$$t_1 = t_0 + t_A = t_0 + \sqrt{\frac{\varepsilon_S p_M V_S}{ad(N_D + p_{SC})}} \qquad [8.42]$$

The hole concentration (p_{SC}) in the space charge layer is a function of time because of the increasing reverse current density. However, its magnitude is typically much smaller than the doping concentration (N_D) in the drift region for 4H-SiC devices. Consequently, the time (t_1) at which the peak reverse recovery current occurs can be computed using:

$$t_1 = \frac{J_{ON}}{a} + \sqrt{\frac{\varepsilon_S p_M V_S}{ad N_D}} \qquad [8.43]$$

This expression indicates that the time for the end of the second phase is reduced with increasing ramp rate and increased with increasing reverse bias supply voltage. The time taken to reach the peak reverse recovery current after the current crosses zero is defined in Fig. 8.8 as t_A. Using this value of time, the peak reverse recovery current can be obtained:

$$J_{PR} = at_A = \sqrt{\frac{a\varepsilon_S p_M V_S}{d(N_D + p_{SC})}} \qquad [8.44]$$

Based upon this expression, it can be concluded that the peak reverse recovery current will increase with increasing ramp rate and increasing reverse bias supply voltage.

It is worth pointing out that electrons are also removed from the vicinity of the interface between the drift region and the N^+ substrate during the second phase of the turn-off process. This is illustrated in Fig. 8.10 by the shaded area marked Q_3. The slope of the carrier profile in this neutral region on the right-hand-side (RHS) is also related to the reverse current density at any time instant because the current is sustained by the diffusion of electrons towards the N^+ substrate:

$$\left(\frac{dn}{dx}\right)_{RHS} = \left(\frac{dp}{dx}\right)_{RHS} = \frac{J_R(t)}{2qD_n} = -\frac{at}{2qD_n} \qquad [8.45]$$

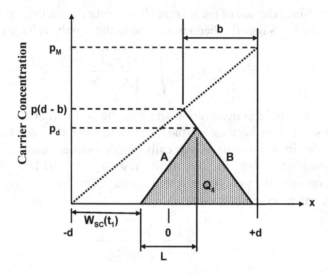

Fig. 8.11 Stored charge within the MPS rectifier at the end of the second phase.

As discussed above, carriers are extracted from the drift region during the second phase by an expanding space charge layer on the left hand side (LHS) and by diffusion of electrons on the right-hand-side. In addition, holes also diffuse towards the space charge layer from the neutral region. In order to account for the reduction in the carrier density due to this process, a linearized carrier profile for the left-hand-side, as shown in Fig. 8.11 by the line marked A, can be used. The positive slope of this line is related to the peak reverse recovery current flowing at the end of the second phase:

$$\left(\frac{dn}{dx}\right)_{LHS} = \left(\frac{dp}{dx}\right)_{LHS} = \frac{J_{PR}}{2qD_p}$$ [8.46]

Using a carrier concentration of zero for this line at $x = [W_{SC}(t_1) - d]$ with the above positive slope yields an equation for line A:

$$P_{LHS}(x) = \frac{J_{PR}}{2qD_p}\left[x + d - W_{SC}(t_1)\right]$$ [8.47]

Similarly, the carrier profile on the right-hand-side of the drift region at the end of the second phase can be derived by using the negative slope given by Eq. [8.48] at time t_1 and the carrier concentration $p(d - b)$:

$$p_{RHS}(x) = p(d-b) - \frac{J_{PR}}{2qD_n}\left[x-(d-b)\right] \qquad \text{[8.48]}$$

The intersection point for these lines provides the maximum carrier density in the drift region at the beginning of the third phase:

$$p_d = \frac{D_n p(d-b)}{D_n + D_p} + \frac{J_{PR}}{2q}\left[\left(\frac{d-b}{D_n + D_p}\right) - \frac{D_n\left[d-W_{SC}(t_I)\right]}{D_p(D_n + D_p)} + \frac{d-W_{SC}(t_I)}{D_p}\right]$$

$$\text{[8.49]}$$

The second term in this expression is much smaller in magnitude than the first term.

During the third phase of the turn-off process, the reverse current reduces at an exponential rate as illustrated in Fig. 8.8 while the voltage supported by the MPS rectifier remains constant at the supply voltage. The stored charge within the drift region after the end of the second phase is illustrated in Fig. 8.11 by the shaded area marked Q_4. This charge must be removed during the third phase of the turn-off process by the diffusion of the free carriers towards the space charge boundary on the left-hand-side and towards the N^+ substrate on the right-hand-side. As in the case of the non-punch-through P-i-N rectifier structure[2], the current in the MPS rectifier reduces is governed by the diffusion of the excess holes remaining in the drift region towards the edges of the space-charge region in a time frame that is much shorter than the recombination lifetime. Since the voltage is supported across the space-charge region during this process, the electric field can be assumed to be small during the diffusion of the excess carriers in the un-depleted region. The drift component of the current in the continuity equation for the excess carriers can therefore be neglected. By using the same approach as used for the non-punch-through P-i-N rectifier, the reverse recovery current for the MPS rectifier during the third phase can be obtained:

$$J_R(t) = J_{PR}\left(\frac{t_1}{t}\right)^3 e^{\frac{L^2}{4D_p}\left(\frac{1}{t}-\frac{1}{t_1}\right)} \frac{\left[1+cotanh\left(L^2/4D_p t\right)\right]}{\left[1+cotanh\left(L^2/4D_p t_1\right)\right]} \qquad \text{[8.50]}$$

Consider the case of a 4H-SiC MPS rectifier designed to support 10,000 volts with a drift region thickness of 80 microns and doping concentration of 5×10^{14} cm^{-3} whose on-state characteristics were discussed in section 8.2.

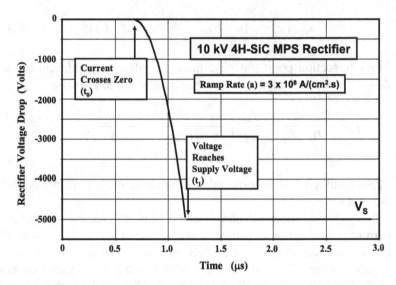

Fig. 8.12 Analytically calculated reverse recovery voltage waveform for the 10 kV 4H-SiC MPS rectifier structure.

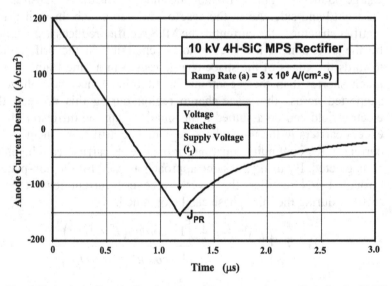

Fig. 8.13 Analytically calculated reverse recovery current waveform for the 10 kV 4H-SiC MPS rectifier structure.

The calculated reverse recovery voltage and current waveforms for the silicon carbide 10 kV MPS rectifier structure are shown in Figs. 8.12 and 8.13 for the case of a ramp rate of 3×10^8 A/cm^2-s. An on-state current density of 200 A/cm^2 was used in this example. The time t_0 at which current crosses zero is 0.667 microseconds. The reverse voltage supported by the silicon carbide MPS rectifier then increases quadratically with time. The time at which the voltage reaches a reverse supply voltage of 5000 volts is indicated in the figure. The time t_1 at which the reverse bias voltage reaches a supply voltage of 5000 volts is 1.18 microseconds for the ramp rate of 3×10^8 A/cm^2-s. This point in the voltage waveforms defines the end of the second phase of the reverse recovery process.

The peak reverse recovery current occurs at the end of the second phase. The peak reverse recovery current densities predicted by the analytical model is 154 A/cm^2 for ramp rate of 3×10^8 A/ cm^2-s. After the second phase, the reverse current decays to zero at an exponential rate that is much faster than the recombination rate dictated by a lifetime of 10 microseconds in the drift region because the model is based upon extraction of the stored charge by diffusion.

8.4.3 Reverse Recovery: Lightly Doped Drift Layer

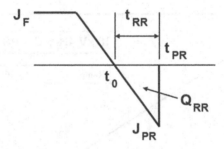

Fig. 8.14 Reverse recovery waveform for the 10 kV 4H-SiC MPS rectifier structure with lightly doped drift layer.

As discussed in chapter 7 for the 4H-SiC rectifier, all the stored charge in the drift region can be removed by the space charge layer well before the voltage reaches the supply voltage if the doping concentration in the drift region is small. This will also occurs for the 4H-SiC MPS rectifier. In this case a simple model for the turn-off can be created by assuming that the charge removed during the reverse recovery is equal to the stored charge. In this model, the current abruptly drops to zero after it reaches the peak value because all the stored charge has been removed at this time.

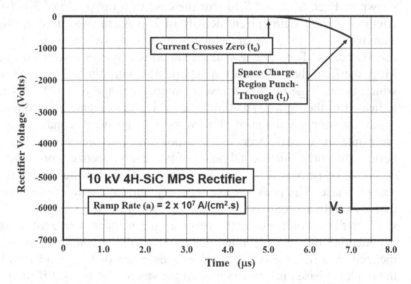

Fig. 8.14 Analytically calculated reverse recovery voltage waveform for the 10 kV 4H-SiC MPS rectifier structure with low drift region doping.

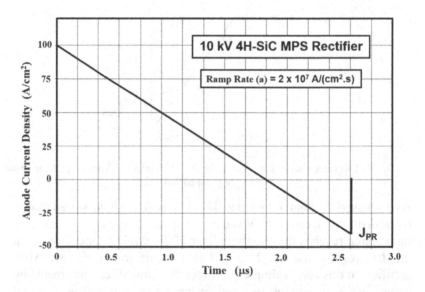

Fig. 8.15 Analytically calculated reverse recovery current waveform for the 10 kV 4H-SiC MPS rectifier structure with low drift region doping.

The charge removed by the reverse recovery current flow is given by:

$$Q_{RR} = \frac{1}{2} J_{PR} t_{RR} = \frac{1}{2} a t_{RR}^2 \quad [8.51]$$

Equating this to the stored charge in the drift region of the MPS rectifier is given by Eq. [8.28] yields:

$$t_{RR} = \sqrt{\frac{2qd\,p_M}{a}} = \sqrt{\frac{d\,L_a\,J_{FC}}{a\,D_n} tanh\left(\frac{2d}{L_a}\right)} \quad [8.52]$$

and

$$J_{PR} = \sqrt{2qad\,p_M} = \sqrt{\frac{ad\,L_a\,J_{FC}}{D_n} tanh\left(\frac{2d}{L_a}\right)} \quad [8.53]$$

The reverse recovery waveforms for the 10 kV 4H-SiC MPS rectifier with a low drift region doping concentration of 1×10^{14} cm^{-3} with a lifetime of 10 microseconds are shown in Figs. 8.14 and 8.15. A ramp rate of 2×10^7 A/cm^2-s was used as in the case of the 4H-SiC P-i-N rectifier turn-off in section 7.5.2 for comparison. In the MPS rectifier, the space charge region spans the entire drift region at a low reverse bias of 620 volts. The time t_{RR} at which this occurs is 2 microseconds according to Eq. [8.52] and the peak reverse recovery current J_{PR} calculated using Eq. [8.53] is 40 A/cm^2. In comparison, the value for t_{RR} and J_{PR} for the 4H-SiC P-i-N rectifier (see section 7.5.2) are 4.5 microseconds and 90 A/cm^2 for a lifetime of 2 microseconds. The on-state voltage drop for the 4H-SiC MPS rectifier is 3.0 volts compared with 3.2 volts for the P-i-N rectifier.

8.5 4H-SiC MPS Rectifier with Low Schottky Barrier Height

In the previous sections, a large barrier height of 2.95 eV was used to illustrate the operating principles of the 4H-SiC MPS rectifier structure. In practice, the metals used as Schottky contacts for 4H-SiC SiC have a barrier height of 1.9 eV at most (reported for Platinum). When the barrier height is reduced, it becomes more difficult to forward bias the P-N junction in the MPS rectifier suppressing the injection of holes into the drift region. This produces a snap-back in the on-state characteristics[2]. This problem can be overcome by making the width of the Schottky contact very small in the 4H-SiC MPS rectifier structure. With a sufficiently small width, the P-N junction depletion width can be sufficient to create a potential barrier under the Schottky contact even at zero bias. This raises the voltage at which current begins to flow through the Schottky which is equivalent to increasing the barrier height. Excellent on-state characteristics can be derived by using this design approach[2].

8.6 4H-SiC MPS Rectifier: Experimental Results

Since the MPS concept is only applicable to 4H-SiC rectifiers with very high blocking voltage capability, there are not many relevant papers in the literature on the device. However, 4H-SiC MPS rectifiers with 10-kV blocking voltage capability have been designed and fabricated at the FREEDM Systems Center using a drift region with doping concentration of 2.7×10^{14} cm^{-3} and thickness of 100 microns[3]. P-i-N and JBS rectifiers were simultaneously fabricated for comparison. The JBS rectifiers had a large spacing of 4 microns between the P-N junctions compared with only 2 micron for the MPS rectifiers. The MPS rectifiers had two designs: a linear cell and a hexagonal cell. The on-state voltage drop for the linear MPS design was very close to that for the P-i-N rectifier at 125 °C with a value of 4.2 volts at an on-state current density of 30 A/cm^2. In comparison, the JBS rectifier had a much higher on-state voltage drop of 8.7 volts. The leakage current for the JBS rectifier was found to be very high, while that for the MPS rectifiers were close to that for P-i-N rectifiers. The reverse recovery charge for the MPS was found to be 60% of that measured for the P-i-N rectifier and only 10% greater than that for the JBS rectifier. These results demonstrate that the 10-kV 4H-SiC MPS rectifier provides a significantly better switching performance when compared with P-i-N rectifiers without higher on-state voltage drop.

The MPS mode of operation has been observed in 4H-SiC JBS rectifiers with low blocking voltages under surge current levels and at elevated temperatures. It has been found that the P-N junctions in the JBS rectifiers begin to inject carriers into the drift region at surge current levels (50x of the on-state current density). This reduces their on-state voltage drop when compared with Schottky rectifiers greatly reducing the power dissipation[4]. The Schottky diodes were destroyed while the MPS rectifiers could survive the surge current pulse.

8.7 Summary

The physics of operation of the MPS rectifier has been analyzed in this chapter. Analytical expressions have been derived for the on-state and blocking state, as well as the reverse recovery transients. At on-state current levels, the injected minority carrier density in the drift region exceeds the relatively low doping concentration required to achieve high breakdown voltages. This high level injection in the drift region modulates its conductivity producing a reduction in the on-state voltage drop. Unlike the P-i-N rectifier, the carrier concentration at the junction is close to zero with a maximum value at the interface between the drift region and the N^+ cathode region. Due to the presence of the Schottky contact with a lower barrier for current flow when compared with the P-N junction, the on-state voltage drop for the MPS rectifier can be smaller than that for the P-i-N rectifier.

The MPS rectifier can support a large voltage in the reverse blocking mode by appropriate choice of the doping concentration and thickness of the drift region. The leakage current in the reverse direction is larger than that for the P-i-N rectifier due to the thermionic current across the Schottky contact. However, this current can be made small by using a large Schottky barrier height and a small area for the Schottky contact.

As in the case of the P-i-N rectifier, the switching of the MPS rectifier from the on-state to the reverse blocking state is accompanied by a significant current flow in the reverse direction. However, the peak reverse recovery current and reverse recovery time are smaller than those observed in the P-i-N rectifier. The reverse recovery charge for the MPS rectifier is much smaller than that for the P-i-N rectifier. This reduces switching losses in the rectifier and the switches in the power circuits. The performance of the MPS rectifier relative to the P-i-N rectifier has been found to improve with increasing temperature[5].

Since a lifetime control process has not matured for silicon carbide bipolar devices, the MPS structure offers an elegant approach to producing high voltage power rectifiers with low on-state voltage as well as much smaller reverse recovery power losses. In addition, the current density flowing through the P-N junction in the silicon carbide MPS rectifier is low because it is a small fraction of on-state current density. This reduced current density at the P-N junction in the silicon carbide MPS rectifier structure suppresses the highly undesirable increase in on-state voltage drop observed during prolonged on-state operation at high current density of silicon carbide P-i-N rectifiers[6].

References

[1] B.J. Baliga, "Analysis of the High-Voltage Merged P-i-N/Schottky (MPS) Rectifier", IEEE Electron Device Letters, Vol. EDL-8, pp. 407-409, 1987.

[2] B.J. Baliga, "Advanced Power Rectifier Concepts", Chapter 7, Springer-Science, New York, 2009.

[3] E.R. Van Brunt, "Development of Optimal 4H-SiC Bipolar Power Diodes for High Voltage High-Frequency Applications", NCSU Ph.D. Thesis, 2012.

[4] B. Heinze *et al.*, "Surge Current Ruggedness of Silicon Carbide Schottky and Merged-PiN-Schottky Diodes" IEEE International Symposium on Power Semiconductor Devices and ICs, Paper WB-P-6, pp. 245-248, 2008.

[5] Y. Jiang *et al.*, "10kV 4H-SiC MPS Diodes for High Temperature Application", IEEE International Symposium on Power Semiconductor Devices and ICs, June 2016.

[6] A. Hefner *et al.*, "Recent Advances in High-Voltage, High-Frequency, Silicon-Carbide Power Devices", IEEE 41st Industrial Application Society Conference, Vol. 1, pp. 330-337, 2006.

Chapter 9

Junction Field Effect Transistors

As already pointed out in earlier chapters, the main advantage of wide band gap semiconductors for power device applications is the very low resistance of the drift region even when it is designed to support large voltages. This favors the development of high voltage unipolar devices which have much superior switching speed than bipolar structures. At early stages of SiC power device development, the inversion layer mobility in the channel of the power MOSFET structure was found to be very low precluding their development. The *Junction Field Effect Transistor (JFET)* and the *Metal Semiconductor Field Effect Transistor (MESFET)* were therefore considered potential candidates as unipolar switches for power applications. These structures have also been called *Static Induction Transistors*[1]. In the case of silicon, the maximum breakdown voltage of JFETs was limited by the increase in the resistance of the drift region[2]. This limitation does not apply to silicon carbide due to the much larger doping concentration within the drift region for high voltage structures. However, as in the case of silicon structures, the normally-on behavior of the high voltage JFETs has been found to be a serious impediment to circuit applications. When powering up any electronic system, it is impossible to ensure that the gate voltage required to block current flow in the JFET is provided to the structure prior to the incidence of the drain voltage on the structure. This situation can result in shoot-through currents between the power rails resulting in destructive failure of the devices. This has discouraged the use of normally-on devices in power electronic applications.

The invention of the *Baliga Pair* circuit configuration[3] enabled achieving a normally-off power switch function with a high speed integral diode that is ideally suitable for H-bridge applications. The Baliga-Pair utilizes a high voltage normally-on silicon carbide JFET or MESFET with a low voltage silicon power MOSFET to achieve a normally-off function.

This chapter reviews the basic principles of operation of the vertical Junction (or Metal-Semiconductor) Field Effect Transistor. These structures must be designed to deplete the channel region by application of a reverse bias voltage to the gate-source junction. To prevent current flow through the device, a potential barrier for electron transport must be created in the channel to suppress electron transport even under large drain bias voltages. Concurrently, the channel must remain un-depleted (preferably at zero gate bias) to allow on-state current flow with low on-state resistance. Unique issues that relate to the wide band gap of silicon carbide must be given special consideration. Experimental results on relevant structures are provided to define the state of the development effort on high voltage silicon carbide JFETs and MESFETs.

9.1 Trench Junction (Metal-Semiconductor) FET Structure

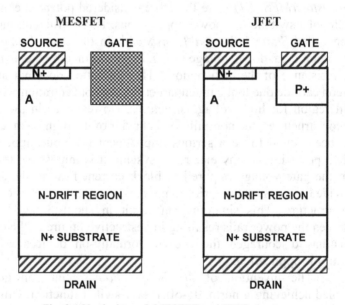

Fig. 9.1 Vertical trench MESFET and JFET structures.

The basic structure of the vertical trench Junction Field Effect Transistor (JFET) and the trench Metal-Semiconductor Field Effect Transistor (MESFET) are shown in Fig. 9.1. The structures contain a drift region (usually N-type) between the drain and the source regions designed to support the desired maximum operating voltage. As discussed in detail later in the chapter, the drift region must be capable of supporting the sum of the drain bias and the reverse gate bias potentials without undergoing breakdown. To prevent current flow under forward blocking conditions, a gate region must be incorporated in the drift region. A P-N junction gate region is used for the JFET structure and a Schottky (metal-semiconductor) contact is used for the MESFET structure. In order to produce a strong potential barrier for suppressing the transport of electrons between the drain and the source in the blocking state, it is preferable to create a gate with vertical sidewalls. For the JFET structure, this can be achieved by utilizing multiple P-type ion implants with increasing energy using a common mask edge. For the MESFET structure, a trench is etched with vertical sidewalls followed by the selective deposition of the Schottky barrier gate metal to fill the trench.

When a negative bias is applied to the gate electrode, a depletion layer extends from the gate into the drift region. With sufficient gate bias, the entire space between the gate regions becomes depleted. The gate bias required to deplete the space between the gate regions is referred to as the *pinch-off voltage*. At a gate bias above the pinch-off voltage, a potential barrier forms under the source region at location A. This barrier suppresses the transport of electrons between the drain and the source. However, the potential barrier is reduced with application of the drain bias. Consequently, as the drain bias increases, drain current flow commences at voltages well below the breakdown voltage capability of the drift region. It is possible to support a larger drain bias voltage before the observation of drain current flow with a larger gate bias because a larger potential barrier is created.

The spacing between the gate regions is usually designed to be more than twice the zero bias depletion width. Consequently, an un-depleted portion of the drift region remains under the source region at zero gate bias. Current flow between the drain and source occurs through this region with the amount limited by the resistance of the channel region (between the gates) and the drift region below the gate. A large un-depleted region is favored for reducing the on-resistance but this reduces the magnitude of the potential barrier when the device operates in the forward blocking mode. Thus, a trade-off between the on-state resistance and blocking characteristics must be made when designing these structures.

9.1.1 Forward Blocking

Since the operating principles for the MESFET and JFET structures are similar, these names will be used interchangeably in this chapter. In a normally-on device structure, an un-depleted portion of the channel exists at zero gate bias. Drain current flow can therefore occur via this un-depleted region at zero gate bias. The forward blocking capability in the JFET structure is achieved by creating a potential barrier in the channel between the gate regions by applying a reverse bias to the gate region. At a gate bias above the pinch-off voltage, the channel becomes completely depleted. As the gate bias is increased above the pinch-off voltage, the potential barrier increases in magnitude. Current transport between the drain and source is suppressed because electrons must overcome the potential barrier. As the drain voltage is increased, the potential barrier reduces allowing injection of electrons over it.

An exponential increase in the drain current is observed in the JFET structure with increasing drain bias with 'triode-like' characteristics. This behavior can be described by[4]:

$$J_D = \frac{qD_n N_D}{L} \sqrt{\frac{q}{\pi kT}} \left(\alpha V_G - \beta V_D\right) \exp\left\{-\left[\frac{q}{kT}\left(\alpha V_G - \beta V_D\right)\right]\right\} \quad [9.1]$$

where α and β are constants that depend upon the gate geometry. In this equation, D_n is the diffusion coefficient for electrons, N_D is the doping concentration in the drift region, L is the gate length (in the direction of current flow), q is the charge of the electron, k is Boltzmann's constant, T is the absolute temperature, V_G is the gate bias voltage and V_D is the drain bias voltage. The exponential variation of drain current with variation of gate and drain bias has been observed in silicon vertical channel JFETs[5].

The JFET structure can exhibit a normally-off behavior up to a limited drain voltage if the space between the gate regions in the JFET structure is reduced so that it becomes completely depleted by the zero bias depletion width. This type of design will exhibit purely triode-like characteristics. On the other hand, if the space between the gate regions is larger than the maximum depletion width at breakdown for the drift region, the structure will exhibit purely pentode-like characteristics. If the space between the gate regions falls between these extremes, the device will exhibit a mixed triode-pentode like characteristics. High voltage normally-on JFETs are usually designed to operate in this mixed triode-pentode mode to obtain a good compromise between low on-state resistance and high blocking voltage capability.

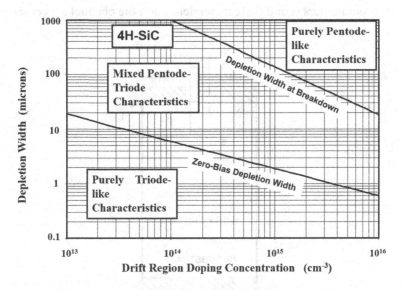

Fig. 9.2 Design space for high voltage 4H-SiC JFETs.

The design space for vertical high voltage JFETs is bounded by the depletion widths at breakdown and the zero-bias depletion width. These boundaries are shown in Fig. 9.2 for the case of 4H-SiC devices. For any given doping concentration, a much larger spacing is required for 4H-SiC devices than silicon devices due to the bigger depletion width at breakdown. In general, a greater latitude exists for the design of 4H-SiC devices to optimize the characteristics.

A good compromise between achieving a low on-state resistance and a good blocking voltage capability for high voltage JFETs requires designing the channel to operate in the mixed pentode-triode regime. At low drain current levels and high drain voltages, these devices exhibit triode-like characteristics. In this mode, it is useful to define a DC blocking gain (G_{DC}) as the ratio of the drain voltage to the gate voltage at a specified leakage current level. A differential blocking gain (G_{AC}) can also be defined as the increase in drain voltage, at a specified leakage current level, for an increase in the gate voltage by 1 volt. From Eq. [9.1], it can be shown that:

$$G_{AC} = \frac{dV_D}{dV_G} = \frac{\alpha}{\beta} \qquad\qquad [9.2]$$

The parameters α and β are dependent upon the channel aspect ratio.

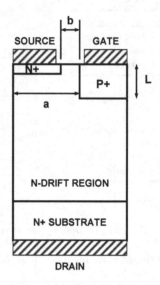

Fig. 9.3 Channel aspect ratio (L/a) for the JFET structures.

The channel aspect ratio is defined as the ratio of the length of the gate (dimension 'L' in Fig. 9.3) in the direction of current flow and the space between the gate regions (dimension 'a' in Fig. 9.3). A large aspect ratio favors obtaining a high blocking gain, which is beneficial for reducing the gate voltage required to block high drain voltages. Theoretical analysis[6] and empirical observations[7] on high voltage silicon JFETs indicate that the blocking gain can be described by:

$$G_{DC} = \frac{V_D}{V_G} = A.\exp\left(B\frac{L}{a} \right) \qquad\qquad [9.3]$$

where A and B are constants. Since the gate junction must support the sum of the negative gate bias voltage and the applied positive drain voltage, the drift region parameters must be chosen to account for the finite blocking gain.

The largest drain voltage that can be supported by the JFET structure before the on-set of significant current flow is determined by several factors. Firstly, it is limited by the intrinsic breakdown voltage capability of the drift region as determined by the doping concentration and thickness. The breakdown voltage can be obtained by using the graphs and equations provided in chapter 3. Secondly, the maximum drain voltage that can be supported without significant current flow can be limited by the applied gate bias and the blocking gain of the structure. In addition, the largest gate bias that can be applied is limited by the on-set of breakdown between the gate and the source regions. The breakdown voltage between the gate and the source regions is determined by the depletion layer punch-through from the gate junction to the highly doped N^+ source region. Since the space between these regions must be kept small in order to obtain a high channel aspect ratio leading to high blocking gain, the electric field between the gate and source regions can be assumed to be uniform. Under this approximation, the gate-source breakdown voltage can be calculated using:

$$BV_{GS} = b.E_C \qquad [9.4]$$

where 'b' is the space between the gate and the source as shown in Fig. 9.3 and E_C is the critical electric field for breakdown. Fortunately, the critical electric field for breakdown in 4H-SiC is large allowing high gate-source breakdown voltages in spite of using small gate-source spacing.

9.1.2 On-State

The on-state current flow pattern for the normally-on JFET design with mixed Pentode-Triode mode of operation is indicated in Fig. 9.4 by the shaded area. The current flows from the source through a uniform cross section between the gate regions with a width (d) determined by the space between the gate regions and the zero-bias depletion width. The current then spreads at a 45 degree angle into the drift region to a depth of 's' under the gate region, and then becomes uniform throughout the cross-section. This current flow pattern can be used to model the on-state resistance:

$$R_{on,sp} = \rho_D (L + W_0) \left(\frac{p}{d} \right) + \rho_D \left(\frac{p(s + W_0)}{p - 2d} \right) \ln \left(\frac{p}{2d} \right)$$
$$+ \rho_D (t - s)$$

[9.5]

where W_0 is the zero bias depletion width.

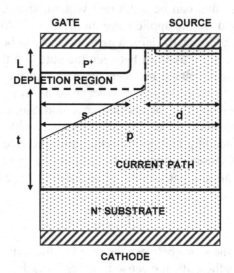

Fig. 9.4 On-state current flow pattern in the 4H-SiC JFET structure.

The specific on-resistance of the JFET can be much larger than the ideal specific on-resistance for any particular blocking voltage capability of the JFET structure. Firstly, this is because the drift region must support the sum of the gate and drain bias voltages. Secondly, it is increased by the contributions from the channel region and the spreading resistance. These contributions become more significant as the space between the gates is reduced. Thus, a compromise must be made between obtaining a low specific on-state resistance and a high blocking gain.

When a negative gate voltage is applied to reverse bias the gate junction, the depletion region extends further into the channel producing an increase in the on-resistance. Further, if the drain voltage becomes comparable to the gate bias voltage, the depletion width at the bottom of the channel on the drain side becomes larger (as determined by V_{GS} plus V_{DS}). This alters the current flow pattern as illustrated in Fig. 9.5. Under the assumption of a field independent mobility and by using the gradual channel approximation, the drain current is determined by[2]:

$$I_D = 2a\rho_D \frac{Z}{L}\left\{V_{Dch} - \frac{2}{3a}\left(\frac{2\varepsilon_S}{qN_D}\right)^{1/2}\left[\left(V_{Dch} + V_G + V_{bi}\right)^{3/2} - \left(V_G + V_{bi}\right)^{3/2}\right]\right\}$$

[9.6]

where Z is the length of the device in the direction orthogonal to the cross-section, and V_{Dch} is the drain voltage at the bottom edge of the channel.

The basic I-V characteristics determined by this equation can be described using three segments. In the first segment, the drain voltage is much smaller than the gate bias voltage. In this case, the channel resistance increases with reverse gate voltage as given by:

$$R_{ch} = \rho_D \frac{L}{2Z}\left[a - \sqrt{\frac{2\varepsilon_S}{qN_D}\left(V_{GS} + V_{bi}\right)}\right]$$

[9.7]

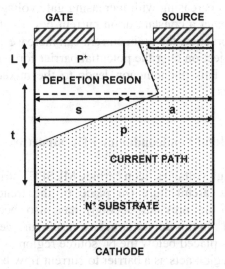

Fig.9.5 Current flow pattern at larger V_{DS} for the JFET structure.

In the second segment, the drain voltage is comparable to the gate voltage. This produces a non-linear characteristic with the resistance increasing with increasing drain bias. Eventually, the entire space between the gate regions becomes depleted at a certain drain bias. This condition is described by:

$$V_P = \left(V_{Dch} + V_{GS} + V_{bi}\right) = \frac{qN_D a^2}{2\varepsilon_S}$$ [9.8]

with V_P defined as the pinch-off voltage. The drain voltage, at which channel pinch-off occurs, decreases linearly with increasing gate voltage.

In the third segment, the drain current becomes constant because the channel is completely pinched off. The saturated drain current is given by:

$$I_D = 2a\rho_D \frac{Z}{L} \left\{ \frac{qN_D a^2}{6\varepsilon_S} - \left(V_{GS} + V_{bi}\right) + \frac{2}{3a}\left(\frac{2\varepsilon_S}{qN_D}\right)^{1/2} \left[\left(V_G + V_{bi}\right)^{3/2}\right] \right\}$$ [9.9]

A family of drain current-voltage curves is formed with the saturated drain current decreasing with increasing gate voltage. Although these equations predict a constant drain current beyond the channel pinch-off point, in practice, the drain current can increase after pinch-off due to electron injection over the potential barrier formed at the bottom of the channel. This phenomenon leads to the mixed Pentode-Triode characteristics.

9.2 Planar Metal-Semiconductor FET Structure

An elegant high voltage silicon carbide MESFET structure that utilizes a planar gate architecture (in place of the trench gate structure discussed in the previous section) has also been proposed and demonstrated[8]. In this device structure, a sub-surface heavily doped P-type region is placed below the N^+ source region as illustrated in Fig. 9.6. The P^+ region acts as a barrier to current flow between the source and drain regions restricting the current to gaps between the P^+ regions. A gate region is placed over these gaps and overlapping the P^+ region to enable control over the transport of current between the drain and source regions. In general, the gate can be constructed as a Metal-Semiconductor contact (to form a MESFET structure), a P-N Junction (to form a JFET structure) or as a Metal-Oxide-Semiconductor sandwich (to form a MOSFET structure). The MESFET structure will be discussed in detail in this section. The operation of the JFET structure is similar but requires taking into account the larger built-in potential of the gate P-N junction. The MOSFET structure is discussed in a subsequent chapter.

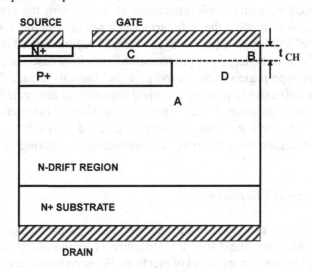

Fig. 9.6 The planar gate MESFET structure.

The sub-surface P$^+$ region in the planar gate MESFET structure can be either connected to the gate or to the source region. If the P$^+$ region is electrically connected to the gate electrode, it collaborates with the metal-semiconductor contact to constrict current flow from the source to the drain region. However, when the gate is reverse biased, a low gate-source breakdown voltage can occur due to the close vertical proximity of the N$^+$ source region and the sub-surface P$^+$ region. It is therefore preferable to connect the P$^+$ region to the source electrode. During the fabrication of the MESFET structure, this can be achieved by interruption of the N$^+$ region in the orthogonal direction to the cross-section shown in Fig. 9.6 and placing a P$^+$ contact region from the surface down to the sub-surface P$^+$ region in these gaps. This approach avoids a potential breakdown problem between the N$^+$ source and P$^+$ sub-surface regions. However, the breakdown voltage between the gate and source electrodes can now occur due to depletion region reach-through from the gate contact to the underlying P$^+$ region. Fortunately, relatively small reverse gate bias voltages are required in the planar MESFET to achieve high drain blocking voltages. The low gate-source breakdown voltage as a result of reach-through is therefore not a serious limitation with proper design of the structure.

The sub-surface P^+ region in the planar MESFET structure can be created by using ion-implantation of boron with the appropriate energy[9]. Alternately, the sub-surface P^+ region can be formed by growth of an N-type epitaxial layer over a P^+ region formed in the drift region by lower energy ion-implantation[10]. In either case, the thickness of the N-type region between the gate and the sub-surface P^+ region must be sufficient to prevent complete depletion at zero gate bias. The doping concentration of the N-type region located between the gate and the sub-surface P^+ region can be increased above that for the N-type drift region if necessary by ion-implantation or during its epitaxial growth.

9.2.1 Forward Blocking

As previously described in this chapter, the forward blocking regime of operation for the MESFET structure is achieved by creating a potential barrier for transport of electrons between the source and drain region by the application of a reverse gate bias. In the planar MESFET structure, this potential barrier is formed in the channel (at location 'C' shown in Fig. 9.6). If the thickness of the channel (shown as t_{CH} in Fig. 9.6) is narrow, a potential barrier can be formed with relatively low reverse gate bias voltages. In addition, the planar MESFET structure contains a second JFET region formed between the adjacent P^+ regions (at location 'D' shown in Fig. 9.6). When the drain bias exceeds the pinch-off voltage for this JFET region, the potential at the surface under the gate becomes isolated from the potential applied at the drain electrode. Consequently, the channel potential barrier is shielded from the drain voltage enabling the support of high drain voltages without the on-set of drain current flow. These features favor producing a very high blocking gain with low reverse gate bias voltages.

9.2.2 On-State Resistance

The planar MESFET can be designed to contain an un-depleted channel region at zero gate bias. Current can then flow between the source and drain regions through the channel and the gap between the P^+ regions down to the N-type drift region. The resistance of these regions must be included in the analysis of the total on-state resistance of the structure.

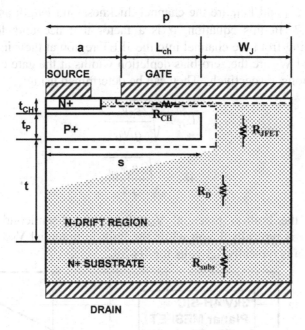

Fig. 9.7 Resistances in a planar gate MESFET structure.

The total specific on-resistance is given by:

$$R_{on,sp} = R_{CH} + R_{JFET} + R_D + R_{subs}$$ [9.10]

where R_{CH} is the channel resistance, R_{JFET} is the resistance of the JFET region, R_D is the resistance of the drift region after taking into account current spreading from the JFET region, and R_{subs} is the resistance of the N^+ substrate. These resistances can be analytically modeled by using the current flow pattern indicated by the shaded regions in Fig. 9.7. In this figure, the depletion region boundaries have also been shown using dashed lines. The drain current flows through a channel region with a small cross-section before entering the JFET region. The current spreads into the drift region from the JFET region at a 45 degree angle and then becomes uniform. The dimension 'a' in Fig. 9.6 is decided by alignment tolerances used during device fabrication. A typical value of 0.5 microns has been assumed for the analysis in this chapter.

The channel resistance is given by:

$$R_{CH} = \frac{\rho_D (L_{CH} + \alpha W_P) p}{(t_{CH} - W_G - W_P)}$$ [9.11]

where t_{CH} and L_{CH} are the channel thickness and length as shown in Fig. 9.7. In this equation, α is a factor that accounts for current spreading from the channel into the JFET region at their intersection. W_G and W_P are the zero-bias depletion widths at the gate contact and P^+ regions, respectively. They can be determined using:

$$W_G = \sqrt{\frac{2\varepsilon_S V_{biG}}{qN_D}} \qquad\qquad [9.12]$$

$$W_P = \sqrt{\frac{2\varepsilon_S V_{biP}}{qN_D}} \qquad\qquad [9.13]$$

where the built-in potential V_{biG} for a metal-semiconductor gate contact is typically 1 volt while the built-in potential V_{biP} for the P^+ junction is typically 3.3 volts for 4H-SiC.

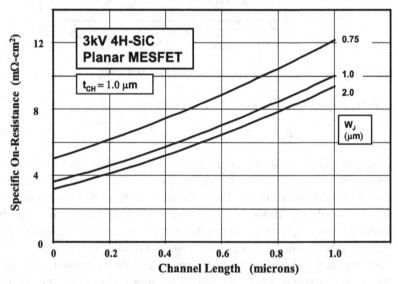

Fig. 9.8 Analytically calculated on-resistance for a 4H-SiC planar MESFET.

The JFET region resistance is given by:

$$R_{JFET} = \rho_D \left(t_{CH} + t_P - W_G \right) \left(\frac{p}{W_J - W_P} \right) \qquad [9.14]$$

where p is the cell pitch. The drift region spreading resistance can be obtained by using:

$$R_D = \rho_D \left(\frac{2p}{W_J - W_P} \right) \ln \left(\frac{2p}{W_J - W_P} \right) + \rho_D (t - s - W_P) \quad [9.15]$$

where t is the thickness of the drift region below the P^+ region and s is the width of the P^+ region.

The contribution to the resistance from the N^+ substrate is given by:

$$R_{subs} = \rho_{subs} \cdot t_{subs} \quad [9.16]$$

where ρ_{subs} and t_{subs} are the resistivity and thickness of the substrate, respectively. A typical value for this contribution is 4×10^{-4} Ω-cm^2.

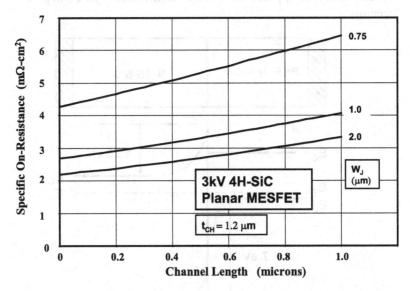

Fig. 9.9 Analytically calculated on-resistance for a 4H-SiC planar MESFET.

Using the analytical expressions, it is possible to model the change in specific on-resistance with alterations of the cell design parameters. The specific on-resistance is most sensitive to variations of the channel length (L_{CH}) and thickness (t_{CH}), as well as the width of the JFET region (W_J). The variation of the specific on-resistance with increasing channel length is shown in Figs. 9.8 and 9.9 for cases of channel thickness of 1 and 1.2 microns, respectively. It can be seen that increasing the channel thickness reduces the specific on-resistance and its dependence on the channel length. Increasing the JFET width beyond 1 micron does not improve the on-resistance significantly

9.3 4H-SiC Experimental Results: Trench Gate Structures

Several approaches can be taken to construct a vertical FET structure that utilizes a potential barrier along the vertical path induced by a gate bias. The gate region in the device can be formed by using ion implantation of P-type dopants to form a JFET structure or by placing a metal contact within a trench etched between the source regions to form a MESFET structure or by forming a hetero-junction gate region within the trench to form a HJFET structure. Prior to the development of a process for formation of heavily doped P-type regions in silicon carbide, it was more practical to construct the hetero-junction gate FET structure.

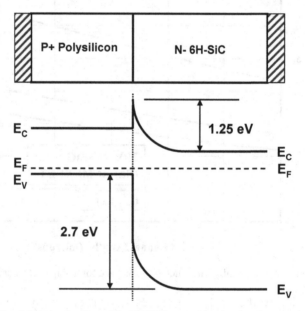

Fig. 9.10 P$^+$ polysilicon/N- 6H-SiC hetero-junction band structure.

The formation of a hetero-junction between P-type polysilicon and N-type silicon carbide was first demonstrated at PSRC in 1996[11]. The band structure for the interface between P$^+$ Polysilicon and N-type 6H-SiC is shown in Fig. 9.10. Good rectification was experimentally confirmed at this hetero-junction[12] allowing consideration of this junction for the gate region in a HJFET structure, shown in Fig. 9.11.

The fabrication process for this structure is simple because the polysilicon gate material can be deposited into the trenches and planarized due to the good selectivity between it and SiC during reactive-ion-etching. The HJFET structure was analyzed in detail in 1996[13] followed by experimental demonstration[14] of a structure with the P^+ polysilicon located within trenches etched between the source regions. Although some gate control was observed, the performance of the structure was poor due to the bad quality of the surface within the trenches after etching.

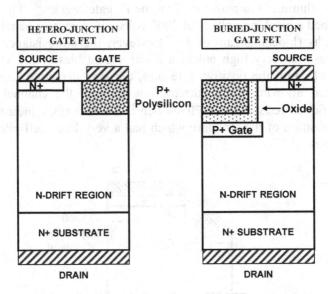

Fig. 9.11 Novel vertical trench gate SiC FET structures.

A trench JFET structure, also illustrated in Fig. 9.11, that contains a MOS sidewall has been proposed and demonstrated[15]. This structure requires ion implantation of the P^+ region at the bottom of the trenches after they have been etched. Since the authors masked the P^+ implant, the process is difficult to implement due to the poor topology for patterning photoresist in the presence of 1 micron deep trenches. In addition, it is not clear that the P-type implant will not occur on the trench sidewalls, an effect disregarded by the authors. The authors deposited an oxide after the P^+ implant and then refilled the trench with polysilicon to create an MOS-structure on the trench sidewalls. The process described by the authors precludes making contact to the P⁺ gate regions along the trenches. Since the contact to the P⁺ gate regions must be located at the periphery of the device, the structure would have poor switching characteristics due to the very high resistance of the P⁺ fingers orthogonal to the cross-section. The authors report obtaining a specific on-resistance of 5 mΩ-cm^2 for a device that is able to block 600 volts using a reverse gate bias of –30 volts.

The vertical JFET structure can also be fabricated using ion implantation to form the P⁺ gate region. In order to obtain a good channel aspect ratio, it is necessary to resort to very high energy (1.3 MeV) aluminum implants to form the P⁺ gate regions[16]. The authors obtained a blocking voltage of 2000 volts with a reverse gate bias of – 50 volts (blocking gain of 40). However, the on-resistance for the devices was very high unless a positive gate bias of 2.5 volts was applied. With the positive gate bias, the gate depletion region was reduced allowing drain current flow through the channel with a specific on-resistance of 70 mΩ-cm^2. These values indicate poor optimization of the structure which had a very large cell pitch of 32 microns.

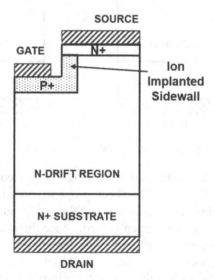

Fig. 9.12 Trench-implanted gate 4H-SiC JFET structure.

A 4H-SiC JFET has been developed by etching vertical trenches followed by ion implantation of the sidewalls to create the P⁺ gate regions[17]. A blocking voltage of 1650-V was obtained using a gate bias of –18 volts (blocking gain of 92) if the channel width is made 0.7 microns. A specific on-resistance of 1.88 $m\Omega$-cm^2 was measured by using a forward gate bias of 3 volts. This results in gate current flow corresponding to the bipolar mode of operation for JFETs. The current gain under these conditions was 100. The on-resistance is very high without the forward bias on the gate.

The JFET structure shown in Fig. 9.12 can also be designed to operate as a normally-off device[18] if the width of the channel is reduced to 0.63 microns. These devices were operated in the bipolar-mode with a gate bias of 5 volts resulting in a current gain of only 10. The difficulties of controlling the width of the trench and the need for bipolar mode of operation makes this type of structure unattractive.

A vertical channel JFET structure has also been developed by ion-implantation of the gate P-regions to obtain a channel length of 2 microns[19]. The device structure is similar to that shown in   Fig. 9.1 with a recessed gate contact. The devices could support 1680-V using a gate bias of –24-V corresponding to a blocking gain of 70. A specific on-resistance of 5.3 $m\Omega$-cm^2 was measured by using a forward gate bias of 2.5 volts. The current gain under these conditions was 16,800. The on-resistance is very high without the forward bias on the gate. The work was extended[20] to achieve a blocking voltage of 2055-V. Attempts to make the device normally-off by the same authors was found to be problematic because the space between the junction had to be reduced to only 0.38 microns leading to very high specific on-resistance[21]. However, a 9-kV nor-mally-on JFET was successfully demonstrated by the authors using e-beam lithography[22]. The device could actually block only 3190 volts with a gate bias of –32-V, a blocking gain of 100. A specific on-resistance of 104 $m\Omega$-cm^2 was measured at zero gate bias.

9.4 4H-SiC Experimental Results: Planar Gate Structures

The planar gate FET structure was originally proposed and patented in 1996 with either a metal-semiconductor, or a junction, or an MOS gate region[8]. These structures have a lateral channel whose thickness and doping level can be controlled to achieve good blocking gain and on-state resistance in the vertical JFET structure. The sub-surface P$^+$ region can be formed by using ion-implantation with appropriate energy to locate the junction below the surface so as to create an un-depleted N-type channel region. An alternative approach is to grow an N-type epitaxial layer over the P$^+$ regions implanted into a substrate to create the N-type channel region. Experimental results on the lateral channel, vertical (planar) MESFET/JFET structures are discussed here.

The first planar MESFET structures were successfully fabricated at PSRC in 1996-97 by performing 380 keV boron ion-implants to form the sub-surface P^+ region[9]. Devices were fabricated from both 6H-SiC and 4H-SiC as the starting material with doping concentration of about 2×10^{16} cm^{-3}. The energy for the boron implants was chosen to locate the center of the P^+ region at about 0.5 microns below the surface[23]. An additional N-type nitrogen implant was used in the channel region to enhance its doping to produce normally-on devices. Contact to the sub-surface P^+ region was made by additional boron implants at lower energy in selective regions within the cell structure. The devices were found to exhibit a low specific on-resistance of about 12 mΩ-cm^2 but had poor gate–drain breakdown voltage of about 50 volts.

The planar JFET structures that were also fabricated at PSRC in 1996-97 using the same process conditions described above had much better blocking capability[24]. In these devices, the P^+ gate region was formed by using a shallow 10 keV boron implant. The 4H-SiC planar JFETs fabricated using a N-type channel implant with a dose of $1-2 \times 10^{13}$ cm^{-2} exhibited a specific on-resistance of only 11-14 mΩ-cm^2. The devices were able to block a drain bias of 1100 volts with a negative gate bias of 40 volts. Excellent gate controlled pentode-like characteristics with drain current saturation was observed in the devices up to the breakdown voltage with a drain current density of 250 A/cm^2.

Lateral channel, vertical JFET structures, fabricated by epitaxial growth of an N-type layer over a P^+ region implanted into the drift region, have been reported by several groups. The first such devices were reported[10] in 1999 by using epitaxial layers capable of supporting 600 V and 1200 V. By growing an epitaxial layer with doping concentration of about 2×10^{16} cm^{-3} with a thickness of 2.5 microns over the sub-surface P^+ region, the devices were able to support a gate bias of more than 10 volts above a channel pinch-off voltage of 40 volts. This was sufficient to allow blocking a drain bias of 1200 volts. The specific on-resistance for the devices was 18, 25, and 40 mΩ-cm^2 for devices capable of supporting 550, 800, and 950 volts, respectively. The performance of these structures was subsequently improved[25] to a blocking voltage of 1800 volts with a specific on-resistance of 24.5 mΩ-cm^2.

A comprehensive study of various gate structural options for 10-kV 4H-SiC normally-on vertical JFETs with a lateral channel region was reported more recently[26]. Four basic lateral channel JFET structures are illustrated in Fig. 9.13. In the SG-JFET structure, the P^+ buried layer is connected to the source region and only the P^+ top layer is used as the gate. In the SBG-JFET structure, the P^+ buried layer and the P^+ top regions are both used as the gate. In the BG-JFET structure, the P^+ top layer is connected to the source and only the P^+ buried layer is used as the gate. In the DG-JFET structure, the P^+ top gate region does not overlap the gap between the P^+ buried layers.

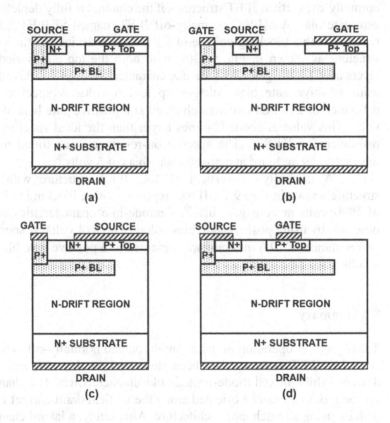

Fig. 9.13 Planar lateral channel gate structures for 4H-SiC JFETs: (a) SG-JFET; (b) SBG-JFET; (c) BG-JFET; and (d) DG-JFET

The above structures were optimized by varying the length of the lateral channel and the gap between the P^+ buried layers. Devices were fabricated using 120 micron thick drift layers with doping concentration of 9×10^{14} cm^{-3} on N^+ substrates to achieve a measured breakdown voltage of 9.4-kV using a gate bias of -18 V. This corresponds to a blocking gain of over 500 which is essential to achive such high blocking voltages. Such high blocking gains can only be achieved by using the lateral channel configuration. A minimum specific on-resistance of 130 mΩ-cm^2 was observed at a gap of 3.5 microns between the P$^+$ buried layers. The measured specific-on resistance is only 2-times larger than the ideal specific on-resistance for a 10-kV 4H-SiC device.

The lateral channel approach can also be used to make normally-off vertical JFET structures if the channel is fully depleted at zero gate bias. A 4H-SiC normally-off JFET, named SEJFET[27], was reported with a blocking voltage of 5.5 kV. This device has the same structure as shown in Fig. 9.13(b) with both the top and buried P$^+$ layers used for the gate. Pentode-like characteristics were observed by using positive gate bias voltages up to 2.6 volts. A specific on-resistance of 218 mΩ-cm^2 was obtained at a positive gate bias of 2.6 volts. This value is about 10-times larger than the ideal specific on-resistance for 4H-SiC. The specific on-resistance was found to be reduced to 69 mΩ-cm^2 at a positive gate bias of 5 volts[28].

A normally-off vertical 4H-SiC JFET structure with the structure shown in Fig. 9.13(d) was reported with a blocking voltage of 1900 volts at zero gate bias[29]. Pentode-like characteristics were observed by using positive gate bias voltages up to 4 volts. A specific on-resistance of 19.6 mΩ-cm^2 was obtained at a positive gate bias of 4 volts.

9.5 Summary

The physics of operation of the normally-on and normally-off vertical JFET/MESFET structure has been described in this chapter. The devices exhibit mixed triode-pentode like characteristics. The channel for these devices can be oriented along the vertical drain current flow path by using a trench gate architecture. Alternately, a lateral channel can be formed by introducing a sub-surface P$^+$ region below a planar gate design. The planar devices have been successfully made using either a deep P$^+$ ion implantation step into the N-drift region or by growth of an N-type epitaxial layer above a previously implanted P$^+$ region in the N-type drift region. Both methods have resulted in devices with low specific on-resistance and good gate controlled current saturation capability with relatively low reverse gate bias voltages. These devices are suitable for utilization in the *Baliga-Pair* configuration discussed in the next chapter.

References

[1] J. Nishizawa, T. Terasaki and J. Shibata, "Field Effect Transistor versus Analog Transistor (Static Induction Transistor)", IEEE Transactions on Electron Devices, Vol. ED22, pp. 185-197, 1975.

[2] B. J. Baliga, "Modern Power Devices", John Wiley and Sons, 1987.

[3] B. J. Baliga, "Silicon Carbide Switching Device with Rectifying Gate", U. S. Patent 5,396,085, Issued March 7, 1995.

[4] P. Plotka and B. Wilamowski, "Interpretation of Exponential type Drain Characteristics of the Static Induction Transistor", Solid State Electronics, Vol. 23, pp. 693-694, 1980.

[5] B. J. Baliga, "A Power Junction Gate Field Effect Transistor Structure with High Blocking Gain", IEEE Transactions on Electron Devices, Vol. 27, pp. 368-373, 1980.

[6] X. C. Kun, "Calculation of Amplification Factor of Static Induction Transistors", IEE Proceedings, Vol. 131, pp. 87-93, 1984.

[7] B. J. Baliga, "High Voltage Junction Gate Field Effect Transistor with Recessed Gates", IEEE Transactions on Electron Devices, Vol. 29, pp. 1560-1570, 1982.

[8] B. J. Baliga, "Silicon Carbide Semiconductor Devices having Buried Silicon Carbide Conduction Barrier Layers Therein", U. S. Patent 5,543,637, Issued August 6, 1996.

[9] P. Shenoy and B. J. Baliga, "The Planar Lateral Channel SiC MESFET", PSRC Technical Report TR-97-038, 1997.

[10] H. Mitlehner *et al.*, "Dynamic Characteristics of High Voltage 4H-SiC Vertical JFETs", IEEE International Symposium on Power Semiconductor Devices and ICs, Abstract 11.1, pp. 339-342, 1999.

[11] P. M. Shenoy and B. J. Baliga, "High Voltage P+ Polysilicon/N- 6H-SiC Heterojunction Diodes", PSRC Technical Report TR-96-050, 1996.

[12] P. M. Shenoy and B. J. Baliga, "High Voltage P+ Polysilicon/N- 6H-SiC Heterojunction Diodes", Electronics Letters, Vol. 33, pp. 1086-1087, 1997.

[13] B. Vijay, K. Makeshwar, P. M. Shenoy and B. J. Baliga, "Analysis of a High Voltage Heterojunction Gate SiC Field Effect Transistor", PSRC Technical Report TR-96-049, 1996.

[14] P. M. Shenoy, V. Bantval, M. Kothandaraman and B. J. Baliga, "A Novel P+ Polysilicon/N- SiC Heterojunction Trench Gate Vertical FET", IEEE International Symposium on Power Semiconductor Devices and ICs, pp. 365-368, 1997.

[15] R. N. Gupta, H.R. Chang, E. Hanna and C. Bui, "A 600 V SiC Trench JFET". Silicon Carbide and Related Materials – 2001, Material Science Forum, Vol. 389-393, pp. 1219-1222, 2002.

[16] H. Onose *et al.*, "2 kV 4H-SiC Junction FETs", Silicon Carbide and Related Materials – 2001, Material Science Forum, Vol. 389-393, pp. 1227-1230, 2002.

[17] Y. Li, P. Alexandrov and J.H. Zhao, "1.88-mΩ.cm2 1650-V Normally-on 4H-SiC TI-JFET", IEEE Transactions on Electron Devices, Vol. 55, pp. 1880-1886, 2008.

[18] J.H. Zhao *et al.*, "4H-SiC Normally-Off Vertical JFET with High Current Density", IEEE Electron Device Letters, Vol. 24, pp. 463-465, 2003.

[19] V. Veliadis *et al.*, "A 1680-V, 54-A Normally-On 4H-SiC JFET with 0.143-cm^2 Active Area", IEEE Electron Device Letters, Vol. 29, pp. 1132-1134, 2008.

[20] V. Veliadis *et al.*, "A 2055-V, 24-A Normally-On 4H-SiC JFET with 6.8-mm^2 Active Area", IEEE Electron Device Letters, Vol. 29, pp. 1325-1327, 2008.

[21] V. Veliadis *et al.*, "Investigation of the Suitability of 1200-V Normally-Off Recessed Implanted Gate SiC VJFETs for Efficient Power Switching Applications", IEEE Electron Device Letters, Vol. 30, pp. 736-738, 2009.

[22] V. Veliadis *et al.*, "A 9-kV Normally-On Vertical Channel SiC JFET for Unipolar Operation", IEEE Electron Device Letters, Vol. 31, pp. 470-472, 2010.

[23] M. S. Janson *et al.*, "Range distributions of Implanted Ions in Silicon Carbide", Silicon Carbide and Related Materials – 2001, Material Science Forum, Vol. 389-393, pp. 779-782, 2002.

[24] P. Shenoy and B. J. Baliga, "A Planar Lateral Channel SiC Vertical High Power JFET", PSRC Technical Report TR-97-036, 1997.

[25] P. Friedrichs *et al.*, "Static and Dynamic Characteristics of 4H-SiC JFETs Designed for Different Blocking Categories", Silicon Carbide and Related Materials – 1999, Material Science Forum, Vol. 338-342, pp. 1243-1246, 2000.

[26] W. Sung *et al.*, "A Comparative Study of Gate Structures for 9.4-kV 4H-SiC Normally-on Vertical JFETs", IEEE Transactions on Electron Devices, Vol. 59, pp. 2417-2423, 2012.

[27] K. Asano *et al.*, "5.5kV Normally-off Low RonS 4H-SiC SEJFET", IEEE International Symposium on Power Semiconductor Devices and ICs, Paper 1.1, pp. 23-26, 2001.

[28] K. Asano *et al.*, "5kV 4H-SiC SEJFET with Low RonS of 69 $m\Omega cm^2$", IEEE International Symposium on Power Semiconductor Devices and ICs, pp. 61-64, 2002.

[29] J.H. Zhao *et al.*, "A Novel High-Voltage Normally-Off 4H-SiC Vertical JFET", Material Science Forum, Vol. 389-393, pp. 1223-1226, 2002.

Chapter 10

The Baliga-Pair (Cascode) Configuration

The ability to produce a high quality interface between silicon carbide and a suitable gate dielectric material was a significant challenge[1]. In addition to a larger density of charge in the oxide and at the interface that causes threshold voltage shift, the inversion layer mobility was found to be very low when compared with silicon. To compound the problem, the conventional silicon power MOSFET structure could not provide the full benefits of the high breakdown field strength of the silicon carbide material because of reliability and rupture problems associated with the enhanced electric field in the gate oxide.

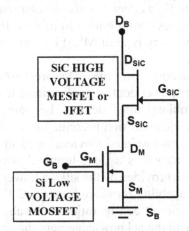

Fig. 10.1 *Baliga-Pair* power switch configuration.

In order to overcome these problems, it was proposed[2] that a normally-on silicon carbide high voltage JFET/MESFET be used together with a low voltage silicon MOSFET to create a configuration with the desired features for a high quality power switch. The basic idea, shown in the patent as Fig. 5, is illustrated in Fig. 10.1. It consists of a high voltage silicon carbide JFET or MESFET structure with its source electrode connected to the drain electrode of a low voltage silicon power MOSFET. An important feature of configuration is that the gate of the silicon carbide device is connected to the source of the silicon power MOSFET which serves as the ground or reference terminal in circuits. Gate signals are exclusively applied to the gate of the silicon power MOSFET. The drain of the silicon carbide device is connected to the load in power circuits as would be done with the drain of silicon power MOSFETs. The third independent claim of the patent states: *A three terminal gate controlled semiconductor switching device, comprising: an insulated-gate field effect transistor having an insulated gate electrode, a first source region and a first drain region; a rectifying-gate field effect transistor having a gate electrode, a second source region and a second drain region, wherein said gate electrode and said second source region are electrically connected to said first source region and first drain region, respectively; a drain contact electrically connected to said second drain region; a source contact electrically connected to said first source region; and wherein said rectifying-gate field effect transistor comprises a MESFET.* The fifth claim extends this configuration to include JFETs. These claims can be seen to cover any type of MESFET structures including GaN HEMT devices.

In an analogy to the *Darlington-Pair* configuration[3] commonly used for power control applications, it was suggested[4] that the proposed combination of devices be named the *Baliga-Pair* configuration. The idea was first presented in 1996 at the Conference on Silicon Carbide and Related Devices[5] held in Kyoto, Japan. The Baliga-Pair configuration is a three-terminal power switch with an MOS-input interface provided by the silicon power MOSFET and high blocking voltage capability provided by the SiC JFET/MESFET.

The same circuit configuration was subsequently called the Cascode circuit[6] with the acknowledgement that it was first disclosed in a textbook in 1996[7]. The term cascode is a misnomer because it was originally coined to describe a two-stage triode vacuum tube amplifier with a second triode serving a load to achieve superior stability[8]. Since this was apparent to the U.S. patent office, the Baliga-Pair circuit was considered novel and distinct from prior art allowing the patent claims to be issued.

This chapter discusses the operating principles of the Baliga-Pair configuration. It is demonstrated that a silicon power MOSFET with low breakdown voltage rating (and hence low specific on-resistance) can be used to control a high voltage, normally-on silicon carbide JFET/MESFET structure. This enables supporting large voltages within the silicon carbide FET while allowing control of the composite switch with signals applied to the MOS gate electrode of the silicon MOSFET. The same type of simple, low cost, integrated control circuits used for silicon power MOSFETs and IGBTs can therefore be utilized for the Baliga-Pair configuration. Since both devices in the configuration are unipolar devices, the Baliga-Pair has very fast switching speed and excellent safe-operating-area. In concert with the low on-resistance for both the FETs, the fast switching speed results in very low overall power dissipation in applications[9]. In addition, this configuration contains an excellent fly-back diode allowing replacement of not only the IGBT but also the fly-back rectifier that is usually connected across it in H-bridge power circuits. From this stand-point, it is preferable to use the silicon carbide MESFET structure due to lower on state voltage drop of the Schottky diode.

10.1 The *Baliga-Pair* Configuration

The Baliga-Pair configuration consists of a low breakdown voltage silicon MOSFET and a normally-on, high voltage silicon carbide JFET/MESFET connected together as shown in Fig. 10.2. Any of the trench gate or planar gate JFET/MESFET structures discussed in the previous chapter can be used to provide the high blocking voltage capability. It is important that the silicon carbide FET structure be designed for normally-on operation with a low specific on-resistance. It is also necessary for the silicon carbide FET to be able to block the drain bias voltage with a gate bias less than the breakdown voltage of the silicon power MOSFET.

The silicon power MOSFET can be either a planar DMOS structure or a trench-gate UMOS structure to provide low specific on-resistance. The source of the silicon carbide FET is connected to the drain of the silicon power MOSFET. Note that the gate of the silicon carbide FET is connected directly to the reference or ground terminal. The path formed between the drain and the gate contact of the silicon carbide FET creates the fly-back diode. The composite switch is controlled by the signal applied to the gate of the silicon power MOSFET. It is preferable to use a silicon power MOSFET with low (< 30-V) breakdown voltage because they have excellent low on-resistance and fast switching capability.

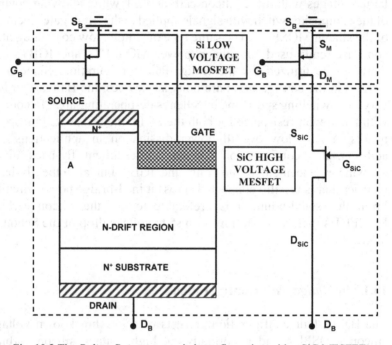

Fig. 10.2 The *Baliga-Pair* power switch configuration with a SiC MESFET.

10.1.1 Voltage Blocking Mode

The composite switch can block current flow when the gate of the silicon power MOSFET is shorted to ground by the external drive circuit. With zero gate bias, the silicon power MOSFET supports any bias applied to its drain terminal (D_M) unless the voltage exceeds its breakdown voltage. Consequently, at lower voltages applied to the drain terminal (D_B) of the composite switch, the voltage is supported across the silicon power MOSFET because the silicon carbide JFET is operating in its normally-on mode. However, as the voltage at the drain (D_M) of the silicon MOSFET increases, an equal positive voltage develops at the source (S_{SiC}) of the silicon carbide FET. Since the gate (G_{SiC}) of the silicon carbide FET is connected to the ground terminal, this produces a reverse bias across the gate-source junction of the silicon carbide FET. Consequently, a depletion layer extends from the gate contact/junction into the channel of the silicon carbide FET. When the depletion region pinches off the channel at location A, further increase in the bias applied to the drain (D_B) of the composite switch is supported across the silicon carbide FET. Since the potential at the source of the silicon carbide FET is then isolated from the drain bias applied to the silicon carbide FET, the voltage across the silicon power MOSFET is also clamped to a value close to the pinch-off voltage of the silicon carbide FET. This feature enables utilization of a silicon power MOSFET with a low breakdown voltage. Such silicon power MOSFETs have very low specific on-resistance with a mature technology available for their production. From this point of view, it is desirable to utilize silicon power MOSFETs with breakdown voltages of below 30 volts.

If the Baliga-Pair is designed to support a drain bias of 3000 volts, the ability to utilize a silicon power MOSFET with a breakdown voltage of 30 volts requires designing the silicon carbide FET so that the channel is pinched-off at a gate bias of below 20 volts. Thus, the blocking gain of the silicon carbide FET should be in excess of 150. This is feasible for both the trench-gate and planar gate architectures for silicon carbide FETs discussed in the previous chapter. As pointed out in that chapter, much larger blocking gains could be achieved with the planar MESFET structure making it an attractive choice for use in the Baliga-Pair configuration.

10.1.2 Forward Conduction Mode

The composite switch shown in Fig. 10.1 can be turned-on by application of a positive gate bias to the gate (G_B). If the gate bias is well above the threshold voltage of the silicon power MOSFET, it operates with a low on-resistance. Under these conditions, any voltage applied to the drain terminal (D_B) produces current flow through the normally-on silicon carbide FET and the silicon MOSFET. Due to the low specific on-resistance of both structures, the total on-resistance of the Baliga-Pair configuration is also very small:

$$R_{on}(Baliga - Pair) = R_{on}(SiliconMOSFET) + R_{on}(SiCFET)$$

[10.1]

Depending up on the size of the two devices, an on-resistance of less than 10 mΩ-cm^2 is feasible even when the switch is designed to support 3000 volts. This indicates that the Baliga-Pair configuration will have an on-state voltage drop of about 1 volt with a nominal on-state current density of 100 A/cm^2 flowing through the devices. This is well below typical values of around 4 volts for an IGBT designed to support such high voltages.

10.1.3 Current Saturation Mode

One of the reasons for the success of the silicon power MOSFET and IGBT in power electronics applications is the gate controlled current saturation capability of these devices. This feature enables controlling the rate of rise of current is power circuits by tailoring the input gate voltage waveform rather than by utilizing snubbers that are required for devices like gate turn-off thyristors. In addition, current saturation is essential for survival of short-circuit conditions where the device must limit the current.

The current saturation capability is inherent in the Baliga-Pair configuration. If the gate voltage applied to the Baliga-Pair configuration is close to the threshold voltage of the silicon MOSFET, it will enter its current saturation mode when the drain bias increases. This produces a constant current through both the silicon MOSFET and the silicon carbide FET while the drain bias applied to the composite switch increases. At lower drain bias voltages applied to the drain terminal (D_B), the voltage is supported across the silicon power MOSFET. As this voltage increases, the channel in the silicon carbide FET gets pinched-off and further voltage is then supported by the silicon carbide FET. Under these bias conditions, both the devices sustain current flow while supporting voltage. The level of the current flowing through the devices is controlled by the applied gate bias. In this sense, the Baliga-Pair behaves like a silicon power MOSFET from the point of view of the external circuit on both the input and output side. This feature makes the configuration attractive for use in power electronic systems because the existing circuit topologies can be used. The safe-operating-area of the composite switch is mainly determined by the silicon carbide FET because it supports a majority of the applied drain voltage. The excellent breakdown strength, thermal conductivity, and wide band gap of silicon carbide ensure good safe-operating-area for the FET structures.

10.1.4 Switching Characteristics

The transition between the on and off modes for the Baliga-Pair configuration is controlled by the applied gate bias to the silicon power MOSFET. During turn-on and turn-off, the gate bias must charge and discharge the capacitance of the silicon power MOSFET. Since silicon power MOSFETs are extensively used for high frequency power conversion, their input capacitance and gate charge have been optimized by the industry[10]. The switching speed of the Baliga-Pair is consequently very high because of the availability of silicon power MOSFETs designed for high frequency applications. The main limitations to the switching speed of the Baliga-Pair will be related to parasitic inductances in the package that could produce high voltage spikes.

10.1.5 Fly-Back Diode

The Baliga-Pair contains an inherent high quality fly-back rectifier. When the drain bias is reversed to a negative value, the gate-drain contact/junction of the silicon carbide FET becomes forward biased. Since the gate of the silicon carbide FET is directly connected to the ground terminal, current can flow through this path when the drain voltage is negative in polarity. From this stand-point, it is preferable to use a metal-semiconductor contact for the gate rather than a P-N junction. The Schottky gate contact provides for a lower on-state voltage drop by proper choice of the work-function for the gate contact. In addition, the Schottky contact has no significant reverse recovery current. This greatly reduces switching losses in both the rectifier and the FETs[9]. Thus, the Baliga-Pair configuration replaces not just the power switch (such as the IGBT or GTO) in applications but also the power rectifier that is normally used across the switch.

Another path for current flow in the Baliga-Pair configuration in the reverse direction is via the body diode of the silicon power MOSFET and the SiC JFET/MESFET. This diode has a low on-state voltage drop corresponding to that for a silicon P-i-N rectifier. The reverse recovery for the body diode of the silicon MOSFET can be improved by electron irradiation[11].

10.2 Numerical Simulation Results

The results of numerical simulations of the Baliga-Pair circuit have been described in detail in previous books[12,13]. The reader is encouraged to read these books for additional knowledge about the operation of this configuration.

10.3 Experimental Results

The attractive features of the Baliga-Pair configuration have been acknowledged by several research groups. The same configuration was referred to as the Cascode arrangement in 1999[6] with the acknowledgement that it was originally proposed and published in 1996[7]. The switching behavior of the Cascode arrangement was reported[14] by using a planar silicon carbide JFET structure with a 50 volt silicon power MOSFET. The authors compared the performance of silicon carbide JFETs with the gate formed using the buried sub-surface P^+ region with a JFET fabricated using a gate formed on the upper surface (similar to the structure discussed in the previous chapter). It was found that the buried gate device had inferior switching performance due to the high resistance in the buried P^+ regions. The turn-off time was limited by the R-C charging time-constant for the buried P^+ regions. In addition, the slow response of this JFET structure resulted in the silicon power MOSFET being driven into avalanche breakdown. These problems were not observed for the surface gate device.

In a subsequent paper[15], the authors stated: *"SiCED favors a combination of a silicon switch and a vertical, normally-on SiC junction field effect transistor"*. Their analysis concluded that the Baliga-Pair configuration is useful upto at least the 4.5 kV range. The excellent performance of the Baliga-Pair circuit has also been confirmed by using numerical simulations and compared with the performance of a silicon carbide MOSFET structure[16]. The authors, who called this a Cascade Configuration, found that the turn-off time for the Baliga-pair configuration was half that for the silicon carbide MOSFET due to the smaller Miller capacitance.

The utility of the reverse conduction path[17] for the Baliga-Pair configuration was reported using 1200-V SiC vertical trench JFET devices. It was demonstrated that the reverse conduction occurs via both paths discussed in section 10.1.5. Switching in a totem-pole circuit demonstrated that the Baliga-Pair configuration can be operated without an additional anti-parallel diode as usually required for silicon power MOSFETs and IGBTs.

It has been found that the switching speed of the SiC/Si cascode circuit is very fast resulting in high [dV/dt] transients. The [dV/dt] can be reduced by (a) using a larger gate drive resistance as in the case of silicon power MOSFETs and IGBTs[18]; (b) by connecting a capacitor between the drain of the SiC JFET and the gate of the Si power MOSFET; or (c) by connecting an R-C circuit between the drain of the SiC JFET and the gate of the SiC JFET. In all the cases, the energy loss during turn-on and turn-off switching events increases approximately inversely as the reduction of the [dV/dt].

Similarly, the [dV/dt] during switching was found to be dependent on the output capacitance of the silicon power MOSFET and could be reduced by adding additional capacitance between its drain and source[19]. This method also results in an increase in the turn-off energy loss. In comparison with 1200-V trench-gate IGBTs, it was found that the SiC JFET/Si MOSFET cascode circuit has very similar power losses during switching.

The switching behavior of the 4H-SiC normally-on JFET/Si power MOSFET cascode configuration has been studied in detail[20]. The turn-on and turn-off switching transients were found to be similar to those for silicon power MOSFETs. A boost converter with efficiency of 98% was constructed at a switching frequency of 100 kHz using these devices.

Excellent turn-on and turn-off switching waveforms were also demonstrated for the 600-V 4H-SiC normally-on JFET/Si power MOSFET cascode configuration[21]. However, the very fast switching transients produced an EMI problem. The [dV/dt] could be reduced using a gate resistor but with significant increase in switching power loss. The losses could be reduced by 45% by adding a resistor in series with the gate of the SiC JFET and reducing the gate resistor. This is because of the large transconductance of the SiC JFET.

The Baliga-Pair configuration has been extended to operation at higher voltages by using the 'super-cascode' concept[22]. In this approach, a single Si power MOSFET is used with multiple SiC JFETs connected in series to support a larger blocking voltage. A voltage clamping diode is required across each of the SiC JFETs to achieve proper voltage sharing. A 6 kV switch was constructed by the use of five 1.2 kV 4H-SiC JFETs demonstrating that no derating of the voltage is required.

10.4 Summary

It has been demonstrated that the Baliga-pair configuration provides an ideal power switch for high voltage power system applications. It can be used in the same manner as IGBTs packaged with anti-parallel rectifiers without alterations of the gate control techniques. The packaging of this combination of two FETs is similar to that for the two chips (IGBT and flyback diode) currently used because the Baliga-Pair contains an inherent fly-back rectifier. Until the development of reliable silicon carbide power MOSFETs with low specific on-resistance, the Baliga-Pair offered a commercially viable near term option for high power electronic systems.

The commercialization of GaN HFETs has been also hampered by their normally-on characteristics. Consequently, the Baliga-Pair configuration has become a popular approach for creating products by many companies[23].

References

[1]　B. J. Baliga, "Critical Nature of Oxide/Interface Quality for SiC Power Devices", Microelectronics Engineering, Vol. 28, pp. 177-184, 1995.

[2] B. J. Baliga, "Silicon Carbide Switching Device with Rectifying Gate", U. S. Patent 5,396,085, Issued March 7, 1995.

[3] S. Darlington, "Semiconductor Signal Translating Device", U. S. Patent 2,663,806, Issued December 22, 1953.

[4] P. M. McLarty, Private Communication, 1995.

[5] B. J. Baliga, "Prospects for Development of SiC Power Devices", Silicon Carbide and Related Materials – 1995, Institute of Physics Conference Series, Vol. 142, pp. 1-6, 1996.

[6] P. Friedrichs *et al.*, "Static and Dynamic Characteristics of 4H-SiC JFETs Designed for Different Blocking Categories", Silicon Carbide and Related Materials – 1999, Material Science Forum, Vol. 338-342, pp. 1243-1246, 2000.

[7] B. J. Baliga, "Power Semiconductor Devices", pp. 418-420, PWS Publishing Company, 1996.

[8] "Cascode", en.wikipedia.org/wiki/Cascode.

[9] B. J. Baliga, "Power Semiconductor Devices for Variable-Frequency Drives", Proceeding of the IEEE, Vol. 82, pp. 1112-1122, 1994.

[10] B. J. Baliga and D. Alok, "Paradigm Shift in Planar Power MOSFET Technology", Power Electronics Technology Magazine, pg. 24-32, November 2003.

[11] B.J. Baliga and J.P. Walden, "Improving the Reverse Recovery of Power MOSFET Integral Diodes by Electron Irradiation", Solid-State Electronics, Vol. 26, pp. 1133-1141, 1983.

[12] B.J. Baliga, "Silicon Carbide Power Devices", World Scientific Publishers, Singapore, 2005.

[13] B.J. Baliga, "Fundamentals of Power Semiconductor Devices", Springer-Science, New York, 2008.

[14] H. Mitlehner *et al.*, "Dynamic Characteristics of High Voltage 4H-SiC Vertical JFETs", IEEE International Symposium on Power Semiconductor Devices and ICs, Abstract 11.1, pp. 339-342, 1999.

[15] P. Friedrichs *et al.*, "Application-Oriented Unipolar Switching SiC Devices", Silicon Carbide and Related Materials – 2001, Material Science Forum, Vol. 389-393, pp. 1185-1190, 2002.

[16] A. Mihaila *et al.*, "Static and Dynamic Behavior of SiC JFET/Si MOSFET Cascade Configuration for High-Performance Power Switches", ", Silicon Carbide and Related Materials – 2001, Material Science Forum, Vol. 389-393, pp. 1239-1242, 2002.

[17] D.C. Sheridan *et al.*, "Reverse conduction properties of Vertical SiC Trench JFETs", IEEE International Symposium on Power Semiconductor Devices and ICs, pp. 385-388, 2012.

[18] D. Aggeler *et al.*, "DV/Dt Control Methods for the SiC JFET/Si MOSFET Cascode", IEEE Transactions on Power Electronics, Vol. 28, pp. 4074-4082, 2013.

[19] R. Pittini, Z. Zhang and M.A.E. Andersen, "SiC JFET Cascode Loss Dependence on the MOSFET Output Capacitance and Performance Comparison with Trench IGBTs", IEEE Applied Power Electronics Conference, pp. 1287-1293, 2013.

[20] A. Rodriquez *et al.*, "Switching Performance Comparison of the SiC JFET and SiC JFET/Si MOSFET Cascode Configuration", IEEE Transactions on Power Electronics, Vol. 29, pp. 2428-2440, 2014.

[21] H. Shimizu *et al.*, "Controllability of Switching Speed and Loss for SiC JFET/Si MOSFET Cascode with External Gate Resistor", IEEE International Symposium on Power Semiconductor Devices and ICs, pp. 221-224, 2014.

[22] J.L. Hostetler *et al.*, "6.5 kV SiC Normally-Off JFETs – Technology Status", IEEE Workshop on Wide Bandgap Power Devices and Applications", pp. 143-146, 2014.

[23] www.transphormusa.com, Product such as TPH3002LD.

Chapter 11

SiC Planar Power MOSFETs

The planar power MOSFET was the first commercially successful unipolar switch developed using silicon technology[1] once the issues related to the metal-oxide-semiconductor interface had been resolved for CMOS technology. In order to reduce cost, the channel in these devices was created by the double diffusion (or D-MOS) process. In the DMOS process, the P-base and N^+ source regions are formed by ion implantations masked by a common edge defined by a refractory polysilicon gate electrode. A drive-in cycle is used after each ion implantation step to move the P-N junction in the lateral direction under the gate electrode. The separation between the N^+/P-base junction and the P-base/N-drift junction under the gate electrode defines the channel. Consequently, the channel length can be reduced to sub-micron dimensions without the need for high resolution lithography. This approach served the industry quite well from the 1970s into the 1990s with silicon planar power MOSFETs still available today for power electronic applications. In the 1990s, the industry borrowed the trench technology originally developed for DRAMs to introduce the UMOSFET structure for commercial applications[2]. This was important for the reduction of the channel and JFET components in the planar MOSFET structure designed for lower (< 30 volts) voltage applications. This UMOSFET structure has also been explored for silicon carbide as described in a subsequent chapter.

In silicon power MOSFETs, the on-resistance becomes dominated by the resistance of the drift region when the breakdown voltage exceeds 200 volts[2]. At high breakdown voltages, the specific on-resistance for these devices becomes greater than 10^{-2} Ω-cm^2 leading to an on-state voltage drop of more than 1 volt at a typical on-state current density of 100 A/cm^2. For this reason, the Insulated Gate Bipolar Transistor (IGBT) was developed in the 1980s to serve medium and high power systems[2]. The superior performance of the IGBT in high voltage applications relegated the silicon MOSFETs to applications with operating voltages below 100 volts. Novel silicon structures that utilize the charge-coupling concept have allowed extending the breakdown voltage of silicon power MOSFETs to the 600 volt range[3]. However, their specific on-resistance is still quite large limiting their use to high operating frequencies where switching losses dominate.

In principle, the much lower resistance of the drift region in silicon carbide should enable development of power MOSFETs with very high breakdown voltages. These devices offer not only fast switching speed but also superior safe operating area when compared to high voltage silicon IGBTs. This allows reduction of both the switching loss and conduction loss components in power circuits[4]. Unfortunately, the power MOSFET structures developed in silicon cannot be directly utilized to form high performance silicon carbide devices. Firstly, the lack of significant diffusion of dopants in silicon carbide prevents the use of the silicon DMOS process. Secondly, a high electric field occurs in the gate oxide of the silicon carbide MOSFET exceeding its rupture strength leading to catastrophic failure of devices in the blocking mode at high voltages. Thirdly, when compared with silicon, the smaller band offset between the conduction band of silicon carbide and silicon dioxide can produce injection of hot carriers into the oxide leading to instability during operation. In addition, the quality of the oxide-semiconductor interface for silicon carbide must be improved to allow good control over the threshold voltage and the channel mobility.

This chapter begins with a review of the basic principles of operation of the SiC planar power MOSFET structure. A SiC planar power MOSFET structure formed by staggering the ion implantation of the P-base and N^+ source regions to create the channel is then described. Next, the problem of exacerbation of reach-through breakdown in silicon carbide is considered. Based upon fundamental considerations, the difference between the threshold voltage for silicon and silicon carbide structures is analyzed. The relatively high doping concentrations and large channel lengths required to prevent reach-through are shown to be serious limitations to obtaining low specific on-resistance in these devices. In addition, the much larger electric field in silicon carbide is shown to lead to a high electric field in the gate oxide. Structures designed to reduce the electric field at the gate oxide by using a shielding region are therefore essential to realization of practical silicon carbide MOSFETs even after the MOS interface quality is improved. These structures are described and analyzed next in the chapter. The description of experimental results on relevant structures is then used to define the state of the development effort on these devices.

11.1 Planar SiC Power MOSFET Structure

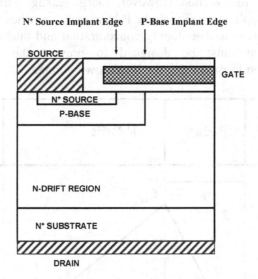

Fig. 11.1 The SiC planar power MOSFET structure.

The basic structure of the SiC planar power MOSFET is shown in Fig. 11.1 together with the location of the ion implantation mask edges. In recognition of the low diffusion coefficients for dopants in silicon carbide, it was proposed[5] that the P-base and N^+ source ion implantations be staggered by using photoresist masks rather that be defined by the gate edge. This approach with staggered P-base and N^+ source implants was subsequently used to fabricate high voltage devices[6]. These devices have been called DIMOSFETs because of the Double-Implant process used for their fabrication. The measured performance of these structures is discussed at the end of the chapter.

11.1.1 Blocking Characteristics

In the forward blocking mode of the SiC planar power MOSFET structure, the voltage is supported by a depletion region formed on both sides of the P-base/N-drift junction. The maximum blocking voltage can be determined by the electric field at this junction becoming equal to the critical electric field for breakdown if the parasitic $N^+/P/N$ bipolar transistor is completely suppressed. This suppression is accomplished by short-circuiting the N^+ source and P-base regions using the source metal as shown on the upper left hand side of the cross-section. However, a large leakage current can occur when the depletion region in the P-base region reaches-through to the N^+ source region. The doping concentration and thickness of the P-base region must be designed to prevent the reach-through phenomenon from limiting the breakdown voltage.

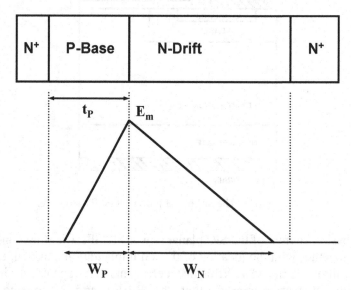

Fig. 11.2 Reach-through in a SiC power MOSFET structure.

Limitation of the blocking voltage for power MOSFETs due to reach-through of the P-base region is a well-known design consideration for silicon power MOSFETs[2]. The problem of reach-through is exacerbated for SiC devices due to the high doping concentration of the N-type drift region and the higher electric field at the P-N junction. The applied drain voltage is supported by the N-drift region and the P-base region with a triangular electric field distribution as shown in Fig. 11.2 if the doping is uniform on both sides. The maximum electric field occurs at the P-base/N-drift junction. The depletion width on the P-base side is related to the maximum electric field by:

$$W_P = \frac{\varepsilon_S E_m}{q N_A} \qquad [11.1]$$

where N_A is the doping concentration in the P-base region. The minimum P-base thickness required to prevent reach-though limited breakdown can be obtained by assuming that the maximum electric field at the P-base/N-drift junction reaches the critical electric field for breakdown when the P-base region is completely depleted:

$$t_P = \frac{\varepsilon_S E_C}{q N_A} \qquad [11.2]$$

where E_C is the critical electric field for breakdown in the semiconductor.

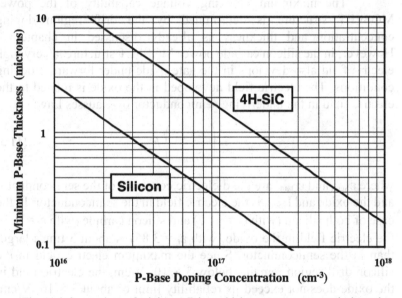

Fig. 11.3 Comparison of minimum P-base thickness to prevent reach-through breakdown in 4H-SiC and Silicon.

The calculated minimum P-base thickness for 4H-SiC power MOSFETs is compared with that for silicon in Fig. 11.3. At any given P-base doping concentration, the thickness for 4H-SiC is about six times larger than for silicon. This implies that the minimum channel length required for silicon carbide devices is much larger than for silicon devices resulting in a big increase in the on-resistance. The enhancement of the on-resistance is compounded by the much lower channel inversion layer mobility observed for silicon carbide when compared with silicon.

The minimum thickness of the P-base region required to prevent reach-through breakdown decreases with increasing doping concentration as shown in Fig. 11.3. The typical P-base doping concentration for silicon power MOSFETs is 1×10^{17} cm^{-3} to obtain a threshold voltage between 1 and 5 volts for a gate oxide thickness of 500 to 1000 angstroms. At this doping level, the P-base thickness can be reduced to 0.5 microns without reach-through limiting the breakdown voltage. In contrast, it is necessary to increase the P-base doping concentration to above 4×10^{17} cm^{-3} for 4H-SiC to prevent reach-through with a 0.5 micron P-base thickness. This higher doping concentration makes the threshold voltage very large for 4H-SiC devices as discussed in the next section.

The maximum blocking voltage capability of the power MOSFET structure is determined by the drift region doping concentration and thickness as already discussed in chapter 3. However, in the silicon carbide power MOSFET structure, a very high electric field also develops in the gate oxide under forward blocking conditions. The electric field developed in the oxide is related to the electric field in the underlying semiconductor by Gausses Law:

$$E_{Oxide} = \left(\frac{\varepsilon_{Semi}}{\varepsilon_{Oxide}} \right) . E_{Semi} \qquad [11.3]$$

where ε_{Semi} and ε_{Oxide} are the dielectric constants of the semiconductor and the oxide and E_{Semi} is the electric field in the semiconductor. In the case of both silicon (with $\varepsilon_r = 11.7$) and silicon carbide (with $\varepsilon_r = 9.7$), the electric field in the oxide (with $\varepsilon_r = 3.85$) is about 3 times larger than in the semiconductor. Since the maximum electric field in the silicon drift region remains below 3×10^5 V/cm, the electric field in the oxide does not exceed its reliability limit of about 3×10^6 V/cm. However, for 4H-SiC, the electric field in the oxide reaches a value of 9×10^6 V/cm when the field in the semiconductor reaches its breakdown strength of about 3×10^6 V/cm. This value not only exceeds the reliability limit but can cause rupture of the oxide leading to catastrophic breakdown. It is therefore important to monitor the electric field in the gate oxide when designing and modeling the silicon carbide MOSFET structures. Structures[7] that shield the gate oxide from high electric field have been proposed and demonstrated to resolve this problem. These structures will be discussed later in this chapter.

11.1.2 On-Resistance

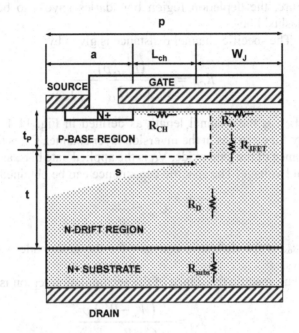

Fig. 11.4 Current flow path and resistances in the planar 4H-SiC power MOSFET.

Current flow between the drain and source can be induced in the power MOSFET structure by creating an inversion layer channel on the surface of the P-base region. The current path is illustrated in Fig. 11.4 by the shaded area. The current flows through the inversion layer channel formed due to the applied gate bias into the JFET region via the accumulation layer formed above it under the gate oxide. It then spreads into the N-drift region at a 45 degree angle and becomes uniform through the rest of the structure. The total on-resistance for the planar power MOSFET structure is determined by the resistance of these components in the current path:

$$R_{on,sp} = R_{CH} + R_A + R_{JFET} + R_D + R_{subs} \qquad [11.4]$$

where R_{CH} is the channel resistance, R_A is the accumulation region resistance, R_{JFET} is the resistance of the JFET region, R_D is the resistance of the drift region after taking into account current spreading from the JFET region, and R_{subs} is the resistance of the N^+ substrate. These resistances can be analytically modeled by using the current flow pattern indicated by the shaded regions in Fig. 11.4. In this figure, the depletion region boundaries have also been shown using dashed lines.

The specific channel resistance is given by:

$$R_{CH} = \frac{(L_{CH} \cdot p)}{\mu_{inv} C_{ox}(V_G - V_T)}$$

[11.5]

where L_{CH} is the channel length as defined in Fig. 11.4, μ_{inv} is the mobility for electrons in the inversion layer channel, C_{ox} is the specific capacitance of the gate oxide, V_G is the applied gate bias, and V_T is the threshold voltage. The specific capacitance can be obtained using:

$$C_{ox} = \frac{\varepsilon_{ox}}{t_{ox}}$$

[11.6]

where ε_{ox} is the dielectric constant for the gate oxide and t_{ox} is its thickness.

The specific resistance of the accumulation region is given by:

$$R_A = \frac{K(W_J - W_P)p}{\mu_a C_{ox}(V_G - V_{TA})}$$

[11.7]

where μ_a is the mobility for electrons in the accumulation layer, C_{ox} is the specific capacitance of the gate oxide, V_G is the applied gate bias, and V_{TA} is the threshold voltage for accumulation (close to zero). The factor K is used to account for two-dimensional current spreading from the channel into the JFET region with a typical value of 0.6 for silicon devices. The same value can also be applied for 4H-SiC power MOSFETs. In this equation, W_P is the zero-bias depletion width at the P-base/N-JFET junction. The JFET region doping concentration ($N_{D,JFET}$) is often increased above the drift region doping concentration for power MOSFETs with larger breakdown voltages. It can be determined using:

$$W_P = \sqrt{\frac{2\varepsilon_s V_{biP}}{q N_{D,JFET}}}$$

[11.8]

where the built-in potential V_{biP} for the P-N junction is typically 3.3 volts for 4H-SiC (can be calculated using Eq. [2.8]).

The specific JFET region resistance is given by:

$$R_{JFET} = \rho_D.t_P\left(\frac{p}{W_J - W_P}\right)$$ [11.9]

where t_p is the depth of the P-base region.

The drift region spreading resistance can be obtained by using:

$$R_D = \rho_D.p.\ln\left(\frac{p}{W_J - W_P}\right) + \rho_D.(t - s - W_P)$$ [11.10]

where t is the thickness of the drift region below the P-base region and s is the width of the P-base region.

The contribution to the resistance from the N^+ substrate is given by:

$$R_{subs} = \rho_{subs}t_{subs}$$ [11.11]

where ρ_{subs} and t_{subs} are the resistivity and thickness of the substrate, respectively. A typical value for the N^+ substrate resistivity for 4H-SiC is 0.020 Ω-cm and its thickness if typically 200 microns leading to a contribution of 4×10^{-4} Ω-cm^2.

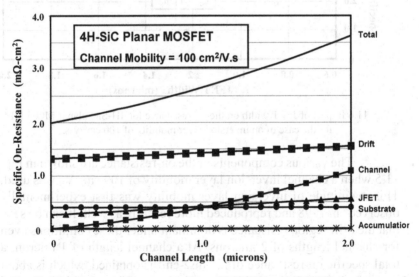

Fig. 11.5 On-Resistance Components for a 4H-SiC Planar MOSFETs for the case of an inversion layer mobility of 100 cm^2/v-s.

The specific on-resistances of 4H-SiC planar MOSFETs with a drift region doping concentration of 1×10^{16} cm^{-3} and thickness of 10 microns can be modeled using the above analytical expressions. The drift region doping corresponds to a device with a breakdown voltage of 1200 volts if the edge termination is limited to 70 % of the ideal breakdown voltage. In all the devices, a gate oxide thickness of 0.1 microns was used. The dimension 'a' in the structure shown in Fig. 11.4 was kept at 1 micron. The accumulation mobility was assumed to be twice the inversion layer mobility and a K-factor of 0.1 was used to provide correlation with the simulations. The P-base was assumed to have a depth of 1 micron and the effective gate drive voltage ($V_G - V_T$) was assumed to be 20 volts.

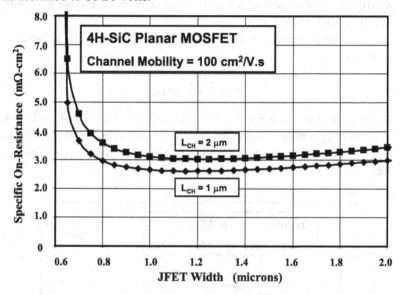

Fig. 11.6 Impact of JFET width on the on-resistance for 4H-SiC planar MOSFETs for the case of an inversion layer mobility of 100 cm²/v-s.

The various components of the on-resistance are shown in Fig. 11.5 when a channel inversion layer mobility of 100 cm²/Vs was used. This magnitude of inversion layer mobility was first experimentally observed[8] in 1998 and reproduced more recently as well. It can be seen that the drift region resistance is dominant under these conditions even for channel lengths of 2 microns. At a channel length of 1 micron, a total specific on-resistance of 2.7 mΩ-cm² is obtained, which is about three times the ideal specific on-resistance. Although this is a good value for a device with the drift regions parameters used for the modeling, the breakdown voltage of the structure is limited by the reach-through problem.

The JFET width in the planar MOSFET structure must be optimized not only to obtain the lowest specific on-resistance but to also control the electric field at the gate oxide interface. The impact of changes to the JFET width on the specific on-resistance is shown in Fig. 11.6 for a structure with the same cell parameters that were given above. It can be observed that the specific on-resistance becomes very large when the JFET width is made smaller than about 0.7 microns. This occurs because the zero bias depletion width for 4H-SiC P-N junctions is about 0.6 microns. When the JFET width approaches this value, the current becomes severely constricted resulting in a high specific on-resistance. As the JFET width is increased, the contribution from its resistance decreases rapidly. However, the increase in cell pitch enhances the specific resistance contributions from the channel resulting in an increase in the total specific on-resistance. A minimum specific on-resistance is observed to occur at a JFET width of about 1 micron.

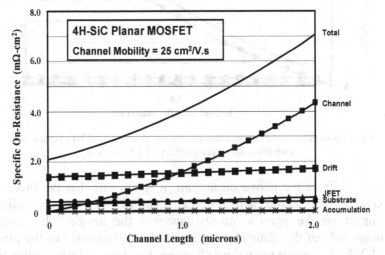

Fig. 11.7 On-resistance components for 4H-SiC planar MOSFETs for the case of an inversion layer mobility of 25 cm²/V-s.

The above results are based upon using a channel mobility of 100 cm²/Vs. However, most groups working on silicon carbide have reported much lower magnitudes for the inversion layer mobility in 4H-SiC structures[9]. Taking this into consideration, it is interesting to examine the impact of reducing the inversion layer mobility as depicted in Figs. 11.7 and 11.8 for the case of mobility values of 25 and 2.5 cm²/Vs. With an inversion layer mobility of 25 cm²/Vs, the channel contribution becomes comparable to the contribution from the drift region. This results in approximately doubling of the specific on-resistance to a value of 4 mΩ-cm² for a channel length of 1 micron. When the inversion layer mobility is reduced to 2.5 cm²/Vs, the channel resistance becomes dominant as shown in Fig. 11.8, with an increase in the specific on-resistance to a value of 18 mΩ-cm² for a channel length of 1 micron.

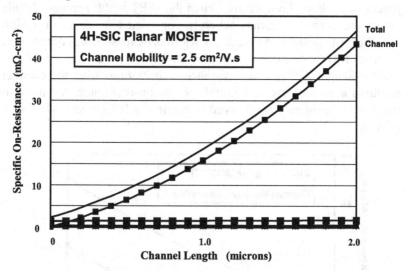

Fig. 11.8 On-resistance components for 4H-SiC planar MOSFETs for the case of an inversion layer mobility of 2.5 cm²/V-s.

From the above discussion, it is obvious that the inversion layer mobility in the channel of the planar MOSFET has a significant impact on the specific on-resistance of the device. The relative magnitude of the channel mobility on the performance of the planar MOSFET depends upon the drift region resistance which is a function of the breakdown voltage of the device. In order to provide a better understanding of this, consider a 4H-SiC planar MOSFET with a JFET region whose doping concentration is held at 1×10^{16} cm⁻³ while the properties of the underlying drift region are adjusted to obtain the desired breakdown voltage. This approach for designing planar MOSFETs with enhanced doping in the JFET region is common practice in silicon technology[2]. The other cell parameters are maintained at the following values: parameter 'a' of 1 micron, P-base depth of 1 micron, gate oxide thickness of 0.05 microns, accumulation layer mobility of 200 cm²/V-s, and channel length of 1 micron. A gate bias voltage of 10 volts above threshold voltage was also assumed.

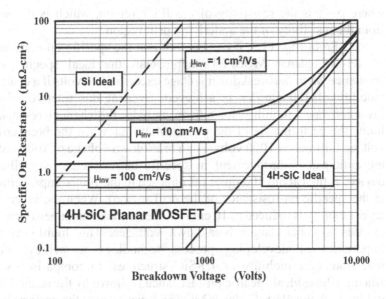

Fig. 11.9 Specific on-resistance for 4H-SiC planar MOSFETs for various breakdown voltages.

The specific on-resistance for the planar 4H-SiC MOSFETs is plotted in Fig. 11.9 as a function of the breakdown voltage for several cases of channel mobility. In performing this modeling, it is important to recognize that the thickness of the drift region (parameter 't' in Fig. 11.4) can become smaller than the cell parameter 's' at lower breakdown voltages (below 700 volts). Under these conditions, the current does not distribute at a 45 degree angle into the drift region from the JFET region. Instead, the current flows from a cross-sectional width of $(W_J - W_P)$ to a cross-section of $(t + W_J - W_P)$. The drift region resistance for these cases can be modeled using:

$$R_D = \rho_D . p . \ln\left(\frac{t + W_J - W_P}{W_J - W_P}\right) \qquad [11.12]$$

In addition, the resistance contributed by the JFET region should be modeled using:

$$R_{JFET} = \rho_{JFET} \cdot t_P \left(\frac{p}{W_J - W_P} \right) \qquad [11.13]$$

where ρ_{JFET} is the resistivity of the JFET region, which is different from the resistivity of the underlying drift region.

From Fig. 11.9, it can be seen that the specific on-resistance of 4H-SiC planar MOSFETs approaches the ideal specific on-resistance when the breakdown voltage exceeds 5000 volts if a channel mobility of 100 cm^2/Vs is achieved. Even at this relatively high inversion layer mobility for silicon carbide, the channel resistance limits the performance of the planar MOSFET when the breakdown voltage falls below 1000 volts. For a breakdown voltage of 1000 volts, the anticipated improvement in specific on-resistance over silicon devices is then about 100 times (as opposed to the 1000x improvement in the specific on-resistance of the drift region). When the inversion layer mobility is reduced to 10 cm^2/Vs, the degradation in performance extends to much larger breakdown voltages. This highlights the importance of developing process technology to achieve high inversion layer mobility in 4H-SiC structures. In comparison with silicon, whose ideal specific on-resistance is shown by the dashed line in Fig. 11.9, the 4H-SiC planar MOSFET can surpass the performance by an order of magnitude at breakdown voltages above 1000 volts if a channel mobility of at least 10 cm^2/Vs is obtained.

11.1.3 Threshold Voltage

The threshold voltage of the power MOSFET is an important design parameter from an application stand-point. A minimum threshold voltage must be maintained at above 1 volt for most system applications to provide immunity against inadvertent turn-on due to voltage spikes arising from noise. At the same time, a high threshold voltage is not desirable because the voltage available for creating the charge in the channel inversion layer is determined by $(V_G - V_T)$ where V_G is the applied gate bias voltage and V_T is the threshold voltage. Most power electronic systems designed for high voltage operation (the most suitable application area for silicon carbide devices) provide a gate drive voltage of 15 volts. Based upon this criterion, the threshold voltage should be kept below 5 volts in order to obtain a low channel resistance contribution.

The threshold voltage can be modeled by defining it as the gate bias at which on-set of *strong inversion* begins to occur in the channel. This voltage can be determined using[10]:

$$V_{TH} = \frac{\sqrt{4\varepsilon_S kTN_A \ln(N_A / n_i)}}{C_{ox}} + \frac{2kT}{q} \ln\left(\frac{N_A}{n_i}\right) \qquad [11.14]$$

where N_A is the doping concentration of the P-base region, k is Boltzmann's constant, and T is the absolute temperature. The presence of positive fixed oxide charge shifts the threshold voltage in the negative direction by:

$$\Delta V_{TH} = \frac{Q_F}{C_{ox}} \qquad [11.15]$$

A further shift of the threshold voltage in the negative direction by 1 volt can be achieved by using heavily doped N-type polysilicon as the gate electrode as routinely done for silicon power MOSFETs.

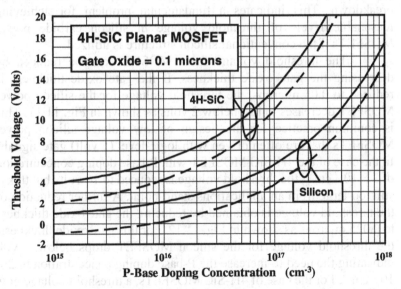

Fig. 11.10 Threshold voltage of 4H-SiC planar MOSFETs.

(Dashed lines include the impact of N+ polysilicon gate and an oxide fixed charge of $2 \times 10^{11} \, cm^{-2}$)

The analytically calculated threshold voltage for 4H-SiC planar MOSFETs is shown in Fig. 11.10 for the case of a gate oxide thickness of 0.1 microns. The results obtained for a silicon power MOSFET with the same gate oxide thickness is also provided in this figure for comparison. In the case of silicon devices, a threshold voltage of about 3 volts is obtained for a P-base doping concentration of 1×10^{17} cm^{-3}. At this doping concentration, the depletion width in the P-base region for silicon devices is less than 0.5 microns, as shown earlier in Fig. 11.3, even when the electric field in the semiconductor approaches the critical electric field for breakdown. This allows the design of silicon power MOSFETs with channel lengths of below 0.5 microns without encountering reach-through breakdown limitations. In contrast, a P-base doping concentration of about 3×10^{17} cm^{-3} is required in 4H-SiC (see Fig. 11.3) to keep the depletion width in the P-base region below 1 micron when the electric field in the semiconductor approaches the critical electric field for breakdown. At this doping concentration, the threshold voltage for the 4H-SiC MOSFET approaches 20 volts. The much larger threshold voltage for silicon carbide is physically related to its larger band gap as well as the higher P-base doping concentration required to suppress reach-through breakdown. This indicates a fundamental problem for achieving reasonable levels of threshold voltage in silicon carbide power MOSFETs if the conventional silicon structure is utilized.

The threshold voltage for a MOSFET can be reduced by decreasing the gate oxide thickness. If the gate oxide thickness is reduced to 0.05 microns, the threshold voltage for the silicon power MOSFET decreases to just below 2 volts as shown in Fig. 11.11. This is common practice for the design of low voltage silicon power MOSFETs that are often driven with logic-level (5 volt) gate signals. In the case of the 4H-SiC MOSFET with P-base doping concentration of 3×10^{17} cm^{-3}, the threshold voltage is reduced to 10 volts but this is still too large from an applications stand-point. A further reduction of the threshold voltage is obtained by reducing the gate oxide thickness to 0.025 microns as shown in Fig. 11.12. For this gate oxide thickness, the threshold voltage for the silicon MOSFET drops below 1 volt indicating the need to increase the P-base doping concentration to 2×10^{17} cm^{-3}. For the case of 4H-SiC MOSFETs, a threshold voltage now becomes about 6 volts which is marginally acceptable. However, operation of very high voltage power MOSFETs with such thin gate oxides may create manufacturing and reliability issues when taking into account the high electric fields under the gate oxide in the semiconductor.

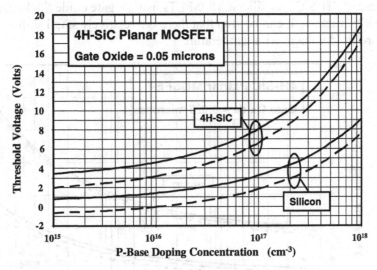

Fig. 11.11 Threshold voltage of 4H-SiC planar MOSFETs.

(Dashed lines include the impact of N^+ polysilicon gate and an oxide fixed charge of $2 \times 10^{11} \, cm^{-2}$)

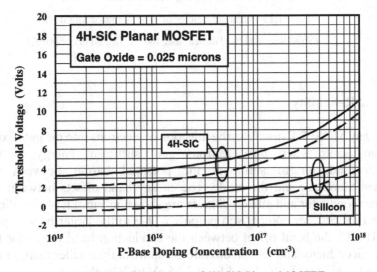

Fig. 11.12 Threshold Voltage of 4H-SiC Planar MOSFETs.

(Dashed lines include the impact of N^+ Polysilicon Gate and an Oxide Fixed Charge of $2 \times 10^{11} \, cm^{-2}$)

The threshold voltage of power MOSFETs decreases with increasing temperature. This variation is shown in Fig. 11.13 for the case of 4H-SiC and silicon MOSFETs using a gate oxide thickness of 0.025 microns. It can be seen that there is a relatively small shift in the threshold voltage with temperature.

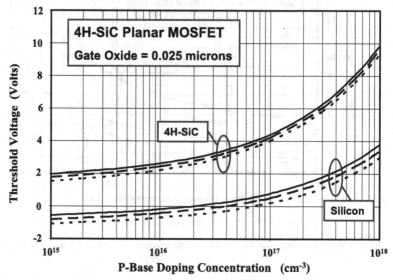

Fig. 11.13 Threshold voltage of 4H-SiC planar MOSFETs.

(Solid line: 300 °K; Dashed line: 400 °K; Dotted line: 500 °K)

11.2 Reliability

In any MOSFET structure, the injection of electrons into the gate oxide can occur when the electrons gain sufficient energy in the semiconductor to surmount the potential barrier between the semiconductor and the oxide. The energy band offsets between the semiconductor and silicon dioxide are shown in Fig. 11.14 for silicon and 4H-SiC for comparison purposes. Due to the larger band gap of 4H-SiC, the band offset between the conduction band edges for the semiconductor and silicon dioxide is significantly smaller than for the case of silicon: i.e. reduced from 3.15 eV to 2.70 eV.

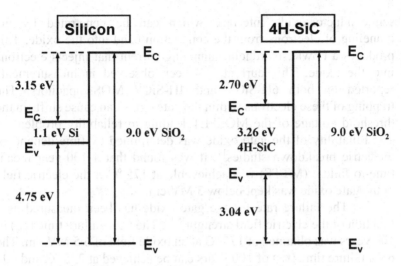

Fig. 11.14 Energy band offsets between oxide and semiconductors.

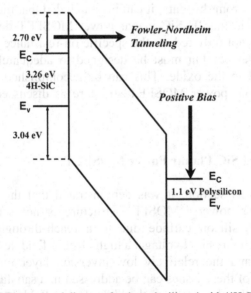

Fig. 11.15 Energy band diagram for the polysilicon/oxide/4H-SiC structure.

The band diagram for the 4H-SiC silicon dioxide interface is shown in Fig. 11.15 when a positive bias is applied to the gate. This bias condition is typical for the on-state mode of operation in power MOSFETs. In this illustration, the gate was assumed to be formed using polysilicon. It can be seen that a narrow barrier is formed at the semiconductor-oxide interface which can be penetrated by the tunneling of electrons from the conduction band into the oxide. This produces a Fowler-Nordheim tunneling current that injects electrons into the oxide. This current has been observed in measurements reported on both 6H-SiC[11] and 4H-SiC[12] MOS-capacitors. The trapping of these electrons within the gate oxide can cause shifts in the threshold voltage of the MOSFET leading to reliability problems[13]. The reliability of the gate oxide was determined by time dependent dielectric breakdown studies[14]. It was found that a 100-year mean-time-to-failure (MTTF) was achievable at 175 °C if the electric field in the gate oxide was kept below 3 MV/cm.

The failure rate for the gate oxide has been measured as a function of the electric field strength[15,16]. The 63% failure time (t_{63}) of 100 years was obtained at 175 °C at an oxide field of 6.5 MV/cm. The 63% failure time (t_{63}) of 100 years can be achieved at 275 °C and 300 °C by reducing the oxide electric field to 4 M/cm and 1.5 MV/cm, respectively. From this data, it can be concluded that the width of the JFET region in the 4H-SiC planar power MOSFET design must be optimized to not only reduce the specific on-resistance as performed for silicon devices but must be designed to adequately reduce the electric field in the oxide. This can be accomplished by using the shielded planar power MOSFET structure as discussed in the next section.

11.3 Shielded SiC Planar Power MOSFET

In the previous section, it was demonstrated that the conventional silicon planar power DMOSFET structure is not satisfactory for utilization in silicon carbide due to a reach-through problem, a relatively high threshold voltage, a high electric field developed in the gate oxide, and the relatively low inversion layer mobility in the channel. All of these issues can be addressed in a satisfactory manner by shielding the channel from the high electric field developed in the drift region. The concept of shielding of the channel region was first proposed[7] at PSRC in the early 1990's with a U.S. patent issued in 1996. The shielding was accomplished by formation of either a P-type region under the channel or by creating a high resistivity conduction barrier region under the channel. The issue of low inversion layer channel mobility was addressed by utilizing an accumulation mode of operation. As discussed in this section, the accumulation mode is also attractive from the point of view of reducing the threshold voltage so as to enhance the channel conductivity.

This section begins with a review of the basic principles of operation of the shielded planar MOSFET structure. The impact of shielding on ameliorating the reach-through breakdown in silicon carbide is described. This shielding approach has more recently been used very effectively for improvement of even silicon low voltage planar power MOSFETs[17]. Based upon fundamental considerations, the difference between the threshold voltage for inversion and accumulation mode silicon carbide structures is then analyzed. The results of two-dimensional numerical simulations of the shielded planar silicon carbide MOSFET structure are included in this section to demonstrate that the shielding concept enables reduction of the electric field developed in the gate oxide leading to the possibility to fully utilize the breakdown field strength of the underlying semiconductor drift region. The shielded accumulation-mode MOSFET structures (named the ACCUFET) discussed in this section have the most promising characteristics for the development of monolithic power switches from silicon carbide.

11.3.1 Shielded Device Structure

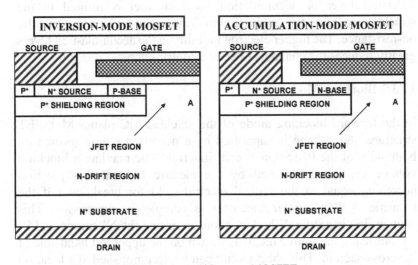

Fig. 11.16 Shielded planar SiC power MOSFET structures.

The basic structures of the shielded SiC planar power MOSFET are shown in Fig. 11.16 with either an inversion layer or an accumulation layer channel. A deep P$^+$ shielding region has been incorporated into the structure shown in Fig. 11.1. In the case of the structure with the inversion layer channel, the P$^+$ shielding region extends under both the N$^+$ source region as well as under the P-base region. It could also extend beyond the edge of the P-base region. In the case of the structure with the accumulation layer channel, the P$^+$ shielding region extends under the N$^+$ source region and the N-base region located under the gate. This N-base region can be formed using an uncompensated portion of the N-type drift region or it can be created by adding N-type dopants near the upper surface with ion implantation or epitaxial growth to independently control its thickness and doping concentration.

The gap between the P$^+$ shielding regions is optimized to obtain a low specific on-resistance while simultaneously shielding the gate oxide interface from the high electric field in the drift region. In both of the structures shown in Fig. 11.16, a potential barrier is formed at location A after the JFET region becomes depleted by the applied drain bias in the blocking mode. This barrier prevents the electric field from becoming large at the gate oxide interface. The potential barrier also presents reach-through breakdown in the P-base region even if it has a low doping concentration to achieve desired threshold voltages.

When a positive bias is applied to the gate electrode, an inversion layer or accumulation layer channel is formed in the structures enabling the conduction of drain current with a low specific on-resistance. The higher electron mobility in the accumula-tion layers greatly reduces the channel resistance contribution.

11.3.2 Blocking Mode

In the forward blocking mode of the shielded SiC planar MOSFET structure, the voltage is supported by a depletion region formed on both sides of the P$^+$ region/N-drift junction. The maximum blocking voltage can be determined by the electric field at this junction becoming equal to the critical electric field for breakdown if the parasitic N$^+$/P/N bipolar transistor is completely suppressed. This suppression is accomplished by short-circuiting the N$^+$ source and P$^+$ regions using the source metal as shown on the upper left hand side of the cross-section. This short circuit can be accomplished at a location orthogonal to the cell cross-section if desired to reduce the cell pitch while optimizing the specific on-resistance. If the doping concentration of the P$^+$ region is high (e.g. 1×10^{18} cm^{-3}), the reach-through breakdown problem discussed in the previous chapter is completely eliminated. In addition, the high doping concentration in the P$^+$ region promotes the depletion of the JFET region at lower drain voltages providing enhanced shielding of the channel and gate oxide.

With the shielding provided by the P$^+$ region, the minimum P-base thickness for 4H-SiC power MOSFETs is no longer constrained by the reach-through limitation. This enables reducing the channel length below the values associated with any particular doping concentration of the P-base region that were shown in Fig. 11.3. In addition, the opportunity to reduce the P-base doping concentration enables decreasing the threshold voltage. The smaller channel length and threshold voltage provide the benefits of reducing the channel resistance contribution.

In the case of the accumulation-mode planar MOSFET structure, the presence of the sub-surface P$^+$ shielding region under the N-base region provides the potential required for *completely depleting* the N-base region if its doping concentration and thickness are appropriately chosen. This enables *normally-off operation* of the accumulation mode planar MOSFET at zero gate bias. It is worth pointing out that this mode of operation is fundamentally different than that of buried channel MOS devices[18]. Buried channel devices contain an un-depleted N-type channel region that provides a current path for drain current flow at zero gate bias. This channel region must be depleted by a negative gate bias creating a normally-on device structure which is not suitable for power electronics applications.

In the accumulation-mode planar MOSFET structure, the depletion of the N-base region is accompanied by the formation of the potential barrier for the flow of electrons through the channel. The channel potential barrier does not have to have a large magnitude because the depletion of the JFET region screens the channel from the drain bias.

The maximum blocking voltage capability of the shielded planar MOSFET structure is determined by the drift region doping concentration and thickness as already discussed in chapter 3. However, to fully utilize the high breakdown electric field strength available in silicon carbide, it is important to screen the gate oxide from the high field within the semiconductor. In the shielded planar MOSFET structure, this is achieved by the formation of a potential barrier at location A by the depletion of the JFET region at a low drain bias voltage.

11.3.3 Inversion Channel Structure Simulations

The results of two-dimensional numerical simulations of the planar 4H-SiC shielded inversion-mode planar MOSFET structure provide insight into its operation. A drift region doping concentration of 1×10^{16} cm^{-3} and thickness of 10 microns was chosen corresponding to a device with a breakdown voltage of 1200 volts if the edge termination is limited to 70% of the ideal breakdown voltage. The baseline device had a gate oxide thickness of 0.04 microns and a P^{+} region depth of 1 micron. The P-base region had a thickness of 0.2 microns with a doping concentration of 9×10^{16} cm^{-3}. The JFET width for the baseline device was 1.5 microns. The cell pitch for the structure, corresponding to the cross-section shown on the left-hand-side of Fig. 11.16, was 3.25 microns.

The blocking capability of the planar 4H-SiC shielded inversion-mode MOSFET was investigated by maintaining zero gate bias while increasing the drain voltage. It was found that the drain current remains below 1×10^{-13} amperes up to a drain bias of 1200 volts. Consequently, the shielding of the P-base region is very effective for preventing the reach-through breakdown problem allowing the device to operate up to the full capability of the drift region.

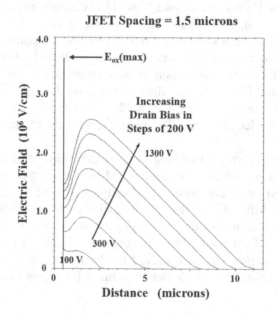

Fig. 11.17 Electric field distribution in the planar 4H-SiC shielded inversion-mode MOSFET structure.

In the planar 4H-SiC shielded inversion-mode MOSFET structure, the largest electric field at the gate oxide interface occurs at the center of the JFET region. The behavior of this electric field with increasing drain bias is shown in Fig. 11.17 for the case of a JFET width of 1.5 microns. From this figure, it is apparent that there is considerable shielding of the gate oxide by the JFET region due to the formation of a potential barrier when this region becomes depleted at lower drain voltages. The maximum electric field occurs inside the semiconductor at a depth of about 2 microns rather than at the surface due to the presence of the P^+ shielding region. The electric field in the oxide reaches a magnitude of 3.5×10^6 V/cm at a drain bias of 1300 volts. This value is sufficiently low to allow reliable operation of the shielded SiC planar power MOSFET structure up to 275 °C as discussed in section 11.2.

11.3.4 Accumulation Channel Structure Simulations

Two-dimensional numerical simulations of the planar 4H-SiC shielded accumulation-mode MOSFET structure can provide valuable insight into its operation in the blocking mode. A drift region doping concentration of 1×10^{16} cm^{-3} and thickness of 10 microns was chosen corresponding to a device with a breakdown voltage of 1200 volts if the edge termination is limited to 70% of the ideal breakdown voltage. The baseline device had the same cell parameters as used for the inversion-mode structure, namely, a gate oxide thickness of 0.04 microns, a P^+ region depth of 1 micron, and a JFET width of 1.5 microns. The N-base region had a thickness of 0.2 microns with a doping concentration of 8.5×10^{15} cm^{-3}. The cell pitch for the structure, corresponding to the cross-section shown on the right-hand-side of Fig. 11.1, was 3.25 microns.

The blocking capability of the planar 4H-SiC shielded accumulation-mode MOSFET was investigated by maintaining zero gate bias while increasing the drain voltage. It was found that the drain current remains below 1×10^{-13} amperes up to a drain bias of 1300 volts. This demonstrates that the depletion of the N-base region by the built-in potential of P^+/N junction creates a potential barrier for electron transport through the channel. In addition, the shielding of the N-base region by the underlying P^+ region is very effective for preventing the reach-through breakdown problem allowing the device to operate up to the full capability of the drift region.

In order to understand the ability to suppress current flow in the accumulation-mode structure in the absence of the P-base region, it is instructive to examine the potential distribution along the channel. The variation of the potential along the channel is shown in Fig. 11.18 for the planar 4H-SiC shielded accumulation-mode MOSFET described above for various drain bias voltages when the gate bias voltage is held at zero volts. It can be seen that there is a potential barrier in the channel with a magnitude of 2 eV at zero drain bias. This barrier for the transport of electrons from the source to the drain is upheld even when the drain bias is increased to 1300 volts. The fundamental operating principle for creating a normally-off device with a depleted N-base region formed using the built-in potential of the P^+/N is demonstrated by this plot.

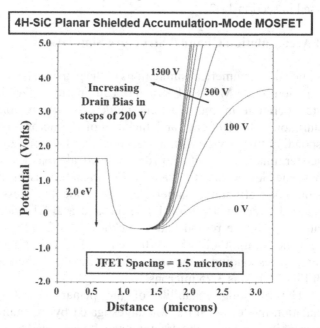

Fig. 11.18 Channel potential barrier in the planar 4H-SiC shielded accumulation-mode MOSFET structure.

For a better perspective on the operation of the accumulation-mode structure, the potential distribution along the channel is provided in Fig. 11.19 for the inversion-mode structure. From this figure, it can be seen that a potential barrier of 2.8 eV in magnitude is created due to the presence of the P-base region. In comparison, the potential barrier in the accumulation-channel structure is smaller at 2.0 eV but still sufficiently large to prevent electron transport across the channel at zero gate bias. The potential barrier is retained intact when the drain bias voltage is increased to 1300 volts in both the inversion and accumulation channel structures. By examining the potential distributions shown in the last two figures, it can be concluded that the accumulation-mode structure operates as well as the inversion-mode structure from the point of view of supporting high drain bias voltages with zero gate bias.

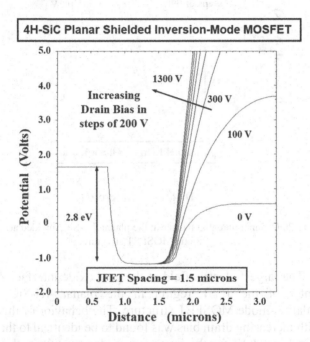

Fig. 11.19 Channel potential barrier in the planar 4H-SiC shielded inversion-mode MOSFET structure.

The normally-off feature can be preserved in the planar 4H-SiC shielded accumulation-mode MOSFET structure even when the channel length is reduced to 0.5 microns. This is demonstrated using the results of numerical simulations. The potential distribution along the channel for the accumulation channel structure with 0.5 micron channel length is shown in Fig. 11.20 for various drain bias voltages. It can be seen that, although the width of the barrier along the channel (x-direction in Fig. 11.20) is smaller than for the structure with the 1 micron channel length, the magnitude of the potential barrier for the transport of electrons from the source to the drain is still maintained at

2.0 eV. This is sufficient for obtained excellent blocking capability in the planar 4H-SiC shielded accumulation-mode MOSFET.

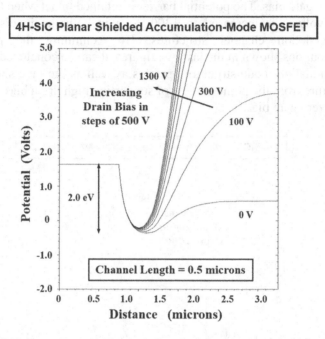

Fig. 11.20 Channel potential barrier in the planar 4H-SiC shielded accumulation-mode MOSFET structure.

The largest electric field at the gate oxide interface occurs at the center of the JFET region in the planar 4H-SiC shielded accumulation-mode MOSFET structure. The behavior of this electric field with increasing drain bias was found to be identical to the profiles shown in Fig. 11.17 for the inversion-mode structure with the same JFET width of 1.5 microns. A further reduction of the electric field in the gate oxide can be achieved by reducing the width of the JFET region. In order to illustrate this, the electric field distribution is shown in Fig. 11.21 for the case of a planar 4H-SiC shielded accumulation-mode MOSFET structure with JFET width of 1 micron. It can be seen that the maximum electric field at the gate oxide interface is reduced to 8.5×10^5 V/cm resulting in a maximum electric field in the gate oxide of only 2.1×10^6 V/cm at a drain bias of 1300 volts. This value is sufficiently low for reliable operation of the structure even at 300 °C especially due to the planar gate structure, where there are no localized electric field enhancements under the gate electrode.

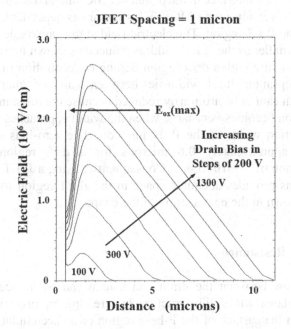

Fig. 11.21 Electric field distribution in the planar 4H-SiC shielded accumulation-mode MOSFET structure.

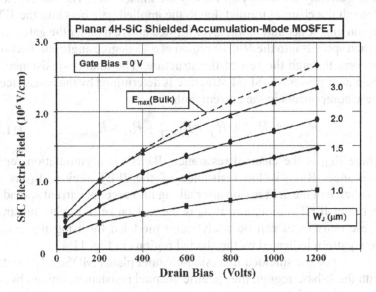

Fig. 11.22 Electric field suppression in the planar 4H-SiC shielded accumulation-mode MOSFET structure.

The magnitude of the electric field developed in the vicinity of the gate oxide interface in the planar 4H-SiC shielded accumulation-mode and inversion-mode MOSFET structure is dependent upon the width of the JFET region. The electric field at the gate oxide interface becomes smaller as the JFET width is reduced as shown in Fig. 11.22. For the structure with a drift region doping concentration of 1×10^{16} cm^{-3}, the optimum JFET width lies between 1 and 1.5 microns. The JFET width cannot be arbitrarily reduced because the resistance of the JFET region becomes very large when the width approaches the zero-bias depletion width of the P$^+$/N junction. The zero-bias depletion width is approximately 0.6 microns for a drift region doping concentration of 1×10^{16} cm^{-3}. Consequently, using a JFET width of 1.5 microns provides adequate space in the JFET region for current flow as shown in the next section of the chapter.

11.3.5 On-Resistance

Current flow between the drain and source can be induced in the shielded planar MOSFET structure by creating an inversion layer channel on the surface of the P-base region or an accumulation layer channel on the surface of the N-base region. The current path is similar to that already shown in Fig. 11.4 by the shaded area. The current flows through the channel formed due to the applied gate bias into the JFET region via the accumulation layer formed above it under the gate oxide. It then spreads into the N-drift region at a 45 degree angle and becomes uniform through the rest of the structure. The total on-resistance for the planar power MOSFET structure is determined by the resistance of these components in the current path:

$$R_{on,sp} = R_{CH} + R_A + R_{JFET} + R_D + R_{subs} \qquad [11.16]$$

where R_{CH} is the channel resistance, R_A is the accumulation region resistance, R_{JFET} is the resistance of the JFET region, R_D is the resistance of the drift region after taking into account current spreading from the JFET region, and R_{subs} is the resistance of the N$^+$ substrate. These resistances can be analytically modeled by using the current flow pattern indicated by the shaded regions in Fig. 11.4.

For the shielded inversion-channel planar MOSFET structure with the P-base region, the specific channel resistance is given by:

$$R_{CH} = \frac{(L_{CH} \cdot p)}{\mu_{inv} C_{ox}(V_G - V_T)}$$ [11.17]

where L_{CH} is the channel length as defined in Fig. 11.4, μ_{inv} is the mobility for electrons in the inversion layer channel, C_{ox} is the specific capacitance of the gate oxide, V_G is the applied gate bias, and V_T is the threshold voltage.

For the shielded accumulation-channel planar MOSFET structure with the N-base region, the specific channel resistance is given by:

$$R_{CH} = \frac{(L_{CH} \cdot p)}{\mu_a C_{ox}(V_G - V_T)}$$ [11.18]

where μ_a is the mobility for electrons in the accumulation layer channel. As discussed later in this chapter, much larger accumulation layer mobility has been experimentally observed in silicon carbide allowing reduction of the specific on-resistance. In addition, the threshold voltage for the accumulation mode is smaller than for the inversion mode allowing further improvement in the channel resistance contribution. The rest of the resistance components in the shielded planar MOSFET structure can be modeled by using the equations provided in section 11.1.2.

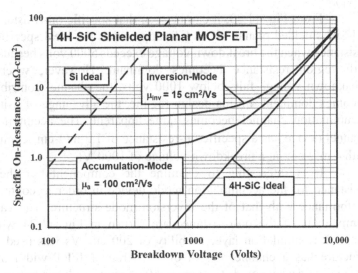

Fig. 11.23 On-resistance for the 4H-SiC shielded planar MOSFETs.

The specific on-resistances of 4H-SiC shielded planar MOSFETs with various breakdown voltages were modeled using the analytical expressions. For comparison of the inversion and accumulation mode structures, a gate oxide thickness of 0.05 microns was used. The dimension 'a' in the structure shown in Fig. 11.4 was kept at 1 micron. The accumulation mobility was assumed to be 100 cm^2/Vs while an inversion layer mobility of 15 cm^2/Vs was used based upon typical values reported in the literature as discussed later in this chapter. A K-factor of 0.1 was used to provide correlation with the simulations. The P^+ region was assumed to have a depth of 1 micron and the effective gate drive voltage ($V_G - V_T$) was assumed to be 10 volts. A channel length of 1 micron and a JFET width (W_J) of 1 micron were used for these structures. The cell pitch was 3 microns. An N^+ substrate contribution of 0.04 mΩ-cm^2 was included.

The specific on-resistance for the shielded planar 4H-SiC MOSFETs is plotted in Fig. 11.23 as a function of the breakdown voltage. As discussed in the previous chapter, in performing this modeling, it is important to recognize that the thickness of the drift region (parameter 't' in Fig. 11.4) can become smaller than the cell parameter 's' at lower breakdown voltages (below 700 volts). Under these conditions, the current does not distribute at a 45 degree angle into the drift region from the JFET region. Instead, the current flows from a cross-sectional width of ($W_J - W_P$) to a cross-section of ($t + W_J - W_P$).

From Fig. 11.23, it can be seen that specific on-resistance of 4H-SiC shielded planar MOSFETs approaches the ideal specific on-resistance when the breakdown voltage exceeds 5000 volts because the drift region resistance becomes dominant. However, when the breakdown voltage falls below 1000 volts, the channel contribution becomes dominant. In this design regime, the advantage of using an accumulation channel becomes quite apparent. The accumulation-channel device has a specific on-resistance of 1.5 mΩ-cm^2 compared with 4 mΩ-cm^2 at a breakdown voltage of 1000-V.

It is interesting to examine the components of the on-resistance for the accumulation-mode device in order to contrast its performance with that of the inversion mode structure. The various components of the on-resistance are shown in Fig. 11.24 when a channel accumulation layer mobility of 200 cm^2/Vs was used. This structure has a channel length of 1 micron, a JFET width of 1.5 microns, and a gate oxide thickness of 0.05 microns. It can be seen that the drift region resistance is dominant under these conditions even for channel lengths of 2 microns. At a channel length of 1 micron, a total specific on-resistance of 2.1 mΩ-cm^2 is obtained, which is about five-times the ideal specific on-resistance for the 1200 volt case. It is worth pointing out that the contribution from the N$^+$ substrate is exceeding that from the channel for this structure. This indicates the need to either reduce the substrate thickness or reduce its resistivity to bring the total specific on-resistance close to that of the drift region.

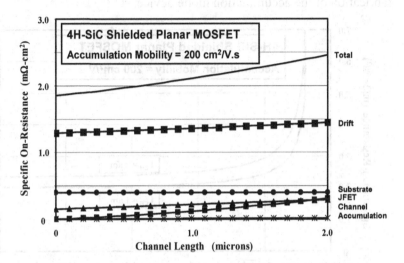

Fig. 11.24 Components of the on-resistance for the 4H-SiC shielded planar MOSFET with accumulation channel.

The JFET width in the planar MOSFET structure must be optimized not only to obtain the lowest specific on-resistance but to also control the electric field at the gate oxide interface. The impact of changes to the JFET width on the specific on-resistance for the accumulation-mode 4H-SiC shielded planar MOSFET structure is shown in Fig. 11.25 for a structure with the same cell parameters that were given above. It can be observed that the specific on-resistance becomes very large when the JFET width is made smaller than about 0.7 microns. This occurs because the zero bias depletion width for 4H-SiC P-N junctions is about 0.6 microns. When the JFET width approaches this value, the current becomes severely constricted resulting in a high specific on-resistance. As the JFET width is increased, the contribution from its resistance decreases rapidly. However, the increase in cell pitch enhances the specific resistance contributions from the channel resulting in a nearly constant total specific on-resistance. A minimum specific on-resistance is observed to occur at a JFET width of above 1.5 microns, which is larger than the value for the planar inversion-mode MOSFETs discussed in section 11.1.2. This is due to the larger mobility in the accumulation channel. The specific on-resistance is observed to be less sensitive to the channel length when compared to the inversion-channel device. This provides more latitude with regard to the design rules used for the fabrication of the accumulation mode device.

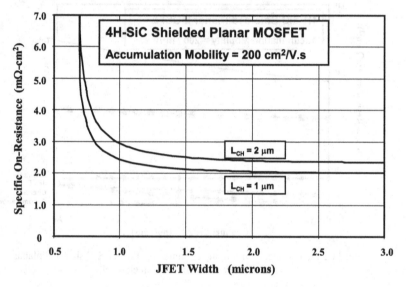

Fig. 11.25 Impact of JFET width on the on-resistance for the 4H-SiC shielded planar MOSFET with accumulation channel.

11.3.6 Threshold Voltage

The threshold voltage of the power MOSFET is an important design parameter from an application stand-point. A minimum threshold voltage must be maintained at above 1 volt for most system applications to provide immunity against inadvertent turn-on due to voltage spikes arising from noise. At the same time, a high threshold voltage is not desirable because the voltage available for creating the charge in the channel is determined by $(V_G - V_T)$ where V_G is the applied gate bias voltage and V_T is the threshold voltage. Most power electronic systems designed for high voltage operation (the most suitable application area for silicon carbide devices) provide a gate drive voltage of 15 volts. Based upon this criterion, the threshold voltage should be kept below 5 volts in order to obtain a low channel resistance contribution.

For the inversion-mode shielded planar MOSFET, the threshold voltage can be modeled by defining it as the gate bias at which on-set of *strong inversion* begins to occur in the channel. This voltage can be determined using Eq. [11.14]. Silicon carbide inversion-channel power MOSFETs have a high threshold voltage, as shown in section 11.1.3, due to much larger band bending required for this wider bandgap semiconductor when compared with silicon.

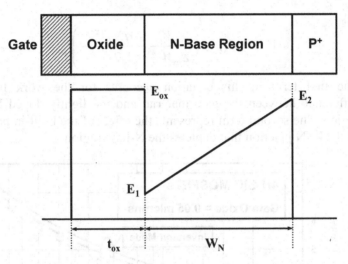

Fig. 11.26 Electric field profile in the gate region for the accumulation-mode MOSFET structure.

The band bending required to create a channel in the accumulation-mode planar MOSFET is much smaller than required for the inversion mode device. This provides the opportunity to reduce the threshold voltage while obtaining the desired normally-off device behavior. A model for the threshold voltage of accumulation-mode MOSFETs has been developed[19] using the electric field profile shown in Fig. 11.26 when the gate is biased at the threshold voltage. In this figure, the electric fields in the semiconductor and oxide are given by:

$$E_1 = \frac{V_{bi}}{W_N} - \frac{qN_D W_N}{2\varepsilon_S}$$

[11.19]

$$E_2 = \frac{V_{bi}}{W_N} + \frac{qN_DW_N}{2\varepsilon_S} \qquad [11.20]$$

$$E_{ox} = \frac{\varepsilon_S}{\varepsilon_{ox}}E_1 \qquad [11.21]$$

Note that this model is based upon neglecting any voltage supported within the P⁺ region under the assumption that it is very heavily doped. Using these electric fields, the threshold voltage is found to be given by:

$$V_{TH} = \phi_{MS} + \left(\frac{\varepsilon_S V_{bi}}{\varepsilon_{ox}W_N} - \frac{qN_DW_N}{2\varepsilon_{ox}}\right)t_{ox} \qquad [11.22]$$

The first term in this equation accounts for the work function difference between the gate material and the lightly doped N-Base region. The second term represents the effect of the built-in potential of the P⁺/N junction that depletes the N-base region.

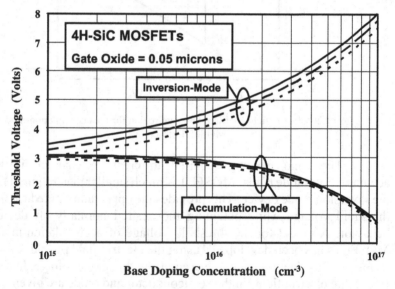

Fig. 11.27 Threshold voltage of 4H-SiC accumulation-mode MOSFETs.

(Solid line: 300 °K; Dash line: 400 °K; Dotted line: 500 °K)

The analytically calculated threshold voltage for 4H-SiC accumulation-mode MOSFETs are provided in Fig. 11.27 for the case of a gate oxide thickness of 0.05 microns, and N-base thickness of 0.2 microns as a function of the N-base doping concentration with the inclusion of a metal-semiconductor work-function difference of 1 volt. For comparison purposes, the threshold voltage for the inversion-mode 4H-SiC MOSFET is also given in this figure for the same gate oxide thickness. A strikingly obvious difference between the structures is a decrease in the threshold voltage for the accumulation-mode structure with increasing doping concentration in the N-base region. This occurs due to the declining influence of the P^+/N junction at the gate oxide interface when the doping concentration of the N-base region is increased. It can also be noted that the temperature dependence of the threshold voltage is smaller for the accumulation-mode structure. Of course, the most important benefit of the accumulation-mode is that lower threshold voltages can be achieved than in the inversion-mode structures. At an N-base doping concentration of 1×10^{16} cm^{-3}, the threshold voltage for the accumulation-mode structure is 2.8 volts. This will be reduced to just below 2 volts in the presence of typical fixed oxide charge in the 4H-SiC/oxide system. This model for the threshold voltage for the accumulation-channel MOSFET structure has been verified by numerical simulations[20].

11.3.7 On-Resistance Simulations

The on-resistance for the planar 4H-SiC shielded accumulation-mode structure was extracted by performing numerical simulations with a gate bias of 10 volts. The lower gate bias voltage, when compared with that used in the previous chapter for the unshielded inversion-mode structure, was chosen because of the smaller threshold voltage of these devices. A gate bias of 10 volts is a typical value for the available gate drive circuits used in medium and high power electronic systems. In addition, due to the improved channel resistance contribution, a channel length of 2 microns was also investigated because this provides greater margin when fabricating the devices.

The specific on-resistance obtained from the numerical simulations performed on planar 4H-SiC shielded accumulation-mode MOSFET structures with various JFET widths is shown in Fig. 11.28 together with the values obtained using the analytical model. These results are based upon using a channel mobility of 90 cm^2/ Vs with a gate oxide thickness of 0.04 microns. A good agreement between the analytical model and the simulations provides further validation of the ability to predict specific on-resistance using the analytical model.

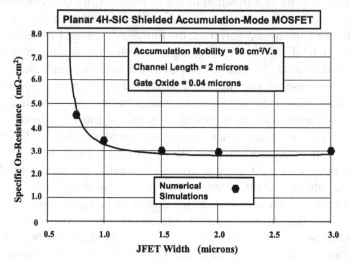

Fig. 11.28 Specific on-resistances for the planar 4H-SiC shielded accumulation-mode MOSFETs.

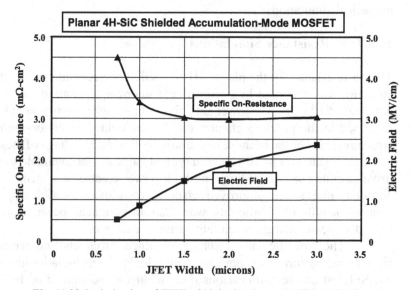

Fig. 11.29 Optimization of JFET width in the planar 4H-SiC shielded accumulation-mode MOSFETs.

The optimization of the JFET width requires simultaneously taking into consideration the specific on-resistance and the electric field developed at the gate oxide interface. The changes in these parameters are provided in Fig. 11.29 for the case of a device with channel length of 2 microns. A gate oxide thickness of 0.04 microns was used here with a channel mobility of 90 cm²/V-s. It can be seen that the electric field in the semiconductor below the gate oxide at the middle of the JFET region can be reduced to less than 1 MV/cm when the JFET width is reduced to 1 micron with a degradation of the specific on-resistance by about 10 percent. This demonstrates that a reliable accumulation-channel shielded 4H-SiC planar power MOSFET structure can be created with a specific on-resistance of about 3 mΩ-cm² for a rated breakdown voltage of 1200 volts.

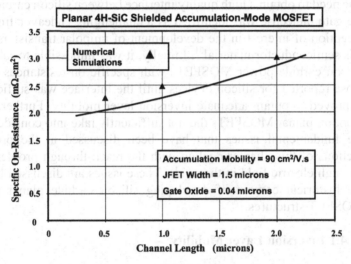

Fig. 11.30 Specific on-resistances for the planar 4H-SiC shielded accumulation-channel MOSFETs.

As discussed in the earlier section, the planar 4H-SiC shielded accumulation-channel MOSFET structure was found to be capable of blocking high drain bias voltages even when the channel length is reduced to 0.5 microns. This creates the opportunity to reduce the specific on-resistance by adjusting the channel length. The results of the specific on-resistance obtained from the numerical simulations performed on planar 4H-SiC shielded accumulation-channel MOSFET structures with various channel lengths are shown in Fig. 11.30 together with the values obtained using the analytical model. These results are based upon using a channel mobility of 90 cm^2/Vs with a gate oxide thickness of 0.04 microns. A good agreement between the analytical model and the simulations provides further validation of the ability to predict specific on-resistance using the analytical model. With a channel length of 0.5 microns, the specific on-resistance for the planar 4H-SiC shielded accumulation-channel MOSFET structure is reduced to 2.3 mΩ-cm^2, which is an extremely low value for a MOSFET capable of supporting 1200 volts.

11.4 Experimental Results: Channel Mobility

The need to obtain a high quality interface between silicon carbide and the gate dielectric was identified as a challenging endeavor from the inception of interest in the development of unipolar transistors from this semiconductor material[21]. Initially, it was impossible to fabricate silicon carbide power MOSFETs with specific on-resistances below those reported for silicon devices until the interface was sufficiently improved to obtain adequate inversion layer mobility. Further, early work on planar MOSFETs did not sufficiently take into consideration the fundamental issues that have been discussed in the previous sections of this chapter dealing with the reach-through problem and the high electric field in the oxide. These issues are discussed here in the historical context of developing silicon carbide planar power MOSFET structures.

11.4.1 Inversion Layer Mobility

Early investigations of the interface between P-type silicon carbide and thermally grown oxide indicated a high density of interface states and positive charge in the oxide. Significant improvement in the interface quality was achieved for 6H-SiC by anneal the thermally grown oxide in a NO ambient[22] leading to a peak inversion layer mobility of 70 cm^2/Vs. However, the inversion layer mobility for 4H-SiC was reported[23,24] to be less than 1 cm^2/Vs.

In 1998, a break-through was achieved at PSRC[25] with the use of deposited oxides on 4H-SiC leading to a reported[26] record high inversion layer mobility of 165 cm^2/Vs. A detailed study[27] of the process steps responsible for producing the improved interface resulting in the high channel mobility was also undertaken at PSRC.

This work produced the first observation of phonon scattering limited inversion layer mobility (which decreased with increasing temperature), and demonstrated that a wet oxide anneal of the deposited oxide was the critical step for producing the high inversion layer mobility. The process proposed and demonstrated at PSRC was subsequently reproduced by the sponsors[28] as well by other research groups[29]. These results indicate that it is possible to achieve sufficiently high inversion layer mobility in 4H-SiC power MOSFETs to obtain low specific on-resistance in high voltage structures.

The low inversion layer mobility observed in 4H-SiC MOSFETs fabricated using thermally grown gate oxides has been traced to the trapping of the carriers at interface states. Hall-effect measurements performed on lateral MOSFETs have demonstrated that the effective Hall mobility for carriers in the inversion layer is high when the trapping effect is taken into consideration[30]. Due to the trapping effect, as recently as 2001, the measured effective mobility in 4H-SiC MOSFETs has been reported[31] to be less than 10 cm^2/V-s when thermal oxidation is used to form the gate oxide. In these devices, the effective mobility was found to increase with temperature which is a signature of trap dominated current conduction in the channel. By using NO ambient anneals of the thermally grown gate oxide, an improvement in the effective channel mobility to 30-35 cm^2/V-s was reported[32] in 2001.

During the last 15 years, considerable effort has been undertaken to understand the interface between SiO_2 and 4H-SiC and correlate this with the inversion layer mobility[33]. The best method for improvement of effective mobility in 4H-SiC for MOSFETs fabricated using thermally grown oxides has been reported to be by post-oxidation annealing in nitric oxide (NO) and nitrous oxide (N_2O). It has been reported that the channel mobility is inversely proportional to the interface state density. A significant reduction in the interface state density has been accomplished by performing annealing at 1175 °C in a nitric oxide (NO) ambient for 2 hours[24,34,35,36]. An effective channel mobility for electrons of 30-35 cm^2/V-s was achieved using this process at the operating gate bias. More recent work[37,38] indicates that charge trapping has been sufficiently suppressed by the nitric oxide annealing leading to effective channel mobility of about 60 cm^2/V-s.

It would be preferable to fabricate 4H-SiC power MOSFETs using ion-implanted P-base regions as in the case of silicon devices. Due to the low diffusion rate for dopants in the 4H-SiC, it is necessary to stagger the edge for the implantation of the P-base and N^+ source regions for the 4H-SiC devices[5,9] to create the DIMOSFET structure. The damage in the ion implanted regions must be removed by annealing followed by surface preparation to reduce interface states. The inversion layer mobility in n-channel lateral MOSFETs fabricated on aluminum implanted layers in 4H-SiC has been reported[39]. The ion-implant dose and energy was selected to achieve a doping concentration of 1×10^{17} cm^{-3} followed by annealing for 10 minutes at 1600 °C. The gate oxide was grown at 1200 °C and then placed in an alumina environment. Peak inversion layer mobility of 100 cm^2/V-s was observed with a high threshold voltage of about 10 volts.

Formation of silicon dioxide on 4H-SiC at a higher temperatures of 1500 °C has been shown to enhance growth rates[40]. The interface state density for this process has been found to be smaller than for oxides typically grown at 1200 °C. This produced an improved field effect mobility of 40 cm^2/V-s for electrons in the inversion layer.

A high peak inversion layer mobility for electrons of 132.6 cm^2/V-s has been obtained by using a gate dielectric stack consisting of 1 nm thick layer of La_2O_3 by Molecular-Beam-Epitaxy followed by a 20 nm layer of SiO_2 by Atomic-Layer-Deposition[41]. The samples were annealed at 900 °C in nitrous oxide by using RTA. The inversion layer mobility reduced sharply with gate bias to 90 cm^2/V-s at a gate voltage of 6 volts. This approach utilizes high-k dielectrics to passivate the SiC surface.

11.4.2 Accumulation Layer Mobility

It has been established that the accumulation layer mobility is significantly greater than the inversion layer mobility in silicon MOSFET structures[42]. This is due to the smaller band bending at the semiconductor surface which reduces surface scattering. The same effect should be applicable to silicon carbide. The first planar silicon carbide accumulation-channel vertical power MOSFET structure, named the planar ACCUFET, was developed at PSRC using 6H-SiC as the semiconductor material[43,44]. The effective accumulation layer mobility extracted[45] from these structures was 120 cm^2/Vs for a structure fabricated using thermally grown gate oxide, which was considerably larger than the best inversion layer mobility (70 cm^2/Vs) reported[8] at that time. Subsequently, 6H and 4H-SiC accumulation-channel planar power MOSFETs were compared using thermally grown gate oxide[46]. The effective mobility for the 4H-SiC structures was found to be much smaller than for the 6H-SiC devices. The control of the threshold voltage for accumulation-mode 4H-SiC lateral

MOSFETs was demonstrated by using counter-doping of the P-shielding layer with nitrogen ion implants[47]. A peak channel field-effect mobility of 35 cm^2/V-s was obtained using this approach.

More recently[48], the accumulation layer mobility has been measured in 4H-SiC MOSFETs using the Hall bar structure to differentiate between the effective mobility, which is limited by charge trapping, and the intrinsic mobility for free carriers in the channel. Note that the gate oxide process used for the fabrication of these MOSFETs was similar to that developed at PSRC[25] with the use of deposited oxides on 4H-SiC leading to a reported[49] record high inversion layer mobility of 165 cm^2/V-s. It was found that the accumulation mobility decreased from 350 cm^2/V-s in the weak accumulation regime of operation to 200 cm^2/V-s in the strong accumulation regime of operation. These results provided the justification for performing the analysis of the planar 4H-SiC shielded accumulation-mode MOSFET structures, as discussed in the previous section, using accumulation mobility values ranging up to 200 cm^2/V-s. However, the effective accumulation-layer mobility was found to be only 3-5 cm^2/V-s indicating substantial trapping of electrons at the gate oxide interface.

Better effective accumulation mobility values have been reported for 4H-SiC MOSFETs by using thermally grown oxide. In 1999, an effective accumulation-layer mobility of 20-30 cm^2/V-s was reported[47] by using nitrogen ion implantation to form the N-base region. Lateral 4H-SiC MOSFETs were reported in 2001 with thermally grown oxide using nitrogen ion implants to create the accumulation-mode device[50]. The nitrogen concentration was kept fixed at 1×10^{17} cm^{-3} and the thickness of the N-type region was varied from 0.15 to 0.20 to 0.25 microns. The threshold voltage was found to reduce with increasing thickness of the N-type layer as predicted by the analytical model in section 11.3.6. A threshold voltage of 0.3 volts was observed for a thickness of 0.2 microns which is in good agreement with the value calculated using the analytical model (see Fig. 11.29). For this case, a peak channel mobility of 140 cm^2/V-s was observed. In 2001, a significantly larger accumulation-layer effective mobility of 200 cm^2/V-s was reported[51] by using a stacked gate oxide process. The stacked gate oxide consisted of a thermally grown layer with thickness of 0.02 microns followed by a NSG-CVD deposited layer with a thickness of 0.03 microns. These results have established the ability to obtain relatively high accumulation-layer mobility in

4H-SiC indicating the possibility for fabrication of high performance vertical power MOSFETs.

In summary, inversion-layer effective mobility values in the range of 10-20 cm^2/V-s have been observed in 4H-SiC MOSFETs while significantly larger effective accumulative-layer mobility values of 100-200 cm^2/V-s have been observed in this material. These values indicate that it is possible to develop high voltage 4H-SiC power MOSFETs using either the inversion-mode or accumulation-mode architecture if the base is shielded from the high electric field in the drift region. The accumulation-mode structure is favored for designs with low channel density where the channel contribution becomes a significant fraction of the total on-resistance.

11.5 Experimental Results: Planar Shielded Power MOSFETs

The need to obtain a high quality interface between silicon carbide and the gate dielectric was identified as a challenging endeavor from the inception of interest in the development of unipolar transistors from this semiconductor material[21]. Initially, it was impossible to fabricate silicon carbide power MOSFETs with specific on-resistances below those reported for silicon devices until the interface was sufficiently improved to obtain adequate inversion layer mobility. Even in recent years, poor inversion layer mobility values (~15 cm^2/V-s) have been reported due to trapping of charge at the oxide-semiconductor interface. To overcome this technological barrier and to obtain a lower threshold voltage, the accumulation-mode structure was proposed[7]. This structure was also designed to shield the gate oxide from high electric fields in the semiconductor drift region. Experimental results on the shielded inversion-channel and accumulation-channel planar power MOSFET structures are discussed in this section.

11.5.1 Inversion-Channel Devices

In recognition of the low diffusion coefficients for dopants in silicon carbide, it was proposed[5] that the P-base and N$^+$ source ion implantations for the planar silicon carbide power MOSFET be staggered by using photoresist masks rather than be defined by the gate edge as conventionally done for silicon devices. The idea to incorporate a sub-surface P-type layer to shield the channel was also proposed at PSRC and subsequently patented[7]. This idea was first successfully demonstrated[52] for 6H-SiC devices in 1997 under the DI-MOSFET moniker. The authors fabricated the devices using multiple energy boron implants to form a P-base region with a box profile at a depth of 1 micron. The energy and dose for the implants was chosen to produce a retrograde doping profile on one of the wafers. This creates a low doped P-base region with an under-lying highly doped P^+ region that shields it from the high electric fields developed in the drift region. Devices were designed with channel length of 2 and 5 microns, and the JFET region width (W_J in Fig. 11.4) was varied from 2.5 to 15 microns. The channel inversion layer mobility was found to be about 20 cm^2/Vs for these 6H-SiC MOSFETs fabricated using a thermally grown oxide. The best specific on-resistance of 130 $m\Omega$-cm^2 was obtained for the smallest JFET width and channel length. The breakdown voltage of the device was found to be 760 volts as limited by reach-through as well as gate oxide rupture leading to catastrophic failure indicating that the design did not provide adequate shielding.

The concept of shielding the P-base region was also reported[53] using a triple ion-implantation process to fabricate 6H-SiC vertical power MOSFETs. The ion-implant steps were designed to produce a low P-base surface concentration of 7×10^{16} cm^{-3} with the doping increasing to 1×10^{18} cm^{-3} at a depth of 0.4 microns. This profile enabled achieving a threshold voltage of 3 volts with a thermally grown gate oxide thickness of 0.036 microns. A breakdown voltage of 1800 volts was obtained for this device by using a drift region doping concentration of 6.5×10^{15} cm^{-3} and thickness of 15 microns. A specific on-resistance of 82 $m\Omega$-cm^2 was observed at a gate bias of 10 volts, which is an order of magnitude better than values that can be obtained with silicon technology. Good dynamic switching performance with ruggedness was confirmed[54] for these MOSFETs.

A DI-MOSFET structure was reported[55] from 4H-SiC in 2001 using 25 micron thick epitaxial layers with doping concentration of 3×10^{15} cm^{-3}. In spite of using the previously discussed deposited oxide process[8], an inversion layer mobility of only 14 cm^2/Vs was observed. Due to the low channel mobility, the devices exhibited a relatively large specific on-resistance of 55 $m\Omega$-cm^2 even when a high gate bias of 25 volts was applied. The device was able to support nearly 2000 volts at zero gate bias. The authors associated the blocking voltage to be limited by open-base bipolar breakdown, namely, the reach-through problem discussed earlier in this chapter. This work was extended[56] to achieve a breakdown voltage of 2400 volts with a specific on-resistance of 42 $m\Omega$-cm^2 in large area (0.1 cm^2) device structures; as well as devices[57] with a breakdown voltage of 10 kV with a specific on-resistance of 236 mΩ-cm². A significant improvement in performance 10-kV 4H-SiC planar inversion-channel power MOSFETs was also reported in 2004 with a specific on-resistance of 123 mΩ-cm² by using a drift region with doping concentration of 8×10^{14} cm⁻³ and thickness of 85 microns[58]. This value of the specific on-resistance is with 2x of the value predicted by the analytical model (see Fig. 11.25).

In 2004, a novel staggered implant process has been proposed to reduce the channel length and applied towards the fabrication of high voltage planar 4H-SiC MOSFETs[59,60]. In this process, oxidation of a sacrificial polysilicon layer is used to form a sidewall spacer that serves as a mask to stagger the P-base and N⁺ source implants to achieve a channel length of less than 0.5 microns. The retrograde ion implant profile discussed earlier was also utilized for these devices to reduce the P-base doping concentration and suppress the reach-through problem. In addition a nitrogen counter-doping ion implant step is used. This may have resulted in accumulation-channel devices although the authors did not indicate this in the paper. Planar 4H-SiC inversion-mode MOSFETs were successfully fabricated with breakdown voltage of 2 kV with a specific on-resistance of 27 mΩ-cm² at a gate bias of 20 volts.

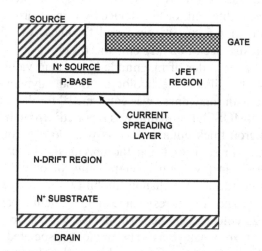

Fig. 11.31 Planar 4H-SiC MOSFET with current spreading layer.

The optimization of a 1-kV DIMOSFET structure was reported in 2007[61]. This structure included a current spreading layer (CSL) as shown in Fig. 11.31. The current spreading layer is formed by epitaxial grown of a more highly doped n-type layer on top of the n-type drift layer. Optimization of the structure indicated that the lowest specific-on-resistance can be achieved by using a CSL doping concentration of 1×10^{17} cm^{-3}. The same doping concentration was found to be suitable for the JFET region. However, it is worth pointing out that the numerical simulations of the structure show a catastrophic drop in breakdown voltage if the JFET is increased beyond 1 micron or when the CSL doping concentration exceeds 2×10^{17} cm^{-3}. This makes the processing of the structure challenging. Nevertheless, the authors were able to obtain functional devices with a specific on-resistance of 6.95 mΩ-cm^2. This value is much larger than the ideal value due to the large channel resistance contribution despite using a self-aligned process to reduce the channel length to 0.3 microns.

With increasing maturity of the 4H-SiC technology to provide 3 inch diameter wafers, large area (0.45 cm $\times$ 0.45 cm chips with 0.15 cm^2 active area) planar power MOSFETs could be fabricated in 2010 with blocking voltage of 1200 volts[62]. These devices could carry 40 amperes in the on-state with a specific on-resistance of 7.1 mΩ-cm^2. This value is close to the analytically computed specific on-resistance for an inversion layer mobility of 15 cm^2/V-s (see Fig. 11.25) indicating that the performance is limited by the channel resistance.

4H-SiC planar power MOSFETs with a blocking voltage of 1500 volts were reported[63] in 2011 with a specific on-resistance of 3.7 mΩ-cm^2. This value is close to the analytically computed specific on-resistance for an inversion layer mobility of 15 cm^2/V-s (see   Fig. 11.25) indicating that the performance is limited by the channel resistance. The devices were fabricated using a self-aligned process to achieve a channel length of 0.5 microns. The authors point out that the sub-threshold slope for the devices is large resulting in a low threshold voltage which must be increased to achieve normally-off operation. Improved high-current devices with blocking voltage of 1600 volts were reported[64] by the authors in 2013 with a specific on-resistance of 7 mΩ-cm^2. These devices had a chip area of 0.56 cm^2 and active area of 0.40 cm^2 allowing a drain current of 150 amperes. The devices were demonstrated to have much smaller switching losses at elevated temperature when compared with silicon IGBTs making them good candidates for high frequency, high power applications.

Further progress towards the development of commercially viable 4H-SiC planar power MOSFETs was reported[65] in 2014 with the fabrication of devices blocking 1700 volts with specific on-resistance of 5.5 mΩ-cm^2. This value is close to the analytically computed specific on-resistance for an inversion layer mobility of 15 cm^2/V-s (see Fig. 11.25) indicating that the performance is still limited by the channel resistance. The devices exhibited a threshold voltage shift of -0.25 volts under a gate bias stress of -15 volts at 225 °C for 20 minutes. The authors claim that this is evidence of stable operation for their devices at 225 °C.

An assessment of the state-of-the-art for 4H-SiC planar power MOSFETs was reported[66] in 2014. The specific on-resistance for devices with various breakdown voltages from this paper are shown in Fig. 11.32 by the triangular symbols. At high breakdown voltages (above 5 kV), the specific on-resistance becomes equal to the ideal value plus the contribution of 0.04 mΩ-cm^2 from the substrate. It is interesting that the specific on-resistance for devices with lower breakdown voltages follows the line for the accumulation-channel device with a channel mobility of 100 cm^2/V-s. This implies that these devices are either utilizing the accumulation-channel concept or are fabricated using a shorter channel length than 1 micron used in the plots. It is worth pointing out that the device blocking voltage rating is about 75% of the breakdown voltage used in the plot.

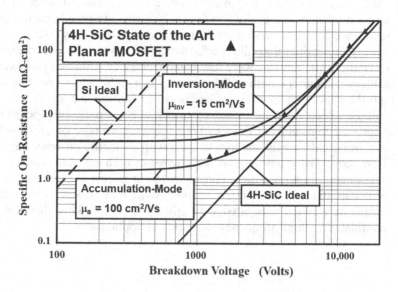

Fig. 11.32 State-of-the-art for planar 4H-SiC MOSFETs in 2014.

11.5.2 Accumulation-Channel Devices

The first planar shielded accumulation-mode MOSFET structure was fabricated[43] at PSRC in 1997 using 6H-SiC. This device, named the planar ACCUFET, was conceived to circumvent the problems observed with obtaining high inversion layer mobility as well as to screen the channel and gate oxide from high electric fields developed in the drift region. The devices were fabricated using a drift region with doping concentration of 1×10^{16} cm^{-3} and thickness of 10 microns, corresponding to a breakdown voltage of 1500 volts. The sub-surface P$^+$ region used to shield the channel and gate oxide was formed by a single boron ion implant with energy of 380 keV and dose of 1×10^{14} cm^{-2}. This produced an N-base region with thickness of about 0.3 microns. The gate oxide with thickness of 125 angstroms was formed using thermal oxidation followed by a re-oxidation anneal.

The un-terminated devices had a breakdown voltage of 350 volts. For the structure with cell pitch of 21 microns (channel length of 2.5 microns and JFET width of 4 microns), a specific on-resistance of 18 mΩ-cm^2 was measured at a gate bias of only 5 volts. This was possible because of the low threshold voltage (~1 volt) and high transconductance of the structure achieved with the small gate oxide thickness. The measured specific on-resistance was within 2.5 times the specific on-resistance of the drift region (8 mΩ-cm^2) despite the rather large cell pitch. In spite of the very thin gate oxide, no evidence of gate rupture was observed in the blocking state due to the shielding by the P$^+$ region. The measured specific on-resistance for these MOSFETs is about 30 times smaller than that for silicon MOSFETs with blocking voltage capability of 1500 volts.

A detailed analysis[67] of the operation and design of the 6H-SiC ACCUFET was published in 1999. In this paper, it was pointed out that a trade-off between reducing the electric field at the gate oxide and minimizing the specific on-resistance must be performed by optimizing the width of the JFET region. It was also found that the specific on-resistance for the ACCUFET increases with increasing temperature. This had not been previously observed in silicon carbide power MOSFETs because the channel conductance improved rapidly with temperature due to trap limited effective inversion-layer mobility. In contrast, the extracted accumulation-layer mobility was found to remain independent of temperature resulting in a positive temperature coefficient for the specific on-resistance due to increase in resistance of the bulk components.

The ACCUFET structure was successfully fabricated using 4H-SiC in 2000 with high blocking voltage capability[68]. These devices were labeled SIAFETs even though the operating principle for the devices was acknowledged by the authors to be identical to that of ACCUFET structures. The authors achieved a blocking voltage capability of 4580 volts with normally-off operation at zero gate bias by using a drift region with a doping concentration of 5×10^{14} cm^{-3} and thickness of 75 microns. These devices exhibited a specific on-resistance of 1200 mΩ-cm^2 in spite of using a gate bias of 40 volts because of a low accumulation-layer mobility of 0.5 cm^2/ V-s. The specific on-resistance was found to reduce by a factor of 6 times by the application of a positive bias to the sub-surface P$^+$ region. Unfortunately, this entails a more complex package and gate drive circuit for the devices that departs from mainstream silicon technology. The performance of this structure, renamed SEMOSFET, was extended[69] to a 5 kV blocking voltage capability with a specific on-resistance of 88 mΩ-cm^2 in the presence of a positive bias of 2 volts applied to the buried P$^+$ region with 20 volts applied to the MOS-gate electrode. Since the bias applied to the buried P$^+$ region was less than the junction potential, these is no minority carrier injection from the P$^+$ region. Consequently, these devices exhibit very fast switching speeds with turn-on and turn-off times of less than 50 nanoseconds while operating at on-state voltage drops slightly less than that for a 4.5 kV IGBT.

4H-SiC ACCUFETs were demonstrated with a blocking voltage of 400 volts and a high specific on-resistance of 90 mΩ-cm^2 in 2000[70] because the accumulation mobility observed for these devices was low (~8 cm^2/V-s). The channel mobility was found to increase with temperature indicating that the channel charge was dominated by traps. A 4H-SiC ACCUFET device was reported[71] in 2002 with breakdown voltage of 600 volts and specific on-resistance of 13 mΩ-cm^2. The authors estimated an accumulation channel mobility of 450 cm^2/V-s validating this approach.

High current 4H-SiC ACCUFETs were demonstrated[72,73] in 2003 with blocking voltages up to 900 volts and drain current of 20 amperes. These devices were fabricated by the grown of an N-type epitaxial layer on top of an ion implanted P$^+$ region to form the N-base region. Devices were found to have a specific on-resistance of 22 mΩ-cm^2 with a breakdown voltage of 550 volts. The authors observed an accumulation channel mobility of 18 cm^2/V-s by using a thermally grown gate oxide followed by N$_2$O anneal.

A double-epitaxial layer process was reported[74,75] in 2004 for the fabrication of an ACCUFET structure called the DEMOSFET by the authors. A highly doped P-type epitaxial layer was first grown to form the P$^+$ shielding regions. This layer was etched to open the JFET regions followed by a second low doped P-type epitaxial layer. Nitrogen ion implantation was used to form the N-base region with a concentration of 1.2 × 10^{17} cm^{-3} and thickness of 0.2 microns. The devices had a threshold voltage of 3 volts and a specific on-resistance of 8.5 mΩ-cm^2. For devices with blocking voltage of 600 volts, the authors reported a high accumulation channel mobility of 105 cm^2/V-s by using a thermally grown gate oxide followed by pyrogenic re-oxidation at 950 °C.

A 4H-SiC ACCUFET structure was reported[76] in 2006 with an epitaxially grown N-type layer to form the N-base region. The doping and thickness of the N-base region were not provided but the threshold voltage of the device was close to zero. A specific on-resistance of 5 mΩ-cm^2 was observed for a gate bias of 20 volts. This is about 10-times larger than the ideal specific on-resistance (see Fig. 11.34). No channel mobility data was provided. The basic structure of this device is provided in Fig. 11.33.

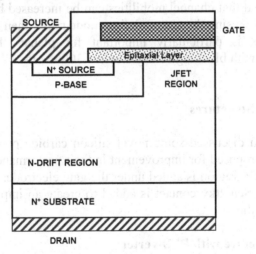

Fig. 11.33 Planar 4H-SiC ACCUFET structure with epitaxial N-base region.

In 2008, the blocking voltage for the 4H-SiC ACCUFETs was increased to 10-kV by using a 100 micron thick epitaxial layer with doping concentration of 5×10^{14} cm^{-3}. These devices also utilized a 0.1 micron thick epitaxially grown N-base region[77,78]. A three-zone, 750 micron wide, JTE edge termination was employed to achieve the breakdown voltage of 10-kV. A specific on-resistance of 170 mΩ-cm^2 was observed for a gate bias of 10 volts. This value is within 3-times the ideal case (see Fig. 11.34) because the channel contribution is becoming a smaller fraction of the total resistance at the high breakdown voltage. The channel mobility was reported to have a value of 15 cm^2/V-s at a gate bias of 15 volts.

The 4H-SiC ACCUFET structure with the epitaxially grown N-base region (as shown in Fig. 11.33) was used to create 40 ampere devices with blocking voltage of 1000 volts[79]. The devices had a specific on-resistance of 3.5 mΩ-cm^2 and a threshold voltage of 2.3 volts. The specific on-resistance is limited by the channel resistance according to the analytical model (see Fig. 11.32). The devices were shown to be suitable for induction cook-top applications at an operating frequency of 30 kHz. The switching loss was reduced from 19 watts for the silicon IGBT to 5 watts with the SiC device.

Based up on the results reported in the literature, it can be concluded that the ACCUFET concept has become popular for creating high voltage 4H-SiC power MOSFETs with low specific on-resistance and low threshold voltages. It has been conclusively demonstrated that channel mobilities can be increased by an order of magnitude by using the accumulation-mode rather than the inversion-mode. This is particularly important for planar 4H-SiC power MOSFETs with blocking voltages below 3000 volts.

11.6 Novel Structures

This section discussed some novel silicon carbide power MOSFET structures proposed for improvement in their performance. In the first structure, a P$^+$ region is added under the gate electrode. In the second structure, a Schottky contact is added to create an improved reverse conducting diode.

11.6.1 Structure with P$^+$ Diverter

A 4H-SiC planar power MOSFET structure containing a P$^+$ diverter region, shown in Fig. 11.34, was reported[80] in 2015. The P$^+$ diverter region is connected to the source electrode (not shown in the figure). The structure was called the CIMOSFET because the authors call the P$^+$ region a Central Implant region. The upper structure for this device is exactly the same as that reported[81] in 1994 for improving the latch-up current and reducing the gate-collector (Miller) capacitance for silicon IGBTs. It was demonstrated that the Miller capacitance could be reduced by a factor of 1.6-times in silicon IGBTs by incorporation of the P$^+$ diverter. In the case of the 4H-SiC devices, the addition of the P+ diverter increased the specific on-resistance from 3 to 5 mΩ-cm2 as expected due to an enhanced JFET effect. However, the Miller capacitance was reduced by a factor of 3 times and the reverse transfer gate charge was reduced by a factor of 2. In addition, the electric field in the oxide was reduced by 2 MV/cm.

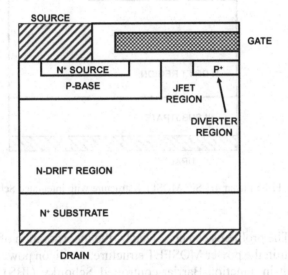

Fig. 11.34 Planar 4H-SiC MOSFET structure with P$^+$ diverter region.

11.6.2 Structure with Integrated Schottky Diode

All power MOSFETs allow current flow in the third quadrant of operation (when a negative drain bias is applied) via the body-diode[2]. In the case of silicon devices, the body diode can be utilized in power electronic circuits by controlling its reverse recovery characteristics by electron irradiation[82]. In principle, the body diode in the 4H-SiC power MOSFET could also be used as an anti-parallel diode. The on-state voltage drop for the 4H-SiC body diode is relatively large (~4 volts) making the on-state power dissipation large. A bigger problem is the generation of stacking faults at basal plane dislocations during bipolar current flow in the on-state. The adverse impact of this on the power MOSFET characteristics will be discussed in the next section.

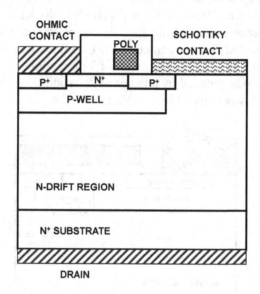

Fig. 11.35 Planar 4H-SiC MOSFET structure with integrated Schottky diode.

The problems can be solved by the incorporation of a Schottky diode within the power MOSFET structure. A silicon power MOSFET with built-in Junction-Barrier-controlled Schottky (JBS) diode that could carry the full rated current was first developed for microprocessor voltage regulator modules[83]. These devices were commercialized as JBSFET products by Silicon Semiconductor Corporation. The incorporation of the JBS diode within a 4H-SiC power MOSFET structure has been reported[84] in 2015. The structure used by the authors is shown in Fig. 11.35. They utilized Nickel for making ohmic contact to the source and Titanium for making the Schottky contact. The on-state voltage drop of the JBS diode was found to be 1.25 volts which is well below the turn-on (or knee) voltage for the body diode. The JBS diode occupied 25% of the active area leading to an increase in the resistance of the MOSFET portion.

11.7 Reliability Issues

Several issues related to long term, stable operation of 4H-SiC power MOSFETs have been described in the literature. Some examples are discussed in this section.

11.7.1 Body-Diode induced Stacking Faults

It is attractive to utilize the body diode in the power MOSFET structure as the anti-parallel diode in power circuits. The body diode behaves as a P-i-N rectifier with high-level injection in the on-state. It has been reported[85] that on-state voltage drop of the body diode increases after it was turned-on for even a short duration of 1 hour in case of thick (100 micron) drift layers for 10-kV devices. The reason for the degradation of the diode characteristics has been found to be the generation of stacking faults at basal-plane-dislocations (BPDs) in the epitaxial layer. The recombination of the injected electrons and holes provides the energy for this process.

It was found that the generation of stacking faults degrades the on-resistance of the 4H-SiC power MOSFET as well. No change in the threshold voltage was observed. It was speculated[85] that the increase in on-resistance is either due to bulk mobility degradation due to additional scattering at the stacking faults or due to reduced concentration of electrons in the drift layer by compensation of dopants at the stacking faults.

This problem is not significant in the case of devices with lower breakdown voltages (600 to 1200 volts) due to the thin drift layers[85]. This has been demonstrated[86] by performing tests on 1200-V 4H-SiC power MOSFETs. The body diode on-state voltage drop increased by 0.8% and the on-resistance of the MOSFET increased by 5.4% after the conduction of the body diode for 1000 hours. In addition, high temperature gate bias (HTGB) stress at a gate bias of − 15 volts and 150 °C led to a threshold voltage shift of −90 mV in 1000 hours and a high temperature gate bias (HTGB) stress at a gate bias of +20 volts and 175 °C led to a threshold voltage shift of 280 mV in 1000 hours. These results indicate that the gate oxide has been sufficiently shielded from the high electric field in the silicon carbide.

11.7.2 Unclamped Inductive Switching Stress

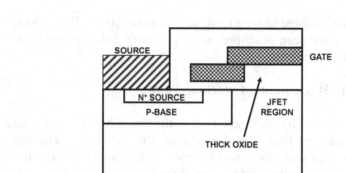

Fig. 11.36 Planar 4H-SiC MOSFET structure with thick gate oxide.

Power MOSFETs operated with inductive loads must tolerate unclamped inductive switching stress (UIL stress) where the energy in the inductance is absorbed by the device while operating under avalanche breakdown. The UIL stress capability for 1.2-kV 4H-SiC power MOSFETs has been studied[87] experimentally and by numerical simulations. The authors observed an increase in leakage current by three orders of magnitude and a reduction in the on-resistance by 4%.

Numerical simulations of the device structure demonstrated that the highest impact ionization in the JFET region occurs near the channel. The authors suggest that hot holes are injected into the gate oxide producing the observed changes. To solve this problem, they have proposed the structure shown in Fig. 11.36 with a thick oxide. Their numerical simulations show a reduction in the impact ionization rate and electric field near the channel with the thick oxide.

11.7.3 Short Circuit Reliability

The short circuit capability for 4H-SiC power MOSFETs is of interest in terms of protection of devices in power electronic applications. 1200-V 4H-SiC power MOSFETs were tested under short circuit conditions with a DC bus voltage of 400 volts[88]. The devices had a short circuit time of 80 microseconds for a gate bias of 10 volts and 50 microseconds for a gate bias of 15 volts which is satisfactory.

A comparison between the planar and planar-shielded 4H-SiC power MOSFET structures was reported[89] in 2015. The authors concluded that the shielding of the gate oxide and the base region produced an improvement in performance under short-circuit conditions.

11.8 Commercial Devices

The commercialization of silicon carbide power devices became feasible with the ability to grow 4H-SiC wafers with diameters of 100 mm. Wafers with diameters of 150 mm are now available from multiple sources. In addition, the ability to grow thick epitaxial layers with doping concentration approaching 1×10^{14} cm^{-3} have made the manufacturing of the very high voltage (> 10-kV) silicon carbide products a possibility. However, companies have targeted the 1200-V and 1700-V markets for their initial products. At this voltage range, the 4H-SiC unipolar devices offer much better switching characteristics than silicon IGBTs while providing a lower on-state voltage drop as well.

The first 4H-SiC products were JBS rectifiers as discussed a previous chapter. Two kinds of silicon carbide switches have been developed as products: JFETs and MOSFETs. The JFETs were introduced earlier because of problems with poor channel mobility in MOSFETs. Recent progress with improving MOS-channels - with advanced gate oxide and interface preparation, and by utilizing the accumulation-mode concept – have made high performance 4H-SiC power MOSFET products feasible. The momentum for power switch product commercialization has now shifted from JFETs to MOSFETs.

11.9 Summary

In this chapter, it has been established that the incorporation of a sub-surface P$^+$ region into the planar MOSFET structure enables shielding the P-base region from reach-through limited breakdown and preventing high electric fields from developing across the gate oxide during the blocking mode. Since relatively low inversion layer mobility has been reported for 4H-SiC MOSFETs, an accumulation-mode structure was proposed with an N-base region that is completely depleted by the built-in potential of the underlying P^+/N junction. This structure takes advantage of the much larger accumulation mobility observed in semiconductors.

The operating principle of the planar shielded MOSFET structures has been reviewed in the chapter providing guidelines for the design of the structures. It has been demonstrated that the JFET width is a critical parameter that controls the electric field at the gate oxide interface as well as the specific on-resistance. Its optimization is important for obtaining high performance devices. In the case of the accumulation-mode structure, the appropriate combination of the doping concentration and thickness of the N-base region must be chosen to ensure that it is completely depleted by the built-in potential of the underlying P^+/N junction. With adequate shielding of the base region, it is found that short channel devices will support high blocking voltages limited only by the properties of the drift region. These devices have excellent safe-operating-area and fast switching speed. This technology has potential for use in systems operating at up to at least 10,000 volts.

Due to the low switching losses for silicon carbide power MOSFETs, they are good competitors to silicon IGBTs in circuits operating at higher frequencies. One example is a 1 MVA solid-state power substation (SSPS) designed using 10-kV SiC modules[90] soft-switching at 20 kHz. Twenty-four MOSFETs and twelve Schottky diodes were used per module to achieve a current handling capability of 120 A per switch. Low voltage silicon Schottky diodes were placed in series with the SiC power MOSFETs to prevent conduction of the body diode. The SSPS converted 13.8-kV to 300-V with 97% efficiency resulting in 75% reduction in weight and 50% reduction in size over the conventional 60-Hz transformer.

4H-SiC power MOSFETs have also been used to create a solid-state transformer (SST) to convert the 3.6-kV distribution grid voltage to 400-V DC and 240/120-V AC voltage[91]. By using 13kV SiC power MOSFETs on the primary side, the topology of the SST is simplified to a single H-bridge AC/DC rectifier, an isolated dual half-bridge (DHB) DC/DC converter, and a DC/AC inverter. The DC link operates at 6-kV inside the SST. The DHB is operated at 15 kHz to reduce the size and weight of the high-frequency transformer. The switching losses in the SiC power MOSFETs are reduced by soft-switching. 1.2 kV SiC power MOSFETs are used on the secondary side to produce the 400-V DC and 220/120V AC outputs. The rectifier stage and the DHB were found to have an efficiency of over 98% while the peak efficiency for the inverter stage was 97%.

In conclusion, silicon carbide power MOSFET have reached a sufficient degree of maturity to find applications at higher frequencies. They can replace silicon IGBTs in applications where the switching losses are dominant to achieve significant reduction of size and weight due to improved efficiency.

References

[1] D. A. Grant and J. Gowar, "Power MOSFETs: Theory and Applications", John Wiley and Sons, 1989.

[2] B. J. Baliga, "Fundamentals of Power Semiconductor Devices", Springer-Science, New York, 2008.

[3] L. Lorenz, G. Deboy, A. Knapp and M. Marz, "COOLMOS – A New Milestone in High Voltage Power MOS", IEEE International Symposium on Power Semiconductor Devices and ICs, Abstract 1.1, pp. 3-10, 1999.

[4] B.J. Baliga, "Power Semiconductor Devices for Variable Frequency Drives", Proceedings of the IEEE, Vol. 82, pp. 1112-1122, 1994.

[5] B. J. Baliga and M. Bhatnagar, "Method of Fabricating Silicon Carbide Field Effect Transistor", U. S. Patent 5,322,802, Issued June 21, 1994.

[6] J. N. Shenoy, J. A. Cooper and M. R. Melloch, "High Voltage Double-Implanted Power MOSFETs in 6H-SiC", IEEE Electron Device Letters, Vol. 18, pp. 93-95, 1997.

[7] B. J. Baliga, "Silicon Carbide Semiconductor Devices having Buried Silicon Carbide Conduction Barrier Layers Therein", U. S. Patent 5,543,637, Issued August 6, 1996.

[8] S. Sridevan and B. J. Baliga, "Inversion Layer Mobility in SiC MOSFETs", Silicon Carbide and Related Materials – 1997, Material Science Forum, Vol. 264-268, pp. 997-1000, 1998.

[9] S-H Ryu *et al.*, "Design and Process Issues for Silicon Carbide Power DiMOSFETs", Material Research Society Symposium Proceeding, Vol. 640, pp. H4.5.1-H4.5.6, 2001.

[10] B. J. Baliga, "Power Semiconductor Devices", Chapter 7, pp. 357-362, PWS Publishing Company, 1996.

[11] D. Alok, P. McLarty and B. J. Baliga, "Electrical Properties of Thermal Oxide Grown on N-type 6H-Silicon Carbide", Applied Physics Letters, Vol. 64, pp. 2845-2846, 1994.

[12] J. B. Casady *et al.*, "4H-SiC Power Devices: Comparative Overview of UMOS, DMOS, and GTO Device Structures", Material Society Research Symposium Proceedings, Vol. 483, pp. 27-38, 1998.

[13] A.K. Agarwal *et al.*, "Temperature Dependence of Fowler-Nordheim Current in 6H and 4H-SiC MOS Capacitors", IEEE Electron Device Letters, Vol. 18, pp. 592-594, 1997.

[14] S. Krishnaswami *et al.*, "Gate Oxide Reliability of 4H-SiC MOS Devices", IEEE International Reliability Physics Symposium, pp. 592-593, 2005.

[15] L. Yu *et al.*, "Oxide Reliability for SiC MOS devices", IEEE International Integrated Reliability Workshop Report, pp. 141-144, 2008.

[16] M.K. Das *et al.*, "SiC MOSFET Reliability Update", Material Science Forum, Vol. 717-720, pp. 1073-1076, 2012.

[17] B. J. Baliga and D. A. Girdhar, "Paradigm Shift in Planar Power MOSFET Technology", Power Electronics Technology Magazine, pp. 24-32, November 2003.

[18] S. T. Sheppard, M. R. Melloch and J. A. Cooper, "Characteristics of Inversion-Channel and Buried-Channel MOS Devices in 6H-SiC", IEEE Transactions on Electron Devices, Vo. 41, pp. 1257-1264, 1994.

[19] N. Thapar and B. J. Baliga, "Analytical Model for the Threshold Voltage of Accumulation Channel MOS-Gated Devices", Solid State Electronics, Vol. 42, pp. 1975-1979, 1998.

[20] B.J. Baliga, "Silicon Carbide Power Devices", World Scientific Publishers, Singapore, 2005.

[21] B. J. Baliga, "Impact of SiC on Power Devices", Proceedings of the 4th International Conference on Amorphous and Crystalline Silicon Carbide", pp. 305-313, 1991.

[22] L. Lipkin and J. W. Palmour, "Improved Oxidation Procedures for Reduced SiO_2/SiC Defects", J. Electronic Materials, Vol. 25, pp. 909-915, 1996.

[23] R. Schorner *et al.*, "Significantly improved performance of MOSFETs on Silicon Carbide using the 15R-SiC Polytype", IEEE Electron Device Letters, Vol. 20, pp. 241-244, 1999.

[24] A.V. Suvorov *et al.*, "4H-SiC Self-Aligned Implant-Diffused Structure for Power DMOSFETs", Materials Research Forum, Vol. 338-342, pp. 1275-1278, 2000.

[25] S. Sridevan and B. J. Baliga, "Lateral N-channel Inversion Mode 4H-SiC MOSFETs", PSRC Technical Report TR-97-019, 1997.

[26] S. Sridevan and B. J. Baliga, "Lateral N-channel Inversion Mode 4H-SiC MOSFETs", IEEE Electron Device Letters, Vol. 19, pp. 228-230, 1998.

[27] S. Sridevan and B. J. Baliga, "Phonon Scattering Limited Mobility in SiC Inversion Layers", PSRC Technical Report TR-98-03, 1998.

[28] D. Alok, E. Arnold and R. Egloff, "Process Dependence of Inversion Layer Mobility in 4H-SiC Devices", Silicon Carbide and Related Materials – 1999, Material Science Forum, Vol. 338-342, pp. 1077-1080, 2000.

[29] K. Chatty *et al.*, "Hall Measurements of Inversion and Accumulation-Mode 4H-SiC MOSFETs", Material Science Forum, Vol. 389-393, pp. 1041-1044, 2002.

[30] N. S. Saks, S. S. Mani and K. Agarwal, Applied Physics Letters, Vol. 76, pp. 2250-2251, 2000.

[31] S.Harada *et al.*, "Temperature Dependence of the Channel Mobility and Threshold Voltage in 4H and 6H-SiC MOSFETs", Material Society Research Symposium Proceedings, Vol. 640, pp. H5.37.1-H5.37.6, 2001.

[32] G. Y. Chung *et al.*, "Improved Inversion Channel Mobility for 4H-SiC MOSFETs following High Temperature Anneals in Nitric Oxide", IEEE Electron Device Letters, Vol. 22, pp. 176-178, 2001.

[33] R.H. Ryu *et al.*, "Critical Issues for MOS Based Power Devices in 4H-SiC" Material Science Forum, Vol. 615-617, pp. 743-748, 2009.

[34] J.R. Williams *et al.*, "Passivation of the 4H-SiC/SiO2 Interface with Nitric Oxide", Material Science Forum, Vol. 389-393, pp. 967-972, 2002.

[35] S. Dhar *et al.*, "Effect of Nitric Oxide Annealing on the Interface Trap Density near the Conduction Band Edge of 4H-SiC at the oxide/(1120) 4H-SiC Interface", Applied Physics Letters, Vl. 84, pp. 1498-1500, 2004.

[36] C-Y. Lu *et al.*, "Effect of Process Variations and Ambient Temperature on Electron Mobility at the SiO2/4H-SiC Interface", IEEE Transactions on Electron Devices, Vol. 50, pp. 1582-1588, 2003.

[37] S. Dhar *et al.*, "Inversion Layer Carrier Concentration and Mobility in 4H-SiC MOSFETs", Journal of Applied Physics, Vol. 108, pp. 054509, 2010.

[38] S. Dhar *et al.*, "Temperature Dependence of Inversion Layer Carrier Concentration and Hall Mobility in 4H-SiC MOSFETs", Material Science Forum, Vol. 717-720, pp. 713-716, 2012.

[39] G. Gudjonsson *et al.*, "High Field-Effect Mobility in n-Channel Si Face 4H-SiC MOSFETs with Gate Oxide Grown on Aluminum Ion-Implanted Material", IEEE Electron Device Letters, Vol. 26, pp. 96-98, 2005.

[40] S.M. Thomas *et al.*, "Enhanced Field Effect Mobility on 4H-SiC by Oxidation at 1500 °C", Journal of the Electron Device Society, Vol. 2, pp. 114-117, 2014.

[41] X. Yang, B. Lee and V. Misra, "High Mobility 4H-SiC Lateral MOSFETs using Lanthanum Silicate and Atomic Layer Deposited SiO2", IEEE Electron Device Letters, Vol. 36, pp. 312-314, 2015.

[42] S. C. Sun and J. D. Plummer, "Electron Mobility in Inversion and Accumulation Layers on Thermally Oxidized Silicon Surfaces", IEEE Transactions on Electron Devices, Vol. 27, pp. 1497-1508, 1980.

[43] P. M. Shenoy and B. J. Baliga, "The Planar 6H-SiC ACCUFET", IEEE Electron Device Letters, Vol. 18, pp. 589-591, 1997.

[44] P.M. Shenoy and B.J. Baliga, "Analysis and Optimization of the Planar 6H-SiC ACCUFET", Solid State Electronics, Vol. 43, pp. 213-220, 1999.

[45] P. M. Shenoy and B. J. Baliga, "High Voltage Planar 6H-SiC ACCUFET", Material Science Forum, Vol. 264-268, pp. 993-996, 1998.

[46] R.K. Chilukuri, P.M. Shenoy and B.J. Baliga, "Comparison of 6H-SiC and 4H-SiC High Voltage Planar ACCUFETs", IEEE International Symposium on Power Semiconductor Devices and ICs, Abstract 6.1, pp. 115-118, 1998.

[47] K. Ueno and T. Oikawa, "Counter-Doped MOSFETs of 4H-SiC", IEEE Electron Device Letters, Vol. 20, pp. 624-626, 1999.

[48] K. Chatty *et al.*, "Accumulation-Layer Electron Mobility in n-Channel 4H-SiC MOSFETs", IEEE Electron Device Letters, Vol. 22, pp. 212-214, 2001.

[49] S. Sridevan and B. J. Baliga, "Lateral N-channel Inversion Mode 4H-SiC MOSFETs", IEEE Electron Device Letters, Vol. 19, pp. 228-230, 1998.

[50] S. Harada *et al.*, "High Channel Mobility in Normally-Off 4H-SiC Buried Channel MOSFETs", IEEE Electron Device Letters, Vol. 22, pp. 272-274, 2001.

[51] S. Kaneko *et al.*, "4H-SiC ACCUFET with a Two-Layer Stacked Gate Oxide", Silicon Carbide and Related Materials – 2001, Material Science Forum, Vol. 389-393, pp. 1073-1076, 2002.

[52] J. N. Shenoy, J. A. Cooper and M. R. Melloch, "High Voltage Double-Implanted Power MOSFETs in 6H-SiC", IEEE Electron Device Letters, Vol. 18, pp. 93-95, 1997.

[53] D. Peters *et al.*, "An 1800V Triple Implanted Vertical 6H-SiC MOSFET", IEEE Transactions on Electron Devices, Vol. 46, pp. 542-545, 1999.

[54] R. Schorner *et al.*, "Rugged Power MOSFETs in 6H-SiC with Blocking Voltage Capability upto 1800V", Silicon Carbide and Related Materials – 1999, Material Science Forum, Vol. 338-342, pp. 1295-1298, 2000.

[55] S-H Ryu *et al.*, "Design and Process Issues for Silicon Carbide Power DiMOSFETs", Material Science Forum, Vol. 640, pp. H4.5.1-H4.5.6, 2001.

[56] S-H. Ryu *et al.*, "Large-Area (3.3 mm × 3.3 mm) Power MOSFETs in 4H-SiC", Silicon Carbide and Related Materials – 2001, Material Science Forum, Vol. 389-393, pp. 1195-1198, 2002.

[57] S-H. Ryu *et al.*, "Development of 10 kV 4H-SiC Power DMOSFETs", Silicon Carbide and Related Materials – 2003, Material Science Forum, Vol. 457-460, pp. 1385-1388, 2004.

[58] S-H. Ryu *et al.*, "10 kV, 123 mΩ-cm^2 4H-SiC Power DMOSFETs", IEEE Electron Device Letters, Vol. 25, pp. 556-558, 2004.

[59] M. Matin, A. Saha and J. A. Cooper, "Self-Aligned Short-Channel Vertical Power DMOSFETs in 4H-SiC", Material Science Forum, Vol. 457-460, pp. 1393-1396, 2004.

[60] M. Matin, A. Saha and J.A. Cooper, "A Self-Aligned Process for High-Voltage Short-Channel Vertical DMOSFETs in 4H-SiC", IEEE Transactions on Electron Devices, Vol. 51, pp. 1721-1725, 2004.

[61] A. Saha and J.A. Cooper, "A 1-kV 4H-SiC Power DMOSFET Optimized doe Low On-Resistance", IEEE Transactions on Electron Devices, Vol. 55, pp. 2786-2791, 2007.

[62] L. Stevanovic *et al.*, "Realizing the Full Potential of Silicon Carbide Power Devices", IEEE Workshop on Controls and Modelling of Power Electronics, pp. 1-6, 2010.

[63] S-H Ryu *et al.*, "3.7 mW-cm2, 1500 V 4H-SiC DMOSFETs for Advanced High Power, High Frequency Applications", IEEE International Symposium on Power Semiconductor Devices and ICs, pp. 227-230, 2011.

[64] L. Cheng *et al.*, "High Performance, Large-Area, 1600 v? 150 A, 4H-SiC DMOSFET for Robust High-Power and High-Temperature

Applications", IEEE International Symposium on Power Semiconductor Devices and ICs, Paper 2-2, pp. 47-50, 2013.

[65] K. Matocha et al., "1700V, 5.5mOhm-cm^2 4H-SiC DMOSFET with Stable 225°C Operation", Material Science Forum, Vol. 778-780, pp. 903-906, 2014.

[66] J. Palmour, "Silicon Carbide Power Device Development for Industrial Markets", IEEE International Electron Devices Meeting, Paper 1, pp. 1.1.1-1.1.8, 2014.

[67] P. M. Shenoy and B. J. Baliga, "Analysis and Optimization of the Planar 6H-SiC ACCUFET", Solid State Electronics, Vol. 43, pp. 213-220, 1999.

[68] Y. Sugawara et al., "4.5 kV Novel High Voltage High Performance SiC-FET (SIAFET)", IEEE International Symposium on Power Semiconductor Devices and ICs, pp. 105-108, 2000.

[69] Y. Sugawara et al., "5.0 kV 4H-SiC SEMOSFET with Low RonS of 88 mOcm2", Material Science Forum, Vol. 389-393, pp. 1199-1202, 2002.

[70] R. Singh, S-H Ryu and J.W. Palmour, "High Temperature, High Current, 4H-SiC Accu-DMOSFET", Material Science Forum, Vol. 338-342, pp. 1271-1274, 2000.

[71] F. Nallet et al., "Very Low R_{ON} measured on 4H-SiC Accu-MOSFET High Power Device", IEEE International Symposium on Power Semiconductor Devices and ICs, pp. 209-212, 2002.

[72] R. Singh et al., "High Channel Density, 20A 4H-SiC ACCUFET with R_{onsp} = 15 mΩ-cm^2", Electronics Letters, Vol. 39, p. 152-153, 2003.

[73] R. Singh et al., "Development of High Current 4H-SiC ACCUFET", IEEE Transactions on Electron Devices, Vol. 50, pp. 471-477, 2003.

[74] S. Harada et al., "8.4 mW-cm2 600-V Double-Epitaxial MOSFETs in 4H-SiC", IEEE Electron Device Letters, Vol. 25, pp. 292-294, 2004.

[75] S. Harada et al., "An Ultra-Low Ron in 4H-SiC Vertical MOSFET: Buried Channel Double-Epitaxial MOSFET", IEEE International Symposium on Power Semiconductor Devices and ICs, pp. 313-316, 2004.

[76] N. Miura et al., "Successful Development of 1.2 kV 4H-SiC MOSFETs with Very Low On-Resistance of 5 mΩ-cm2", IEEE International Symposium on Power Semiconductor Devices and ICs, pp. 1-4, 2006.

[77] R.S. Howell *et al.*, "A 10-kV Large-Area 4H-SiC Power DMOSFET with Stable Subthreshold behavior independent of Temperature", IEEE Transactions on Electron Devices, Vol. 55, pp. 1807-1815, 2008.
[78] R.S. Howell *et al.*, "Comparison of Design and Yield for Large Area 10-kV 4H-SiC DMOSFETs", IEEE Transactions on Electron Devices, Vol. 55, pp. 1816-1823, 2008.
[79] M. Kitabatake *et al.*, "4H-SiC DIMOSFET Power Device for Home Appliances", IEEE International Power Electronics Conference, pp. 3249-3253, 2010.
[80] Q. Zhang *et al.*, "Latest Results on 1200 V 4H-SiC CIMOSFETs with $R_{sp,on}$ of 3.9 mΩ-cm2 at 150 °C", IEEE International Symposium on Power Semiconductor Devices and ICs, pp. 89-92, 2015.
[81] N. Thapar and B.J. Baliga, "A New IGBT Structure with a wider Safe Operating Area (SOA)", IEEE International Symposium on Power Semiconductor Devices and ICs, Paper 4.3, pp. 177-182, 1994.
[82] B.J. Baliga and J.P Walden, "Improving the Reverse Recovery of Power MOSFET Integral Diodes by Electron Irradiation", Solid State Electronics, Vol. 26, pp. 1133-1141, 1983.
[83] B.J. Baliga and D. Alok, "Paradigm Shift in Planar MOSFET Technology", Power Electronics Technology Magazine, pp. 24-32, November 2003.
[84] C-T. Yen *et al.*, "1700V/30A 4H-SiC MOSFET with Low Cut-in Voltage Embedded Diode and Room Temperature Boron Implanted Termination", IEEE International Symposium on Power Semiconductor Devices and ICs, pp. 265-268, 2015.
[85] A. Agarwal *et al.*, "A New Degradation Mechanism in High-Voltage SiC Power MOSFETs", IEEE Electron Device Letters, Vol. 28, pp. 587-589, 2007.
[86] B. Hull *et al.*, "Reliability and Stability of SiC Power MOSFETs and Next Generation SiC MOSFETs", IEEE Workshop on Wide Bandgap Power Devices and Applications", pp. 139-142, 2014.
[87] S. Liu *et al.*, "Investigations on Degradation and Optimization of 1.2kV 4H-SiC MOSFET under repetitive Unclamped Inductive Switching Stress", IEEE International Symposium on Power Semiconductor Devices and ICs, pp. 205-208, 2015.
[88] X. Huang *et al.*, "Short-circuit Capability of 1200V SiC MOSFET and JFET for fault Protection", IEEE Applied Power Electronics Conference, pp. 197-200, 2013.

[89] T-T. Nguyen *et al.*, "Gate Oxide Reliability Issues of SiC MOSFETs under Short-Circuit Operation", IEEE Transactions on Power Electronics, Vol. 30, pp. 2445-2455, 2015.

[90] M.K. Das *et al.*, "10kV, 120A SiC Half H-Bridge Power MOSFET Modules Suitable for High Frequency, Medium Voltage Applications", IEEE Energy Conversion Congress and Exposition, pp. 2689-2692, 2011.

[91] F. Wang *et al.*, "A 3.6kV High Performance Solid State Transformer Based on 13kV SiC MOSFET", IEEE Energy Conversion Congress and Exposition, pp. 4553-4560, 2014.

Chapter 12

SiC Trench-Gate Power MOSFETs

The silicon trench-gate power MOSFET was developed in the 1990s by borrowing the trench technology originally developed for DRAMs. Before the introduction of the trench-gate structure, it was found that the ability to reduce the specific on-resistance for silicon power MOSFETs was constrained by the poor channel density and the JFET region resistance[1]. Although the use of more advanced lithographic design rules enabled a steady reduction of the specific on-resistance in the 1970s and early 1980s, the improvements to the DMOSFET structure were saturating[2]. The trench-gate or UMOSFET structure enabled significant increase in the channel density and elimination of the JFET resistance contribution resulting in a major enhancement in low voltage (<50 volt) silicon power MOSFET performance. The first trench-gate devices[3,4] were fabricated in the mid-1980s and shown to have significantly lower specific on-resistance than the DMOSFET structure. However, problems with controlling the quality of the trench surface and oxide reliability problems needed to be solved before the introduction of commercial devices. Eventually, this technology overtook the planar DMOSFET technology in the 1990s and has now taken a dominant position in the industry for serving portable appliances, such as laptops, PDAs, etc.

In the silicon power DMOSFET structure, the channel and JFET resistances were found to become the dominant components when the breakdown voltage was reduced below 50 volts because of the low resistance of the drift region. Similarly, due to the much lower specific on-resistance for the drift region in silicon carbide, the channel and JFET contributions become dominant for breakdown voltages below 5000 volts, especially if the channel mobility is poor. This has motivated the development of the trench-gate silicon carbide power MOSFETs. In addition, the P-base region for the trench-gate device could be fabricated using epitaxial growth which was at a more advanced state than the ability to use ion implantation to create P-type layers in silicon carbide. This allowed the fabrication of these structures before the planar structures. However, their performance was limited by issues unique to SiC created by simply replicating the silicon device structure[5].

This chapter reviews the basic principles of operation of the trench-gate MOSFET structure. The specific on-resistance for this structure is shown to be significantly lower than that for the DMOSFET structure. However, the gate oxide in the UMOSFET structure is exposed to the very high electric field developed in the silicon carbide drift region during the blocking-mode. This is a major limitation to adopting the basic UMOSFET structure from silicon to silicon carbide. Two fundamental approaches to solving this problem are discussed: (a) the use of high-k dielectrics as the gate insulator and (b) the shielding of the gate oxide from the high electric field in the SiC drift region.

Structures designed to reduce the electric field at the gate oxide by using a shielding region are essential to realization of practical silicon carbide MOSFET structures[6]. In this chapter, analytical models of the basic and shielded trench-gate silicon carbide MOSFET structure are provided followed by the description of experimental results on relevant structures to define the state of the development effort on these devices.

12.1 Basic Device Structure

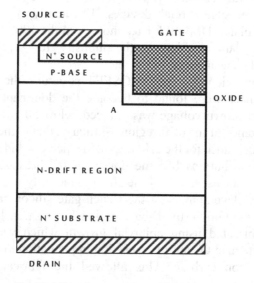

Fig. 12.1 Trench-gate power MOSFET structure.

The basic structure of the trench-gate power MOSFET is shown in Fig. 12.1. The structure can be fabricated by either the epitaxial growth of the P-base region over the drift region or by introducing the P-type dopants using ion-implantation. The first 4H-SiC UMOSFET structure was fabricated by the epitaxial growth of the P-base region due to problems with activation of P-type ion implanted dopants in silicon carbide[5]. However, this requires either removal of the P-type layer on the edges of the structure to form a mesa edge termination[5] or multiple trench-isolated guard rings[7]. The trench-gate structure can be fabricated by using reactive-ion etching of the SiC layers to form the U-shaped trenches. The gate oxide was then created by thermal oxidation followed by refilling the trench with polysilicon as done for silicon UMOSFET structures.

Note that the P-base region is short-circuited to the N^+ source region by the source metal. Although this is routinely done in silicon devices using a common ohmic contact metal for both the N^+ and P-type regions, this has been difficult to achieve in silicon carbide. It is usual to use different metals to form the ohmic contact to silicon carbide, typically with titanium for N-type regions and aluminum for P-type regions. In this case, it may be preferable to short circuit the source-base junction at a location orthogonal to the cross-section shown in the figure. This is also advantageous for reducing the cell pitch which results in a larger channel density.

12.1.1 Blocking Characteristics

When the trench-gate power MOSFET is operating in the forward blocking mode, the voltage is supported by a depletion region formed on both sides of the P-base/N-drift junction. The maximum blocking voltage can be determined by the electric field at this junction becoming equal to the critical electric field for breakdown if the parasitic $N^+/P/N$ bipolar transistor is completely suppressed. This suppression is accomplished by short-circuiting the N^+ source and P-base regions using the source metal as shown on the upper left hand side of the cross-section. However, a large leakage current can occur when the depletion region in the P-base region reaches-through to the N^+ source region. The doping concentration and thickness of the P-base region must be designed to prevent the reach-through phenomenon from limiting the breakdown voltage. The physics governing the reach-through process is identical to that already described in chapter 11 for the planar MOSFET structure. Consequently, the design rules provided earlier with the aid of Fig. 11.3

can be applied to the trench-gate structure as well. This implies that the minimum channel length required for the silicon carbide trench-gate devices is much larger than for silicon resulting in a substantial increase in the on-resistance. The degradation of the on-resistance is compounded by the lower channel inversion layer mobility observed for silicon carbide.

The minimum thickness of the P-base region required to prevent reach-through breakdown decreases with increasing doping concentration as shown in Fig. 11.3. For 4H-SiC, it is necessary to increase the P-base doping concentration to above 2×10^{17} cm^{-3} to prevent reach-through with a 1 micron P-base thickness. This higher doping concentration makes the threshold voltage prohibitively large as already discussed in chapter 11.

The maximum blocking voltage capability of the trench-gate MOSFET structure is determined by the drift region doping concentration and thickness as already discussed in chapter 3. However, in the trench-gate MOSFET structure, the gate extends into the drift region exposing the gate oxide to the high electric field developed in the drift region under forward blocking conditions. For 4H-SiC, the electric field in the oxide reaches a value of 9×10^6 V/cm when the field in the semiconductor reaches its breakdown strength of about 3×10^6 V/cm. This value not only exceeds the reliability limit but can cause rupture of the oxide leading to catastrophic breakdown because the problem is exacerbated by electric field enhancement at the corners of the trenches at location A in the figure. Novel structures[6] that shield the gate oxide from high electric field have been proposed and demonstrated to resolve this problem. These structures are discussed in a subsequent section.

12.1.2 On-Resistance

Current flow between the drain and source of the trench-gate power MOSFET can be induced by creating an inversion layer channel on the surface of the P-base region. The current path is illustrated in   Fig. 12.2 by the shaded area. The current flows from the source region into the drift region through the inversion layer channel formed on the vertical side-walls of the trench due to the applied gate bias. It then spreads into the N-drift region from the bottom of the trench at a 45 degree angle and becomes uniform through the rest of the structure.

The total on-resistance for the trench-gate power MOSFET structure is determined by the resistance of these components in the current path:

$$R_{on,sp} = R_{CH} + R_D + R_{subs} \qquad [12.1]$$

where R_{CH} is the channel resistance, R_D is the resistance of the drift region after taking into account current spreading from the channel, and R_{subs} is the resistance of the N^+ substrate. These resistances can be analytically modeled by using the current flow pattern indicated by the shaded regions in Fig. 12.2.

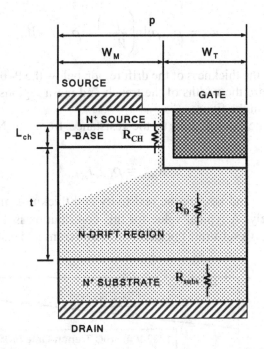

Fig. 12.2 Current flow path in the trench-gate 4H-SiC power MOSFET.

The specific channel resistance is given by:

$$R_{CH} = \frac{L_{CH} \cdot p}{\mu_{inv} C_{ox} (V_G - V_T)} \qquad [12.2]$$

where L_{CH} is the channel length determined by the depth of the P-base and the N^+ source regions as shown in Fig. 12.2, μ_{inv} is the mobility for electrons in the inversion layer channel, C_{ox} is the specific capacitance of the gate oxide, V_G is the applied gate bias, and V_T is the threshold voltage. The specific capacitance can be obtained using:

$$C_{ox} = \frac{\varepsilon_{ox}}{t_{ox}}$$ [12.3]

where ε_{ox} is the dielectric constant for the gate oxide and t_{ox} is its thickness.

The drift region spreading resistance can be obtained by using:

$$R_D = \rho_D \cdot p \cdot \ln\left(\frac{p}{W_T}\right) + \rho_D \cdot (t - W_M)$$ [12.4]

where t is the thickness of the drift region below the P- base region and W_T, W_M are the widths of the trench and mesa regions, respectively, as shown in the figure.

The contribution to the resistance from the N⁺ substrate is given by:

$$R_{subs} = \rho_{subs} \cdot t_{subs}$$ [12.5]

where ρ_{subs} and t_{subs} are the resistivity and thickness of the substrate, respectively. A typical value for this contribution is 4×10^{-4} Ω-cm² based up on a substrate thickness of 200 microns and resistivity of 0.02 Ω-cm.

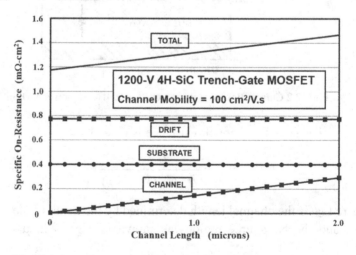

Fig. 12.3 On-resistance components for 4H-SiC trench-gate MOSFETs.

The specific on-resistances of 1200-V rated 4H-SiC trench MOSFETs were modeled using the above analytical expressions. An ideal breakdown voltage of 1670 volts is obtained for a 4H-SiC drift region with doping concentration of 1×10^{16} cm^{-3}. The device breakdown voltage is then 1336 volts if the edge termination limits the breakdown to 80% of the ideal value. This provides some margin for a 1200-V rated device. At this blocking voltage, the depletion width is 11.35 microns. A drift region thickness (t) of 12 microns was therefore selected. A gate oxide thickness of 0.05 microns was used. The mesa width (W_M) and trench width (W_T) for the structure shown in Fig. 12.2 were kept at 0.25 microns resulting in a pitch (p) of 0.5 microns. The gate bias voltage (V_G) was assumed to be 15 volts with a threshold voltage (V_T) of 5 volts.

The various components of the on-resistance are shown in Fig. 12.3 with increasing channel length for the case of a channel inversion layer mobility of 100 cm^2/V-s. As expected, only the channel resistance increases with increasing channel length. The other components are unaffected by the increase in channel length because the cell pitch remains unaltered. It can be seen that the drift region resistance is dominant in this case. A total specific on-resistance of 1.25 mΩ-cm^2 is obtained for a channel length of 0.5 microns, which is within two times the ideal specific on-resistance of the drift region.

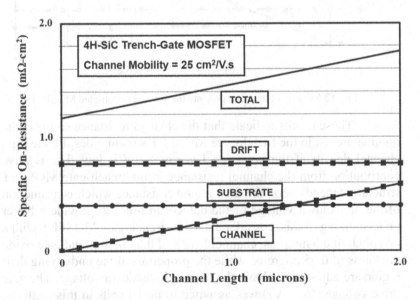

Fig. 12.4 On-resistance components for a 4H-SiC trench-gate MOSFETs.

The above results are based upon using a channel mobility of 100 cm²/Vs. Since most groups working on silicon carbide have reported much lower magnitudes for the inversion layer mobility in 4H-SiC structures[8], the impact of reducing the inversion layer mobility for the trench-gate MOSFET structure is depicted in Figs. 12.4 and 12.5 for the case of mobility values of 25 and 2.5 cm²/Vs. With an inversion layer mobility of 25 cm²/Vs, the specific on-resistance increasing to a value of 1.315 mΩ-cm² for a channel length of 0.5 micron. When the inversion layer mobility is reduced to 5 cm²/Vs, the channel resistance becomes equal to that of the drift region, as shown in Fig. 12.5, with an increase in the specific on-resistance to a value of 1.90 mΩ-cm² for a channel length of 0.5 micron.

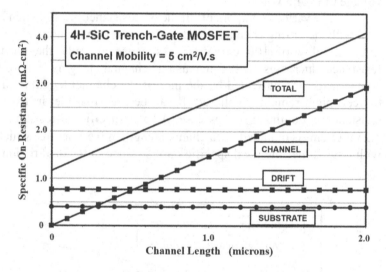

Fig. 12.5 On-resistance components for 4H-SiC trench-gate MOSFETs.

These results indicate that the channel resistance can become dominant even in the trench-gate MOSFET structure despite the high channel density if the inversion layer mobility is low. The relative contribution from the channel resistance in the trench-gate MOSFET structure depends upon the drift region resistance which is a function of the breakdown voltage of the device. In order to provide a better understanding of this, consider a 4H-SiC trench-gate MOSFET with a cell pitch of 0.5 microns, channel length of 0.5 microns and gate oxide thickness of 0.05 microns, while the properties of the underlying drift region are adjusted to obtain the desired breakdown voltage. The gate drive voltage ($V_G - V_T$) was assumed to be 10 volts in this analysis. The specific on-resistance for this trench-gate 4H-SiC MOSFET structure is plotted in Fig. 12.6 as a function of the breakdown voltage using a channel mobility of 25 cm^2/V-s. In performing this modeling, it is important to recognize that the thickness of the drift region (parameter 't' in Fig. 12.2) can become smaller than the mesa width (W_M) at lower breakdown voltages (below 600 volts). Under these conditions, the current does not distribute at a 45 degree angle into the drift region from the bottom of the trench. Instead, the current flows from a cross-sectional width of (W_T) to a cross-section of ($p = W_M + W_P$). The drift region resistance for these cases can be modeled using:

$$R_D = \rho_D . p . \ln\left(\frac{p}{W_T}\right)$$ [12.6]

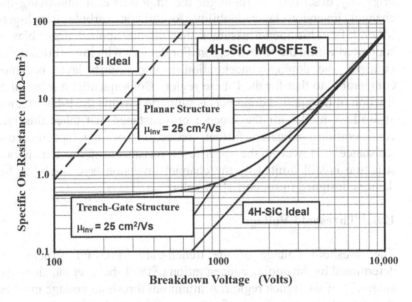

Fig. 12.6 Specific on-resistance for 4H-SiC trench-gate MOSFETs.

From Fig. 12.6, it can be concluded that the specific on-resistance of 4H-SiC trench-gate MOSFETs approaches the ideal specific on-resistance when the breakdown voltage exceeds 5000 volts even with a channel mobility of 25 cm^2/V-s. However, the channel resistance limits the performance of the trench-gate 4H-SiC MOSFET structure when the breakdown voltage falls below 2000 volts. For the case of a breakdown voltage of 1000 volts, the anticipated improvement in specific on-resistance over silicon power MOSFETs is then about 100 times (as opposed to the 2000x improvement in the specific on-resistance of the drift region). For a better perspective, the performance of the planar 4H-SiC inversion-mode MOSFET structure is also shown in the figure. The planar MOSFET analysis was based upon using the same channel length, gate oxide thickness, and channel mobility of 25 cm^2/V-s as the trench-gate structure. A cell-pitch of 3 microns was used for the planar structure. It can be seen that the trench-gate structure offers a substantial (4-fold) reduction of the specific-on-resistance due to the higher channel density. These results highlight the need for development of aggressive processing techniques, like those routinely used for low voltage silicon power UMOSFET structures, to achieve a small cell pitch in the trench-gate 4H-SiC MOSFET structure.

The on-resistance model presented in this section was originally described[9] to highlight the importance of improving the channel inversion layer mobility for silicon carbide trench-gate MOSFETs. This model assumes that all the applied drain bias is supported within the N-drift region. For devices with lower breakdown voltages, the doping concentration in the N-drift layer becomes comparable to that for the P-base region. Consequently, a substantial fraction of the applied drain bias is supported within the P-base region as well. A model for the specific on-resistance that takes this into consideration indicates further reduction of the specific on-resistance[10]. However, the specific on-resistance of the trench-gate structure is still limited by the channel inversion layer mobility for devices designed to support 1000 volts.

12.1.3 Threshold Voltage

The threshold voltage of the trench-gate MOSFET structure is determined by the doping concentration of the P-base region along the sidewalls of the trench region. A minimum threshold voltage must be maintained at above 3 volts for most system applications to provide immunity against inadvertent turn-on due to voltage spikes arising from noise. At the same time, a high threshold voltage is not desirable because the voltage available for creating the charge in the channel inversion layer is determined by ($V_G - V_T$) where V_G is the applied gate bias voltage and V_T is the threshold voltage. Most power electronic systems designed for high voltage operation (the most suitable application area for silicon carbide devices) provide a gate drive voltage of only up to 15 volts. Based upon this criterion, the threshold voltage should be kept below 5 volts in order to obtain a low channel resistance contribution.

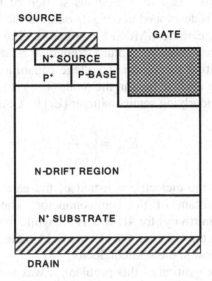

Fig. 12.7 Enhanced trench-gate power MOSFET structure.

The threshold voltage for the trench-gate inversion-mode MOSFET structure can be modeled using the same physics described in chapter 11 for the planar gate inversion-mode MOSFET structure. The threshold voltage for 4H-SiC devices can be obtained using the graphs provided in that chapter. Based upon that analysis, it is preferable to use a gate oxide thickness of 0.05 microns and a P-base doping concentration of 1×10^{17} cm^{-3} to obtain a threshold voltage of about 5 volts. However, the low P-base doping concentration can lead to a reach-through induced breakdown problem as discussed in chapter 11.

A structural enhancement that can circumvent this problem is shown in Fig. 12.7. A P$^+$ region is incorporated into the trench-gate MOSFET structure under the source region to suppress the extension of the depletion region in this portion of the cell structure. A relatively thin, lightly doped P-base region is formed adjacent to the trench sidewalls to determine the threshold voltage. It has been found that the presence of the P$^+$ region also suppresses reach-through within the adjacent P-base region allowing the structure to support high drain voltages upto the breakdown voltage capability of the drift region.

12.2 High Permittivity Gate Insulator

As already discussed in a previous section of this chapter, a high electric field is developed in the gate oxide at the bottom of the trench in the silicon carbide UMOSFET structure. The electric field in the gate oxide can exceed its rupture strength at relatively low drain bias voltages limiting the blocking voltage capability of the trench-gate structure. The electric field in the oxide (E_{OX}) is related to the electric field in the underlying semiconductor (E_S) by Gausses Law:

$$E_{OX} = \left(\frac{\varepsilon_S}{\varepsilon_{OX}} \right).E_S \qquad [12.7]$$

where ε_{OX} is the dielectric constant of the gate oxide and ε_S is the dielectric constant of the semiconductor. The relative dielectric constant (permittivity) for 4H-SiC is 9.7 while that for silicon dioxide is 3.85. Using these values, the electric field in the gate oxide is a factor of 2.5 times that in the semiconductor.

In recognition of this problem, it was proposed[11] that silicon dioxide be replaced by a gate insulator with high permittivity – preferably at least 10 times that of free space. With the increased dielectric constant for the gate insulator, the electric field developed in the gate dielectric would become closer to that developed in the semiconductor[12]. This implies that the maximum electric field in the gate dielectric can be reduced to 3×10^6 V/cm, which should be satisfactory for reliable operation. This approach can take advantage of the large effort being undertaken to develop high dielectric constant insulators for mainstream silicon DRAM chips.

A simple one-dimensional analysis of the metal-oxide-semiconductor structure provides interesting insight into the benefits of using high permittivity gate insulators in silicon carbide MOSFETs. Since the highest electric field occurs at the flat bottom of the trench region in the UMOSFET structure, a one dimensional analysis is a reasonable approximation for analytical purposes.

The electric field profile in the MOS gate stack is illustrated in Fig. 12.8 for three cases of dielectric constants when it is supporting the same voltage (e.g. 1200 volts). In the first two cases with permittivity of 3.85 (corresponding to silicon dioxide) and 7.7 (corresponding to silicon nitride), it is assumed that the field distribution is constrained by reaching a maximum electric field of 3 $\times 10^6$ V/cm in the gate dielectric. This value was chosen based upon the maximum electric field in oxides for reliable operation. Under this constraint, the maximum electric field in the silicon carbide becomes much lower than its breakdown field strength (1.2×10^6 V/cm for the permittivity of 3.85 and 2.4×10^6 V/cm for the permittivity of 7.7). However, when the permittivity is increased to 15.4, the electric field distribution becomes constrained by the maximum allowable electric field in the silicon carbide (3.3×10^6 V/cm for 4H-SiC at a doping concentration of about 1×10^{16} cm^{-3}). Under this limitation, the electric field in the oxide is reduced to about 2×10^6 V/cm for the permittivity of 15.4.

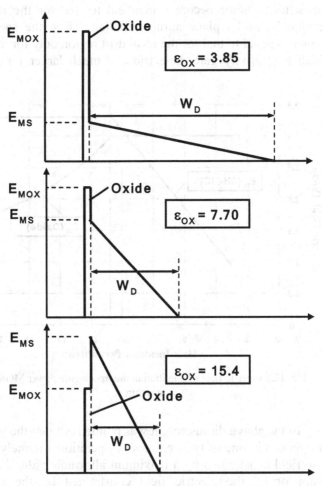

Fig. 12.8 Electric field distribution in the trench-gate power MOSFET structure.

12.2.1 Specific On-Resistance

The changes in the electric field distribution have a strong impact on the doping concentration and thickness of the drift region required to support the drain bias (e.g. 1200 volts) as indicated in Fig. 12.8. For the lower dielectric constants, a much thicker drift region with lower doping concentration (reflected in the smaller slope of the electric field profile in the semiconductor) is required because of the reduced maximum electric field, when compared with the properties of the ideal drift region. For the high dielectric constant case, the electric field in the semiconductor becomes identical to that for the ideal one-dimensional parallel plane abrupt junction. Thus, the specific on-resistance is equal to that for the ideal drift region only for the case of the high permittivity gate dielectric and much larger for the other cases.

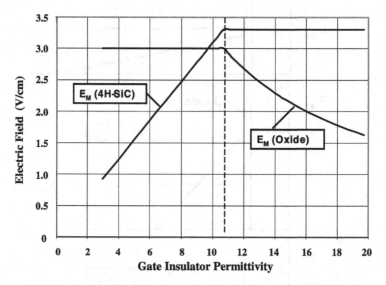

Fig. 12.9 Electric field distribution in the trench-gate power MOSFET structure.

In the above discussion, it was pointed out that the gate MOS stack operates in one of two regimes of operation – namely with the electric field constrained by a maximum allowable value in the gate insulator or by the electric field constrained by the maximum allowable value in the semiconductor. The results of the analytically calculated electric fields in the gate insulator and the semiconductor under these constraints are shown in Fig. 12.9. It can be seen that the transition occurs at a permittivity given by:

$$\varepsilon_{OX} = \left(\frac{E_{MS}(4H - SiC)}{E_{MOX}} \right).\varepsilon_S \qquad [12.8]$$

where $E_{MS}(4H\text{-}SiC)$ is the critical electric field for breakdown in 4H-SiC and E_{MOX} is the maximum electric field in the dielectric for reliable operation. For the values for these parameters described above, the transition occurs at a permittivity of 10.7. This is the minimum permittivity needed to reduce the specific on-resistance of the drift region to that for the ideal drift region.

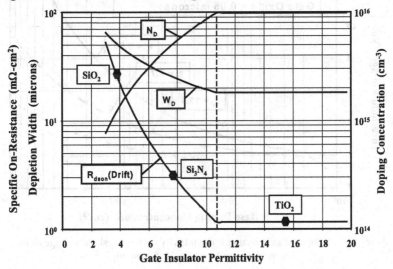

Fig. 12.10 Impact of gate insulator permittivity for the trench-gate power MOSFET structure.

The permittivity of the gate insulator has a strong influence on the doping concentration and thickness of the drift region. This is shown in Fig. 12.10. The doping concentration of the drift region increases and its thickness decreases as the permittivity is increased up to a value of 10. This results in a drastic reduction of the specific on-resistance of the drift region. As examples, points are shown for selected insulators on the graph. In the case of the commonly used silicon dioxide gate insulator, the specific resistance of the drift region is degraded by a factor of 25 times. If silicon nitride is used as the gate insulator, the drift region resistance becomes only 3 times larger than the ideal case. With the use of a much higher dielectric constant of 15 with titanium dioxide as the insulator, the ideal drift region resistance can be obtained together with reduced electric field in the insulator. Other high-k dielectrics that could be potential candidates based on their applications in silicon CMOS technology include aluminum oxide and hafnium oxide[13].

12.2.2 Threshold Voltage

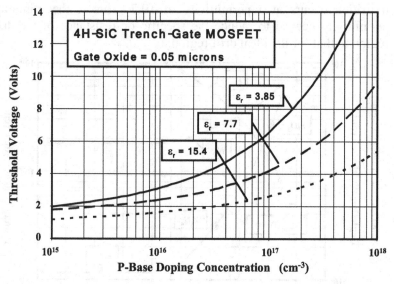

Fig. 12.11 Impact of gate insulator permittivity on the threshold voltage of the trench-gate power MOSFET structure.

A larger dielectric constant for the gate insulator is also beneficial for achieving a lower threshold voltage with any given insulator thickness. This is demonstrated in Fig. 12.11 for the case of a gate oxide thickness of 0.05 microns. The calculations of the threshold voltage were performed using the same assumptions regarding a work-function difference of 1 volt and the presence of a fixed oxide charge of 2×10^{11} cm^{-2} used in chapter 11 (see Fig. 11.11). At a P-base doping concentration of 3×10^{17} cm^{-3}, it can be seen that the threshold voltage is reduced from 10 volts for the silicon dioxide case (permittivity of 3.85) to 6 volts for the silicon nitride case (permittivity of 7.7), and even further to about 3 volts for a permittivity of 15.4. This is favorable for increasing the charge induced in the channel for a typical gate bias of 15 volts resulting in reducing the specific on-resistance.

12.2.3 Reliability

The analysis described above highlights the benefits of using gate insulators with high permittivity. However, it must be pointed out that the breakdown strength of insulators has been empirically found to reduce with increasing permittivity. Consequently, care must be taken to ensure that the gate insulator with high permittivity will operate in a reliable manner even when the electric field in it has been reduced as described above. In addition, the band offset between the gate insulator and silicon carbide should be examined to make sure that the Fowler-Nordheim tunneling described in chapter 11 does not create instabilities.

The high temperature reliability of various gate dielectrics has been investigated[14]. The study considered silicon nitride (7.5), aluminum nitride (8,4), aluminum oxy-nitride (12.4), titanium oxide (30-40), and tantalum oxide (25) due to their larger permittivity's when compared to silicon dioxide (3.85) as given in brackets. Experimental investigations were performed with the silicon nitride grown by low pressure chemical vapor deposition, the aluminum nitride layers grown by metal-organic chemical vapor deposition, and aluminum oxy-nitride (ALO:N) by thermal oxidation of AlN. The maximum breakdown electric field strength was found to be 10 MV/cm for SiO_2, 5.8 MV/cm for Si_3N_4, 1 MV/cm for AlN, and 4.8 MV/cm for AlO:N. The most favorable gate dielectric was found to be silicon nitride. In addition, a oxide-nitride-oxide sandwich was reported to have good breakdown strength. The inversion layer mobility for this sandwich was reported to be as high as 40 cm^2/V-s. The high temperature reliability of this dielectric sandwich was found to be 100-times superior to that for deposited silicon dioxide.

12.3 Shielded Trench-Gate Power MOSFET Structure

In the previous section, it was demonstrated that the performance of the trench-gate silicon carbide power MOSFET structure is severely compromised by the development of a high electric field in the gate oxide during the blocking mode of operation. This problem occurs because the trench penetrates the P-base region exposing the gate oxide at the bottom of the trench to the high electric field in the silicon carbide drift region. The electric field in the gate oxide reaches its rupture strength well before the electric field in the semiconductor approaches its breakdown field strength. Consequently, in order to operate at any given blocking voltage, the drift region doping concentration has to be reduced and its thickness increased until the specific on-resistance becomes 25 times larger than that for the ideal drift region. This problem inhibited the performance of the first trench-gate 4H-SiC power MOSFETs.

In order to suppress the development of high electric fields in the gate oxide, a trench-gate power MOSFET structure has been proposed[6] with a shielding region incorporated at the bottom of the trench. The basic principles of operation of this shielded trench-gate MOSFET structure are discussed in this section. The impact of the JFET region, formed by the incorporation of the shielding region, on the specific on-resistance for this structure is analyzed here. It is demonstrated that the JFET region resistance can be reduced by enhancement of the doping concentration between the shielding regions while retaining a small cell pitch to obtain a high channel density as first disclosed in the patent[6].

12.3.1 Device Structure

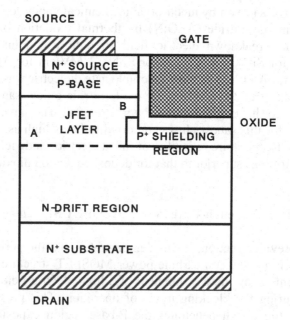

Fig. 12.12 Shielded trench-gate power MOSFET structure.

The basic structure of the shielded trench-gate power MOSFET is shown in Fig. 12.12. The shielding region consists of a heavily doped P-type (P$^+$) region located at the bottom of the trench. The P$^+$ shielding region is connected to the source electrode at a location orthogonal to the device cross-section shown above. The structure then behaves like a monolithic version of the *Baliga-Pair* configuration discussed in chapter 8 with both the JFET and MOSFET formed in the same semiconductor, namely, silicon carbide. From this point of view, the portion of the N-drift region located just below the P-base region serves as both the drain of the MOSFET and the source of the JFET. For this reason, both the *Baliga-Pair* configuration and the monolithic option were patented together[6].

The shielded trench-gate power MOSFET structure can be fabricated by using the same process used for the conventional trench-gate device structure with the addition of an ion implantation step to form the P$^+$ shielding region. The ion implant used to form the P$^+$ shielding region must be performed after etching the trenches. It is worth pointing out that this implantation step should not dope the sidewalls of the trenches. This can be accomplished by doing the ion implant orthogonal to the wafer surface. Alternately, a conformal oxide can be deposited on the trench sidewalls and removed from the trench bottom using anisotropic reactive ion etching to selectively expose the trench bottom to the P-type dopant during the ion implantation. Another option to reduce the introduction of P-type doping on the trench sidewalls is by using a relatively low P-type doping concentration for the shielding region. The shielding is effective as long as the gate oxide is buffered from the high electric field in the N-drift region.

Although the P-type shielding region can be confined to just the bottom of the trench, it is preferable that it overlaps the trench corner as illustrated in Fig. 12.12. The overlap can be a natural outcome of the straggle in the ion implant or the removal of some of the masking oxide on the trench sidewall if its profile is slightly tapered.

12.3.2 Blocking Characteristics

The shielded trench-gate power MOSFET operates in the forward blocking mode when the gate electrode is shorted to the source by the external gate drive circuit. At low drain bias voltages, the voltage is supported by a depletion region formed on both sides of the P-base/ N-drift junction. Consequently, the drain potential appears across the MOSFET located at the top of the structure. This produces a positive potential at location 'A' in Fig. 12.12, which reverse biases the junction between the P^+ shielding region and the N-drift region because the P^+ shielding region is held at zero volts. The depletion region that extends from the P^+/N junction pinches off the JFET region producing a potential barrier at location 'A'. The potential barrier shields the P-base region from any additional bias applied to the drain electrode. Consequently, a high electric field can develop in the N-drift region below the P^+ shielding region while the electric field at the P-base region remains low. This has the beneficial effects of mitigating the reach-through of the depletion region within the P-base region and in keeping the electric field in the gate oxide low at location 'B' where it is exposed to the N-drift region.

The maximum blocking voltage of the shielded trench-gate power MOSFET is determined by the properties of the drift region. This allows reduction of the drift region resistance close to that of the ideal case. In addition, the reduction of the electric field in the vicinity of the P-base region allows reduction of the channel length as well as the gate oxide thickness. This is beneficial for further reduction of the device on-state resistance.

It is worth pointing out that the P^+ shielding region must be adequately short-circuited to the source terminal in order for the shielding to be fully effective. The location of the P^+ region at the bottom of the trench implies that contact to it must be provided at selected locations orthogonal to the cross-section of the device shown in Fig. 12.12. Since the sheet resistance of the ion implanted P^+ region can be quite high, it is important to provide the contact to the P^+ region frequently in the orthogonal direction during chip design. This must be accomplished without significant loss of channel density if low specific on-resistance is to be realized.

12.3.3 Forward Conduction

In the shielded trench-gate MOSFET structure, current flow between the drain and source can be induced by creating an inversion layer channel on the surface of the P-base region along the trench sidewalls. The current flows from the source region into the drift region through the inversion layer channel formed on the vertical side-walls of the trench due to the applied gate bias. It must then flow from point 'B' in the cross-section through the first JFET region which constricts the current into location 'A' shown in the cross-section. The current then spreads into the N-drift region at a 45 degree angle and becomes uniform through the rest of the structure.

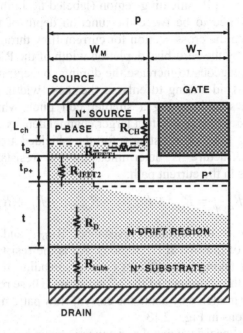

Fig. 12.13 Current flow path in the shielded trench-gate 4H-SiC power MOSFET.

The current flow path is illustrated in Fig. 12.13 by the shaded area together with the zero-bias depletion boundaries of the junctions indicated by the dashed lines. It can be seen that the addition of the P^+ shielding region introduces *two* JFET regions into the basic trench-gate MOSFET structure. The first one, labeled R_{JFET1} in the figure, is formed between the P-base region and the P^+ shielding region with the current constricted by their zero-bias depletion boundaries. The spacing between these regions (labeled t_B in the figure) must be chosen to prevent it from becoming completely depleted. For a typical N-drift region doping concentration of 1×10^{16} cm^{-3}, corresponding to a parallel-plane breakdown voltage of 1670 volts, the zero-base depletion width is about 0.6 microns. In this case, the spacing (t_B) would have to be about 1.5 microns to ensure the existence of an un-depleted path for the transport of electrons. This spacing can be reduced if the doping concentration in the JFET region is selectively increased[6] when compared with the N-drift region as indicated in Fig. 12.12.

The second JFET region, labeled R_{JFET2} in the figure, is formed between the P^+ shielding regions. Its resistance is determined by the thickness of the P^+ shielding region (labeled t_{P+} in the figure), which can be assumed to be twice the junction depth of the P^+ shielding region. Since the cross-section for current flow through this region is constricted by the zero-bias depletion width of the P^+/N junction, it is again advantageous to increase the doping concentration in the JFET region[6] to avoid having to enlarge the mesa width. A smaller mesa width allows maintaining a smaller cell pitch which reduces the specific on-resistance due to a larger channel density.

The total on-resistance for the shielded trench-gate power MOSFET structure is determined by the resistance of all the components in the current path:

$$R_{on,sp} = R_{CH} + R_{JFET1} + R_{JFET2} + R_D + R_{subs} \qquad [12.9]$$

where R_{CH} is the channel resistance, R_{JFET1} and R_{JFET2} are the resistances of the two JFET regions, R_D is the resistance of the drift region after taking into account current spreading from the channel, and R_{subs} is the resistance of the N^+ substrate. These resistances can be analytically modeled by using the current flow pattern indicated by the shaded regions in Fig. 12.13.

The specific channel resistance is given by:

$$R_{CH} = \frac{(L_{CH} \cdot p)}{\mu_{inv} C_{ox}(V_G - V_T)} \qquad [12.10]$$

where L_{CH} is the channel length determined by the width of the P-base region as shown in Fig. 12.13, μ_{inv} is the mobility for electrons in the inversion layer channel, C_{ox} is the specific capacitance of the gate oxide, V_G is the applied gate bias, and V_T is the threshold voltage. The specific capacitance can be obtained using:

$$C_{ox} = \frac{\varepsilon_{ox}}{t_{ox}} \qquad [12.11]$$

where ε_{ox} is the dielectric constant for the gate oxide and t_{ox} is its thickness.

The specific resistance of the first JFET region can be calculated using:

$$R_{JFET1} = \rho_{JFET} \cdot p \cdot \left(\frac{x_{P+} + W_P}{t_B - 2W_P} \right) \qquad [12.12]$$

where ρ_{JFET} is the resistivity of the JFET region, x_{P+} is the junction depth of the P^+ shielding region, and W_P is the zero-bias depletion width *in the JFET region*. The resistivity and zero-bias depletion width used in this equation must be computed using the enhanced doping concentration of the JFET region. The specific resistance of the second JFET region can be calculated using:

$$R_{JFET2} = \rho_{JFET} \cdot p \cdot \left(\frac{t_{P+} + 2W_P}{W_M - x_{P+} - W_P} \right) \qquad [12.13]$$

The drift region spreading resistance can be obtained by using:

$$R_D = \rho_D \cdot p \cdot \ln \left(\frac{p}{W_M - x_{P+} - W_P} \right) + \rho_D \cdot (t - W_T - x_{P+} - W_P)$$

$$[12.14]$$

where t is the thickness of the drift region below the P^+ shielding region and W_T, W_M are the widths of the trench and mesa regions, respectively, as shown in the figure.

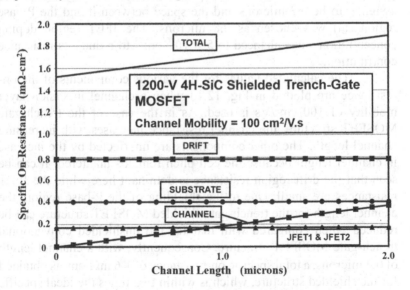

Fig. 12.14 On-resistance components for a 4H-SiC shielded trench-gate MOSFET structure.

The contribution to the resistance from the N$^+$ substrate is given by:

$$R_{subs} = \rho_{subs} \cdot t_{subs} \qquad\qquad [12.15]$$

where ρ_{subs} and t_{subs} are the resistivity and thickness of the substrate, respectively. A typical value for this contribution is 4×10^{-4} Ω-cm^2.

The specific on-resistances of 1200-V rated 4H-SiC shielded trench MOSFETs were modeled using the above analytical expressions. An ideal breakdown voltage of 1670 volts is obtained for a 4H-SiC drift region with doping concentration of 1×10^{16} cm^{-3}. The device breakdown voltage is then 1336 volts if the edge termination limits the breakdown to 80% of the ideal value. This provides some margin for a 1200-V rated device. At this blocking voltage, the depletion width is 11.35 microns. A drift region thickness (t) of 12 microns was therefore selected. A gate oxide thickness of 0.05 microns was used. The mesa width (W_M) and trench width (W_T) for the structure shown in Fig. 12.2 were kept at 1.0 microns and 0.25 microns resulting in a pitch (p) of 1.25 microns. The wider mesa region for the shielded trench-gate structure is required to allow for the depletion region from the shielding region. This enlarges the cell pitch. The gate bias voltage (V_G) was assumed to be 15 volts with a threshold voltage (V_T) of 5 volts. The junction depth of the P$^+$ shielding region was assumed to be 0.2 microns and the space between it and the P-base region (t_B) was chosen as 0.8 microns. The JFET region doping concentration was enhanced to 5×10^{16} cm^{-3} to reduce its resistance contribution.

The calculated values for the various components of the on-resistance are plotted in Fig. 12.14 when a channel inversion layer mobility of 100 cm^2/Vs is used. As in the case of the trench-gate MOSFET structure, the channel resistance increases with increasing channel length. The other components are unaffected by the increase in channel length because the cell pitch remains unaltered. It can be seen that the drift region resistance is dominant here while the JFET resistances are small. Due to the shielding of the P-base region, the channel length for the trench-gate shielded MOSFET structure can be reduced when compared with that for the unshielded conventional trench-gate MOSFET structure. Consequently, with a channel length of 0.4 microns, a total specific on-resistance of 1.6 mΩ-cm^2 is obtained for the shielded structure, which is within two times the ideal specific on-resistance of the drift region.

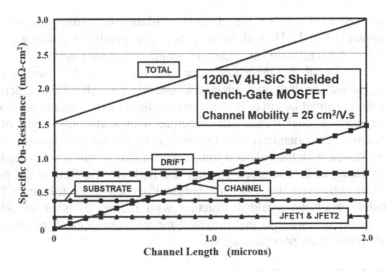

Fig. 12.15 On-resistance components for a 4H-SiC shielded trench-gate MOSFET.

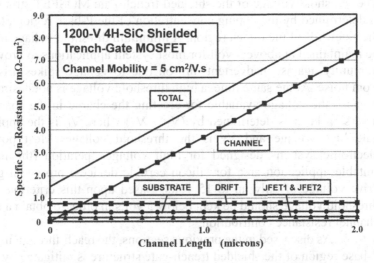

Fig. 12.16 On-resistance components for a 4H-SiC shielded trench-gate MOSFET.

The impact of reducing the inversion layer mobility for the shielded trench-gate MOSFET structure is depicted in Figs. 12.15 and 12.16 for the case of mobility values of 25 and 5 cm²/V-s. With an inversion layer mobility of 25 cm²/Vs, the channel contribution remains smaller than the contribution from the drift region. This results in the specific on-resistance increasing to a value of 1.8 mΩ-cm² for a channel length of 0.4 microns. When the inversion layer mobility is reduced to 5 cm^2/Vs, the channel resistance becomes dominant as shown in Fig. 12.16, with an increase in the specific on-resistance to a value of 3.0 mΩ-cm^2 for a channel length of 0.4 microns.

In all the inversion layer mobility cases, the specific on-resistances of the shielded trench-gate 4H-SiC MOSFET structure with a channel length of 0.4 microns are found to be close to those for the conventional trench-gate structure with a channel length of 1 micron. This demonstrates the ability to obtain low specific on-resistance with the shielded trench-gate structure while resolving the problems of P-base reach-through and high electric field in the gate oxide observed for the conventional trench-gate structure. The behavior of the shielded structure with respect to other blocking voltages is similar to that provided for the conventional-trench gate structure in the previous section.

12.3.4 Threshold Voltage

The threshold voltage of the shielded trench-gate MOSFET structure is determined by the doping concentration of the P-base region along the sidewalls of the trench region. A minimum threshold voltage must be maintained at above 3 volt for most system applications to provide immunity against inadvertent turn-on due to voltage spikes arising from noise. At the same time, a high threshold voltage is not desirable because the voltage available for creating the charge in the channel inversion layer is determined by $(V_G - V_T)$ where V_G is the applied gate bias voltage and V_T is the threshold voltage. Most power electronic systems designed for high voltage operation (the most suitable application area for silicon carbide devices) provide a gate drive voltage of only up to 15 volts. Based upon this criterion, the threshold voltage should be kept below 5 volts in order to obtain a low channel resistance contribution.

As discussed in the previous sections, the reach-through in the P-base region of the shielded trench-gate structure is mitigated by the reduced potential under the P-base/N-drift junction. This allows decreasing the P-base doping concentration to reduce the threshold voltage. In addition, a P$^+$ region can be incorporated under the source region to suppress the extension of the depletion region in this portion of the cell structure, as previously described in this chapter. A relatively thin, lightly doped P-base region is then formed adjacent to the trench sidewalls to determine the threshold voltage. The presence of the P$^+$ region also suppresses reach-through within the adjacent

P-base region allowing the structure to support high drain voltages up to the breakdown voltage capability of the drift region.

12.4 Experimental Results

The lack of significant diffusion of dopants in silicon carbide motivated the investigation of trench-gate structures before planar-gate structures. The need to obtain a high quality interface between silicon carbide and the gate dielectric was identified as a challenging endeavor from the inception of interest in the development of unipolar transistors from this semiconductor material[15]. In addition, the problem of high electric fields developed across the gate oxide, especially at the trench corners, was identified as a limitation to achieving high electric fields within the semiconductor[10]. These issues were indeed found to be a major limitation on the performance of the first trench-gate 4H-SiC power MOSFET structures.

The first reported[5] trench-gate power MOSFETs fabricated using 4H-SiC were limited to a blocking voltage of 260 volts due to rupture of the gate oxide. The specific on-resistance of 18 mΩ-cm^2, measured at a gate bias of 22 volts, was inferior to that of silicon power MOSFETs because of a large cell pitch and low channel mobility ($\sim$ 10 cm^2/V-s).

Subsequently, a trench-gate 4H-SiC power MOSFET structure with breakdown voltage of 1100 volts was reported[16] by using a drift region with thickness of 12 microns and doping concentration of 1×10^{15} cm^{-3}. The low doping concentration in the drift region was required to keep the maximum electric field in the silicon carbide to only 1×10^6 V/cm to prevent oxide rupture. This is consistent with the analysis in the previous section of this chapter. These devices exhibited a very high specific on-resistance (over 1 Ω-cm^2) at room temperature due to the poor channel mobility ($\sim$1.5 cm^2/V-s). The specific on-resistance reduced to about 180 mΩ-cm^2 (using data in Fig. 3 in the publication) due to an increase in the channel mobility. In these devices, a threshold voltage of 5 volts was obtained by using a low P-base doping concentration of 6.5×10^{16} cm^{-3} with a gate oxide thickness of 0.1 microns. However, this necessitated increasing the channel length to 4 microns to prevent reach-through problems. The large channel length, in conjunction with the poor channel mobility, was responsible for the poor (worse than silicon power MOSFETs) specific on-resistance of the devices.

A trench-gate 4H-SiC power MOSFET was reported[17] in 1998 with a breakdown voltage of 1.4 kV. The maximum electric field in the gate oxide was reduced to 2.5 MV/cm by reducing the maximum electric field in the silicon carbide to only 1.3 MV/cm. This required reducing the doping concentration of the drift region to 1×10^{15} cm^{-3} and increasing its thickness to 25 microns. This approach is consistent with Fig. 12.10 producing a huge increase in the specific on-resistance to 30 mΩ-cm^2 from an ideal value of less than 1 mΩ-cm^2. Moreover, the threshold voltage for the devices was reported to be 24-28 volts resulting in a measured specific on-resistance of 311 mΩ-cm^2 despite a gate bias of 40 volts.

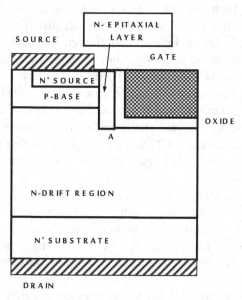

Fig. 12.17 Epitaxial-channel accumulation-mode trench-gate power MOSFET structure.

In chapter 11, an accumulation-mode planar MOSFET structure was described as an approach to reduce the specific on-resistance due to an increase in the channel mobility. An accumulation-mode structure that combines a trench-gate structure with a depleted N-base region located adjacent to a P-N junction has been proposed[18] and experimentally demonstrated. This structure, shown in Fig. 12.17, contains an N-base region that is epitaxially grown on the trench sidewall surface. The authors observed an accumulation layer mobility of 108 cm^2/V-s for this structure resulting in a specific on-resistance of 11 $m\Omega$-cm^2 at a gate bias of 10 volts. The breakdown voltage of 500 volts for this structure is still limited by the development of high electric field in the gate oxide.

Experimental results for inversion-mode trench-gate 4H-SiC power MOSFETs with a P^+ shielding region incorporated at the bottom of the trench were first reported[19] in 2002. These devices had all the features previously described in 1995[6]. The structures were fabricated using 50 micron thick N-type drift regions with doping concentration of 8.5 × 10^{14} cm^{-3}. An N-type layer with doping concentration of 2 × 10^{17} cm^{-3} was grown on the drift region to provide the enhanced doping in the JFET region. The P-base region was formed by growth of a 1 micron thick P-type layer with doping concentration of 2 × 10^{17} cm^{-3} followed by the growth of an N^+ source region with doping concentration of 1 × 10^{19} cm^{-3}. The P^+ shielding regions were formed (after etching the trenches) with a junction depth of 0.8 microns by using aluminum ion implantation with a dose of 4 × 10^{13} cm^{-2}. The authors used two sacrificial thermal oxidation steps to remove any surface residue remaining from the implant anneal step. This may have also removed any P-type doping on the trench sidewalls. A relatively thick gate oxide of 0.275 microns was prepared by the thermal oxidation of a deposited layer of polysilicon into the trench to avoid the non-uniformity of thermally grown oxide. The thick gate oxide resulted in a high threshold voltage of 40 volts. The devices exhibited a breakdown voltage of 3000 volts by the use of a JTE edge termination. A specific on-resistance of 120 $m\Omega$-cm^2 was observed at a gate bias of 100 volts. The relatively high value is due to the poor channel density in the cell design, the large gate oxide thickness, and the low inversion layer channel mobility (2 cm^2/V-s) observed by the authors. However, these results demonstrated the ability to support high voltage in the drift region without encountering gate oxide rupture in a trench-gate device. The authors subsequently reported[20] the fabrication of devices using 115 micron thick N-type drift regions with doping concentration of 7.5 × 10^{14} cm^{-3}. These devices exhibited a specific on-resistance of 228 $m\Omega$-cm^2 at a gate bias of 40 volts. The epitaxial layer was stated to be capable of supporting 14 kV although the actual measured breakdown voltage was only 5 kV.

The accumulation-mode trench-gate MOSFET structure with an epitaxially grown N-base region on the trench sidewalls (see Fig. 12.17) has been supplemented with shielding provided by the P-type region implanted at the bottom of the trenches to prevent gate oxide rupture[21]. These devices were fabricated using 10 micron thick N-type drift regions with doping concentration of 2.5 × 10^{15} cm^{-3} to obtain a breakdown voltage of 1400 volts. The rest of the process was similar to that described in the previous paragraph. A gate oxide thickness of 0.13 microns was formed by thermal oxidation of a polysilicon layer deposited in the trenches. A specific on-resistance of 16 mΩ-cm^2 was observed at a gate bias of 40 volts. This improved specific on-resistance was correlated with an accumulation layer mobility of 9-30 cm^2/V-s by the authors. Accumulation-mode shielded trench-gate MOSFETs fabricated from 4H-SiC were also reported[19] with breakdown voltage of 3360 volts. These devices had a specific on-resistance of 199 mΩ-cm^2 at a gate bias of 100 volts, which was worse than that for the inversion-mode structures. In addition, the leakage current for the accumulation-mode structure was reported to be about 100 times worse than for the inversion-mode structure.

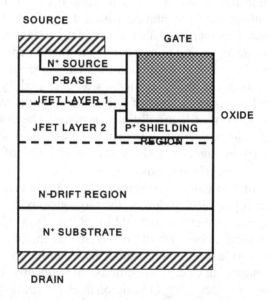

Fig. 12.18 Shielded trench-gate power MOSFET structure with two JFET layers.

An improvement in performance of the shielded trench-gate 4H-SiC power MOSFET was achieved by using two JFET layers[22] as illustrated in Fig. 12.18. The goal was to reduce the JFET resistance (R_{JFET1}) by enhancing the doping concentration between the P-base region and the P$^+$ shielding regions by using JFET layer 1 and to reduce the JFET resistance (R_{JFET2}) between the P$^+$ shielding regions by using JFET layer 2. The authors used a JFET layer 1 doping concentration of 0.5-1 $\times$ 10^{17} cm^{-3} with a thickness of 0.5 microns and a JFET layer

2 doping concentration of 1-2 × 10^{16} cm^{-3} with a thickness of 3 microns. A blocking voltage of 1600V was observed for the case of a drift layer with doping concentration of 3-5 × 10^{16} cm^{-3} with a thickness of 25 microns. The JFET layers must be removed from the edge termination region to allow application of the JTE region. A specific on-resistance of 50 mΩ-cm^2 was observed for devices when a gate bias of 60 volts was applied. The high gate bias was required due to the large gate oxide thickness of 0.16 microns. An important conclusion made in this work was that the inversion layer mobility is 20-times larger when the trench is oriented orthogonal to the wafer flat because of conduction along the (112-0) plane as opposed to the (1100) plane when the trenches are oriented parallel to the flat.

It is necessary to connect the shielding P$^+$ region to the source contact to achieve the desired shielding of the gate oxide. The contact to the shielding P$^+$ region must be performed at selected locations orthogonal to the cross-section shown in Fig. 12.18. Various layout designs for this connection have been explored[23]. It was found that at least one ground for every 9 cells was required to get good SC-SOA.

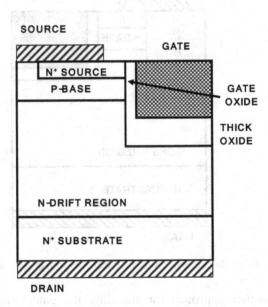

Fig. 12.19 Trench-gate power MOSFET structure with thick trench bottom oxide.

The concept of using a thick oxide at the bottom of the trench, as shown in Fig. 12.19, has been adopted for silicon power MOSFETs to reduce the gate-drain capacitance and charge[24,25]. This idea has been extended to 4H-SiC trench-gate power MOSFETs to also reduce the electric field in the gate dielectric[26]. In the case of a device with breakdown voltage of 1400-V based on a drift region doping concentration of 6×10^{15} cm^{-3} with a thickness of 13 microns, the electric field in the thick bottom oxide (thickness not provided) was found to be reduced to 5.0 MV/cm compared with 7.8 MV/cm for case with thin bottom oxide. Although this is insufficient for reliable operation, the authors found that it prevented catastrophic oxide breakdown. The edge termination for the devices were floating field rings created by etching trenches through the epitaxially grown P-base layer. This concept was first proposed[27] and patented in 1993. A specific on-resistance of 4.6 mΩ-cm^2 was observed at a gate bias of 20-V for devices with gate oxide thickness of 0.075 microns.

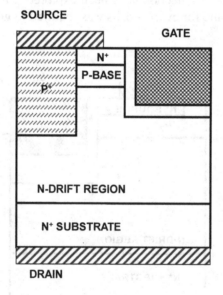

Fig. 12.20 Trench-gate power MOSFET structure with deep P$^+$ shielding region.

Another approach for shielding the gate oxide from high electric fields generated in the silicon carbide drift region is to incorporate a deep P$^+$ region in the mesa region[28] as illustrated in Fig. 12.20. The deep P$^+$ regions create a JFET under the trench with a potential barrier that reduces the electric field at the bottom of the trenches. Numerical simulations demonstrated that the electric field in the oxide could be reduced to less than 3 MV/cm by making the P^+ region 1.2 microns deeper than the bottom of the gate trench. A specific on-resistance of 1 mΩ-cm^2 was projected for a 660-V device and 7 mΩ-cm^2 for a 3300-V device. This structure was experimentally proven[29] to yield 4H-SiC devices with specific on-resistance of 0.79 mΩ-cm^2 for a breakdown voltage of 600 volts. 1200-V devices with this structures were subsequently reported[30] with specific on-resistance of 4 mΩ-cm^2 at a gate bias of 18 volts. Devices with an active area of 0.1 cm^2 were demonstrated to withstand avalanche stress test up to a current of 40 amperes.

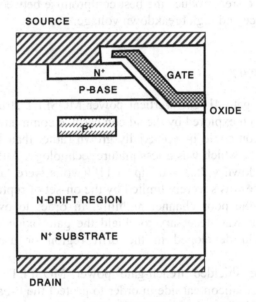

Fig. 12.21 V-groove trench-gate power MOSFET structure with P^+ shielding region.

The trench 4H-SiC power MOSFET structure shown in   Fig. 12.21 has been developed using V-groove etching by thermochemical etching in a chlorine ambient[31]. The process exposes the (0-33-8) surface of 4H-SiC which the authors report have superior MOS interface properties. The device fabrication consists of first growing the n-type drift layer with doping concentration of 4.5 × 10^{15} cm^{-3} with a thickness of 12 microns. The buried P^+ regions are then formed by ion implantation of Aluminum followed by growth of a second epitaxial layer with doping concentration of 7 × 10^{15} cm^{-3} with a thickness of 3 microns. The P-base and N^+ source regions are created by Al and P ion implantation to obtain a channel length of 0.6 microns on the trench sidewalls. The P^+ regions protect the gate oxide from high electric fields generated in the drift region. Devices with the buried P+ regions displayed a breakdown voltage of 1700-V compared with only 575-V without them due to rupture of the gate oxide. A specific on-resistance of 3.6 mΩ-cm^2 was observed at a gate bias of 18 volts with the P^+ buried grid versus 3.1 mΩ-cm^2 without the P^+ buried grid. The channel inversion layer mobility along the trench sidewalls was reported[32] to be 80 cm^2/V-s. A hexagonal layout design for these devices was optimized[33] by varying the area occupied by the P^+ buried regions. It was found that the designs where the P^+ regions occupy 30% of the active area produce the best compromise between low specific on-resistance and high breakdown voltage.

12.5 Summary

The trench-gate 4H-SiC vertical power MOSFET structure was the first approach explored by the silicon carbide community because the P-base region could be epitaxially grown rather than formed by ion implantation, which was a less mature technology. Although devices with breakdown voltages of up to 1100 volts were fabricated, their performance was severely limited by the on-set of rupture of the gate oxide and the poor channel mobility. In order to overcome these problems, it was necessary to shield the gate oxide from the high electric field developed in the drift region or use high-k gate dielectrics.

The shielded trench-gate power MOSFET structure was proposed for silicon carbide in order to protect the P-base region and the gate oxide from the high electric field generated in the drift region during the blocking mode. The shielding can be provided by the addition of a P-type region located at the bottom of the trenches. This P-type shielding region must be grounded by connection to the source electrode. Its presence creates two JFET regions in the trench gate structure, which can increase the on-resistance unless the doping concentration in the vicinity of the trenches is enhanced. With the enhanced doping concentration, it has been found that the full blocking voltage capability of the drift region in 4H-SiC can be utilized with low electric field at the P-base region and the gate oxide. This provides a very attractive power MOSFET structure that can be made from 4H-SiC with specific on-resistance approaching the ideal specific on-resistance of the drift region if a channel mobility of over 25 cm^2/Vs is achieved.

Several other methods for shielding the P-base region and the gate oxide have been proposed and demonstrated. The most promising among them is to use a P^+ region located in the mesa which is deeper than the gate trench region. Another approach is to make use of buried P^+ regions formed by the growth of an epitaxial layer in which the junctions and gate structure can be formed. It can be combined with making a V-groove shaped trench gate structure to yield good device breakdown voltage and specific on-resistance. These innovations allow producing 4H-SiC power MOSFETs that are superior in performance to the planat-gate devices.

References

[1] B. J. Baliga, "Fundamentals of Power Semiconductor Devices", Chapter 6, pp. 279-503, Springer-Science, 2008.

[2] B. J. Baliga, "Evolution of MOS-Bipolar Power Semiconductor Technology", Proceeding of the IEEE, Vol. 74, pp. 409-418, 1988.

[3] D. Ueda, H. Takagi and G. Kano, "A New Vertical Power MOSFET Structure with Extremely Reduced On-Resistance", IEEE Transactions on Electron Devices, Vol. 32, pp. 2-6, 1985.

[4] H-R. Chang *et al.*, "Ultra-Low Specific On-Resistance UMOSFET", IEEE International Electron Devices Meeting, Abstract 28.3, pp. 642-645, 1986.

[5] J. W. Palmour *et al.*, "4H-Silicon Carbide Power Switching Devices", Silicon Carbide and Related Materials – 1995, Institute of Physics Conference Series, Vol. 142, pp. 813-816, 1996.

[6] B. J. Baliga, "Silicon Carbide Switching Device with Rectifying Gate", U. S. Patent 5,396,085, Issued March 7, 1995.

[7] B. J. Baliga, "Silicon Carbide Power MOSFET with Floating Field Ring and Floating Field Plate", U. S. Patent 5,233,215, Issued August 3, 1993.

[8] S-H. Ryu *et al.*, "Design and Process Issues for Silicon Carbide Power DiMOSFETs", Material Research Society Symposium Proceeding, Vol. 640, pp. H4.5.1-H4.5.6, 2001.

[9] B. J. Baliga, "Critical Nature of Oxide/Interface Quality for SiC Power Devices", INFOS'95, Paper 1.1, June 1995. Published in Microelectronics Journal, Vol. 28, pp. 177-184, 1995.

[10] M. Bhatnagar, D. Alok and B. J. Baliga, "SiC Power UMOSFET: Design, Analysis, and Technological Feasibility", Silicon Carbide and Related Materials – 1993, Institute of Physics Conference Series, Vol. 137, pp. 703-706, 1994.

[11] S. Sridevan, P. K. McLarty and B. J. Baliga, "Silicon Carbide Switching Devices having Near Ideal Breakdown Voltage Capability and Ultra-Low On-State Resistance", U. S. Patent 5,742,076, Issued April 21, 1998.

[12] S. Sridevan, P.K. McLarty and B.J. Baliga, "Analysis of Gate Dielectrics for SiC Power UMOSFETs", IEEE International Symposium on Power Semiconductor Devices and ICs, pp. 153-156, 1997.

[13] G.D. Wilk, R.M. Wallace and J.M. Anthony, "High-k Gate Dielectrics: Current Status and materials Properties Considerations", Journal of Applied Physics, Vol. 89, pp. 5243-5275, 2001.

[14] L.A. Lipkin and J.W. Palmour, "Insulator Investigation on SiC for Improved Reliability", IEEE Transactions on Electron Devices, Vol. 46, pp. 525-532, 1999.

[15] B. J. Baliga, "Impact of SiC on Power Devices", Proceedings of the 4th International Conference on Amorphous and Crystalline Silicon Carbide", pp. 305-313, 1991.

[16] A. K. Agarwal *et al.*, "1.1 kV 4H-SiC Power UMOSFETs", IEEE Electron Device Letters, Vol. 18, pp. 586-588, 1997.

[17] Y. Sugawara and K. Asano, "1.4kV 4H-SiC UMOSFET with ow Specific On-Resistance", IEEE International Symposium on Power Semiconductor Devices and ICs, pp. 119-122, 1998.

[18] K. Hara, "Vital Issues for SiC Power Devices", Silicon Carbide and Related Materials – 1997, Materials Science Forum, Vols. 264-268, pp. 901-906, 1998.

[19] Y. Li, J.A. Cooper and M.A. Capano, "High Voltage (3kV) UMOSFETs in 4H-SiC", IEEE Transactions on Electron Devices, Vol. 49, pp. 972-975, 2002.

[20] Y. Sui, T. Tsuji and J. A. Cooper, "On-State Characteristics of SiC Power UMOSFETs on 115 micron Drift Layers", IEEE Electron Device Letters, Vol. 26, pp. 255-257, 2005.

[21] J. Tan, J.A. Cooper and M.R. Melloch, "High Voltage Accumulation-Layer UMOSFETs in 4H-SiC", IEEE Electron Device Letters, Vol. 19, pp. 487-489, 1998.

[22] Q. Zhang *et al.*, "1600V 4H-SiC UMOSFETs with Dual Buffer Layers", IEEE International Symposium on Power Semiconductor Devices and ICs, pp. 159-162, 2005.

[23] R. Tanaka *et al.*, "Impact of Grounding the Bottom Oxide Protection Layer on the Short-Circuit Ruggedness of 4H-SiC Trench MOSFETs", IEEE International Symposium on Power Semiconductor Devices and ICs, pp. 75-78, 2014.

[24] B.J. Baliga, "Advanced Power MOSFET Concepts", Springer-Science, New York, 2010.

[25] M. Darwish *et al.*, "A New Power W-Gated Trench MOSFET with High Switching performance", IEEE International Symposium on Power Semiconductor Devices and ICs, pp. 100-104, 2003.

[26] H. Takaya *et al.*, "A 4H-SiC Trench MOSFET with Thick Bottom Oxide for improving characteristics", IEEE International Symposium on Power Semiconductor Devices and ICs, pp. 43-46, 2013.

[27] B.J. Baliga, "Silicon Carbide Power MOSFET with Floating Field Ring and Floating Field Plate", U.S. Patent 5,233,215, Issued August 3, 1993.

[28] S. Harada *et al.*, "Determination of Optimum Structure of 4H-SiC Trench MOSFET", IEEE International Symposium on Power Semiconductor Devices and ICs, pp. 253-256, 2012.

[29] T. Kimoto, H. Yoshioka and T. Nakamura, "Physics of SiC MOS Interface and Development of Trench MOSFETs", IEEE Workshop on Wide Band Gap Power Devices and Applications, pp. 135-138, 2013.

[30] R. Nakamura *et al.*, "1200V 4H-SiC Trench Gate Devices", Power Conversion and Intelligent Motion Conference, pp. 441-447, 2014.

[31] K. Wada *et al.*, "Fast Switching 4H-SiC V-groove Trench MOSFETs with Buried P+ Structure", IEEE International Symposium on Power Semiconductor Devices and ICs, pp. 225-228, 2014.

[32] Y. Mikamura *et al.*, "Novel Designed SiC Devices for High Power and High Efficiency Systems", IEEE Transactions on Electron Devices, Vol. 62, pp. 382-389, 2015.

[33] K. Uchida *et al.*, "The Optimized Design and Characterization of 1200 V/2.0 mΩ-cm^2 4H-SiC V-groove Trench MOSFETs", IEEE International Symposium on Power Semiconductor Devices and ICs, pp. 85-88, 2015.

Chapter 13

GaN Vertical Power HFETs

The development of vertical GaN power MOSFETs with structures similar to those of silicon or 4H-SiC devices has been hampered until recently by the lack of availability of high quality GaN substrates. Bulk GaN substrates have now become available with diameters of up to 3 inches[1]. Vertical power MOSFETs can be conceived in GaN with the same architecture as discussed in the previous chapters on 4H-SiC devices. However, it is advantageous to utilize the two-dimensional electron gas in the AlGaN/GaN heterojunction to improve the performance of the GaN devices beyond that possible with silicon or 4H-SiC devices. These structures are discussed in this chapter.

13.1 Planar GaN Power HFET Structure

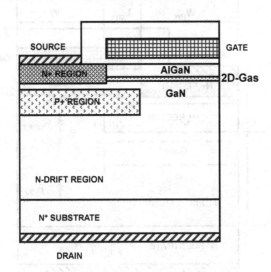

Fig. 13.1 Vertical GaN planar power MOSHFET structure.

The basic structure[2] of the vertical GaN planar power HFET is shown in Fig. 13.1. It contains buried P-type GaN regions formed by the growth of a P-type GaN layer using magnesium as the dopant. A hole is cut through the P-type epitaxial layer followed by the growth of an N-type GaN layer. The thin AlGaN layer is grown on the GaN layer to form the heterojunction and create a 2-dimensional electron gas (2D-Gas) in the GaN at the interface. The gate stack is formed by deposition of an oxide followed by polysilicon. The N-type source regions are formed by ion implantation of silicon after patterning the polysilicon. The source metal contacts are usually Ti/Al.

The formation of the p-type buried GaN regions can also be performed by using ion-implantation of magnesium. However, the damage produced by the ion implantation can lead to degradation of the AlGaN/GaN interface and reduction of the mobility and charge in the 2D-electron gas layer[3]. Even as-grown epitaxial p-type GaN buried regions have a very high resistivity which must be reduced by annealing in a N_2/O_2 ambient.

13.1.1 On-Resistance

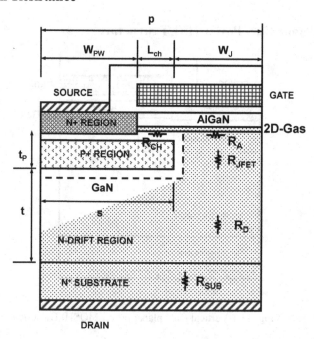

Fig. 13.2 Current flow path in the vertical planar GaN power MOSHFET.

A 2D-gas of electrons is created in the GaN planar HFET structure with no bias applied to the gate electrode. Consequently, current can flow from the drain to the source via the shaded path indicated in Fig. 13.2. The device, therefore, has normally-on characteristics. The current flows from the N^+ source region via the 2D-gas in the channel region and spreads into the JFET region via the "accumulation region" also comprised of the 2D-gas layer. After the current flows vertically down in the JFET region, it spreads into the N-drift region at a 45 degree angle and then becomes uniform through the rest of the structure.

The total on-resistance for the planar GaN power MOSHFET structure is determined by the resistance of the components in the current path shown in Fig. 13.2:

$$R_{on,sp} = R_{CH} + R_A + R_{JFET} + R_D + R_{subs} \qquad [13.1]$$

where R_{CH} is the channel resistance, R_A is the "accumulation region" resistance, R_{JFET} is the resistance of the JFET region, R_D is the resistance of the drift region after taking into account current spreading from the JFET region, and R_{subs} is the resistance of the N^+ substrate. These resistances can be analytically modeled by using the current flow pattern indicated by the shaded regions in Fig. 13.2. In this figure, the depletion region boundaries have also been shown using dashed lines.

In the GaN MOSHFET structure, the conduction in the channel region occurs via the 2D-gas layer. As previously discussed in chapter 2, the typical sheet resistance (ρ_{2-DEG}) for the 2D-gas layer reported in the literature is 300 Ω/sq. The specific channel resistance is given by:

$$R_{CH} = \rho_{2-DEG} \cdot L_{CH} \cdot p \qquad [13.2]$$

where L_{CH} is the channel length and p is the cell pitch as defined in Fig. 13.2.

In the GaN MOSHFET structure, the conduction in the "accumulation region" also occurs via the 2D-gas layer. The specific resistance of the accumulation region is given by:

$$R_A = K \cdot \rho_{2-DEG} \cdot W_J \cdot p \qquad [13.3]$$

where W_J is the width of the JFET region as defined in Fig. 13.2. The factor K is used to account for two-dimensional current spreading from the channel into the JFET region with a typical value of 0.6.

The specific JFET region resistance is given by:

$$R_{JFET} = \rho_D \cdot (t_P + W_0) \cdot \left(\frac{p}{W_J - W_0} \right) \qquad [13.4]$$

where t_p is the depth of the P^+ region below the 2D-gas layer. In this equation, W_0 is the zero-bias depletion width at the P^+/N JFET junction. It can be determined using:

$$W_0 = \sqrt{\frac{2\varepsilon_S V_{biP+}}{qN_D}} \qquad [13.5]$$

where the built-in potential V_{biP+} for the P^+/N junction is typically 3.4 volts for GaN (can be calculated using Eq. [2.8]).

The drift region spreading resistance can be obtained by using:

$$R_D = \rho_D \cdot p \cdot \ln \left(\frac{p}{W_J - W_0} \right) + \rho_D \cdot (t - s - W_0) \qquad [13.6]$$

where t is the thickness of the drift region below the P-base region and s is the width of the P-base region.

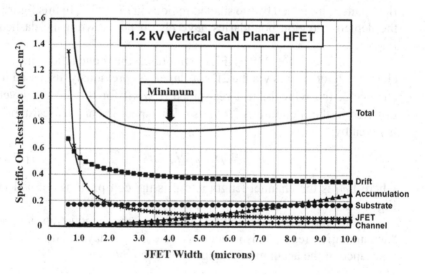

Fig. 13.3 Impact of JFET width on the specific on-resistance for the vertical GaN planar MOSHFET.

The contribution to the resistance from the N^+ substrate is given by:

$$R_{subs} = \rho_{subs} \cdot t_{subs} \qquad [13.7]$$

where ρ_{subs} and t_{subs} are the resistivity and thickness of the substrate, respectively. Freestanding GaN substrates with resistivity of 8.5×10^{-3} Ω-cm have been reported[4]. In the case of a substrate thickness of 200 microns, a specific resistance contribution of 1.7×10^{-4} Ω-cm^2 is obtained

The specific on-resistances of the vertical GaN planar MOSHFETs with a drift region doping concentration of 2×10^{16} cm^{-3} and thickness of 10 microns can be modeled using the above analytical expressions. The drift region doping corresponds to a device with a breakdown voltage of 1200 volts if the edge termination is limited to 70% of the ideal breakdown voltage. The dimension 's' in the structure shown in Fig. 13.2 was kept at 4 micron based up on a gate window of 2 microns and a channel length of 1 micron. The sheet resistance of the 2D-electron gas layer was assumed to be 300 Ω/sq. A K-factor of 0.6 was used for current spreading in the accumulation layer. The P$^+$ region was assumed to have a depth (t_P) of 1 micron.

The various components of the on-resistance are shown in Fig. 13.3. The JFET width in the vertical GaN planar HFET structure was adjusted to obtain the lowest specific on-resistance. It can be observed that the specific on-resistance has a minimum value when the JFET width is 4 microns. When the JFET width is increased beyond this point, the contribution from the "accumulation region" starts to produce a larger specific on-resistance for this structure. It is worth pointing out that the channel resistance contribution is very small because it is formed using the 2D-electron gas with a very low sheet resistance. The minimum specific on-resistance predicted by the analytical model is 0.74 mΩ-cm^2. This value is 5-times larger than the ideal specific on-resistance of the drift region for a 1200-V GaN device. It is about 3-times smaller than that of the accumulation-mode 4H-SiC shielded gate planar power MOSFET structure and 6-times smaller than that of the inversion-mode 4H-SiC shielded gate planar power MOSFET structure. However, the specific on-resistance for the trench-gate SiC power MOSFET is close to that of the vertical GaN planar MOSHFET.

The main issue with the vertical GaN planar MOSHFET structure is that it is a normally-on or depletion-mode device. A negative gate bias must be used to operate the device in the blocking mode. Consequently, it must be used in a Baliga-Pair configuration (see chapter 10) to create a normally-off power switch that is acceptable for power circuits. It is possible to obtain a positive threshold voltage by using a P-type GaN gate region or by addition of Flourine ions to the oxide. These structures are discussed in the experimental results section.

13.1.2 Switching Performance

The vertical GaN planar MOSHFET structure discussed in the previous section has normally-on characteristics. It can be used to create the Baliga-pair circuit with a silicon power MOSFET. This configuration has been used to analyze its switching performance based on a hybrid circuit model[5]. The behavior was found to be very similar to that previously reported by performing numerical simulations for the Baliga-pair with a SiC JFET[6,7]. In the case of the Baliga-pair with the vertical GaN planar MOSHFET, the turn-on time was found to be about 10 ns and the turn-off time was found to be 45 ns. The authors concluded that the switching energy loss for the Baliga-pair with a vertical GaN planar HFET is 6-times smaller than for the Baliga-pair with the SiC JFET and the stand alone SiC power MOSFET.

13.2 Experimental Results

The first report[2] of a vertical GaN planar HFET structure on a freestanding GaN substrate was published in 2007. The device was fabricated by growth of an n-type drift region on the GaN substrate with a doping concentration of 1×10^{16} cm^{-3} and thickness of 3 microns. This produces a punch-through design during the blocking state according to the analysis in chapter 3. The authors did not report the breakdown voltage and showed characteristics to a drain bias of only 10 volts. A magnesium doped p-type GaN layer with thickness of 0.1 microns was grown on the n-type drift region and apertures were etched in it using Chlorine based reactive ion etching. A 0.3 micron thick n-type GaN layer was grown over the aperture followed by the AlGaN layer to create the 2D-electron gas. The device had a MOS gate structure using high temperature deposited oxide as the gate dielectric and N$^+$ polysilicon as the gate electrode as shown in Fig. 13.1. The devices had normally-on characteristics with a threshold voltage of − 16 volts. The specific on-resistance of 2.6 mΩ-cm2 was measured at zero gate bias. The authors stated that better voltage blocking may be possible by increasing the thickness of the p-type GaN current blocking layer.

An improved vertical GaN planar MOSHFET structure was reported[8] with the structure shown in Fig. 13.4. An AlN nitride layer was added over the p-type GaN layer to prevent the magnesium from being incorporated into the second n-type GaN layer. In addition, a undoped GaN (u-GaN) layer was grown over the AlN layer to prevent etching of the AlN layer during the wet cleaning process before the growth of the second GaN n-type layer. The threshold voltage was controlled by adjusting the thickness of the second n-type GaN layer. The threshold voltage decreased from –20 volts to   –10 volts when the thickness was reduced from 150 to 50 nm. A gate oxide thickness of 500 angstroms was used. The large negative threshold voltage and leakage current through the p-type GaN layer did not allow measurement of the breakdown voltage according to the authors.

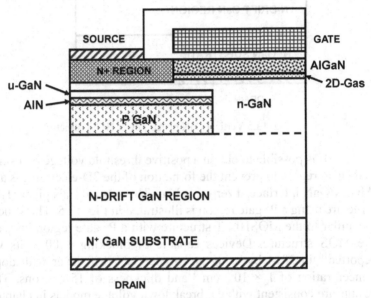

Fig. 13.4 Improved vertical GaN planar power MOSHFET structure.

The vertical GaN planar HFET structure has also been fabricated by using magnesium ion-implantation to form the p-type GaN buried layers[9]. In this process, the magnesium ion-implantation was performed using an SiO_2 mask. The ion-implant was activated at 1280 °C in ammonia and nitrogen ambient. Devices with various JFET region width were fabricated. The specific on-resistance reduced rapidly with increasing JFET width to a value of 4 microns and then continued to decrease up to a JFET width of 10 microns. The best devices had a breakdown voltage of 250 volts with a specific on-resistance of 2.2 mΩ-cm^2.

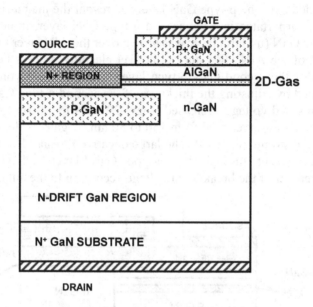

Fig. 13.5 Vertical GaN planar power HFET structure.

It is possible to obtain a positive threshold voltage by using P$^+$ GaN gate region to prevent the formation of the 2D-electron gas at the AlGaN/GaN interface at zero gate bias. The vertical GaN planar HFET structure using a P$^+$ gate region is illustrated in Fig. 13.5. The structure is similar to the MOSHFET structure with a P$^+$ gate region instead of the MOS structure. Devices capable of blocking 1500 volts were reported[10] in 2014 by using an n-type GaN drift layer with doping concentration of 1×10^{16} cm^{-3} and thickness of 15 microns. These results are consistent with the breakdown voltage models in chapter 3. The paper states: *"The thickness and doping of the p-GaN layer in the channel region and the charge in the AlGaN/GaN interface were designed to result in a positive threshold voltage."* However, a threshold voltage of only 0.5 volts was achieved, which is too low for power electronic applications. The specific on-resistance was reported as 2.2 mΩ-cm^2 at a positive gate bias of over 3 volts. A gate voltage of over 3 volts leads to substantial gate current flow due to injection of holes from the gate junction. The reported specific on-resistance for the device is about 3-times larger than the minimum value for an optimized design according to the analytical model. This indicates room for further improvement on the performance of this vertical GaN power HFET structure.

In order for GaN power devices to be used in commercial and industrial power electronics, it is necessary to bring the cost of the GaN devices into parity with the silicon IGBTs. A target of $ 0.1 per ampere has been set by the ARPA-E SWITCHES (Strategies for Wide Bandgap Inexpensive Transistors for Controlling High-Efficiency Systems) for 1.2-kV devices[11]. This effort is funding the development of the structures discussed in the chapter.

13.3 Summary

Most of the effort on GaN power devices has been devoted to lateral device structures fabricated on silicon substrates to reduce the cost. Recently, free-standing GaN substrates have become available allowing the development of vertical GaN power HFETs. The analysis of the specific on-resistance for these devices in this chapter demonstrates very promising values superior to that for the SiC power MOSFETs because of utilizing the 2D-electron gas to reduce the channel and accumulation layer resistances. Challenges to building devices with buried P$^+$ regions to achieve high blocking voltages have been overcome with demonstration of low specific on-resistances. The devices have a normally-on characteristics suitable for use in a Baliga-pair configuration.

References

[1] T. Paskova, D.A. Hanser and K.R. Evans, "GaN Substrates for III-Nitride Devices", Proceedings of the IEEE, Vol. 98, pp. 1324-13338, 2010.

[2] M. Kanechika *et al.*, "A Vertical Insulated Gate AlGaN/GaN Heterojunction Field-Effect Transistor", Japanese Journal of Applied Physics, Vol. 46, pp. L503-L505, 2007.

[3] R. Yeluri *et al.*, "Design, Fabrication, and Performance Analysis of GaN Vertical Electron Transistors with a Buried P-N Junction", Applied Physics Letters, Vol. 106, pp. 183502-1 – 183502-5, 2015.

[4] K. Kojima *et al.*, "Low-resistivity m-plane Freestanding GaN Substrate with very low point defect concentrations grown by Hydride Vapor Phase Epitaxy on a GaN Seed Crystal synthesized by the Ammonothermal Method", Applied Physics Express, Vol. 8, pp. 095501-1 – 095501-4, 2015.

[5] D. Ji and S. Chowdhury, "A Discussion on the DC and Switching Performance of a Gallium Nitride CAVET for 1.2 kV Application", IEEE Workshop on Wide bandgap Power Devices and Applications, pp. 174-179, 2015.

[6] B.J. Baliga, "Silicon Carbide Power Devices", World Scientific Press, Singapore, 2005.

[7] B.J. Baliga, "Fundamentals of Power Semiconductor Devices", Springer-Science, New York, 2008.

[8] M. Sugimoto *et al.*, "Vertical Device operation of AlGaN/GaN HEMTs on free-standing n-GaN Substrates", IEEE Power Conversion Conference, pp. 368-372, 2007.

[9] S. Chowdhury *et al.*, "CAVET on Bulk GaN Substrates achieved with MBE-regrown AlGaN/GaN Layers to Suppress Dispersion", IEEE Electron Device Letters, Vol. 33, pp. 61-63, 2012.

[10] H. Nie *et al.*, "1.5-kV and 2.2 mW-cm2 Vertical GaN Transistors on Bulk-GaN Substrates", IEEE Electron Device Letters, Vol. 35, pp. 939-941, 2014.

[11] T.D. Heidel and P. Gradzki, "Power Devices on Bulk Gallium Nitride Substrates: An Overview of ARPA-E's SWITCHES Program", IEEE International Electron Devices Meeting, pp. 2.7.1-2.7.4, 2014.

Chapter 14

GaN Lateral Power HFETs

The ability to grow gallium nitride layers on silicon substrates with sufficiently low defect density despite the large lattice mismatch has been a technological break-through that promises device cost reduction for wide bandgap power devices. These devices require a lateral configuration like the RESURF devices used to make silicon high voltage integrated circuits[1]. The prospects for lateral GaN Hetrojunction-Field-Effect-Transistor (HFET) power devices have been reviewed in several papers over the years[2,3,4,5]. One of the major advantages of the lateral GaN HFET structure is the excellent conductivity of the 2-dimensional electron gas formed at the interface between AlGaN and GaN. This has allowed rapid progress with demonstrating high voltage lateral GaN HFETs with low specific on-resistance. The physics of operation of these devices is discussed in this chapter. Analytical models are developed for performing evaluation of the performance of the devices and comparing them with silicon and silicon carbide devices.

One of the major challenges with the lateral GaN HFET structure has been its normally-on (depletion-mode) behavior, which is unacceptable for power electronics applications. Many companies use the normally-on lateral GaN HFETs in the Baliga-Pair (or cascode) configuration to create an enhancement-mode switch. As discussed in this chapter, many device structural innovations have also been proposed and demonstrated to make normally-off (enhancement-mode) devices.

Another issue that has plagued these devices is the phenomenon of dynamic on-resistance (or current collapse). The physical mechanisms responsible for this behavior are discussed in this chapter together with methods used to suppress it. This is critical to reducing power switching losses in high frequency circuits.

In addition, it has been shown that the full exploitation of the merits of lateral GaN HFETs will be possible only with the integration of multiple devices and their gate drivers. These monolithic circuits are also described here.

14.1 Lateral GaN Power HFET Structure

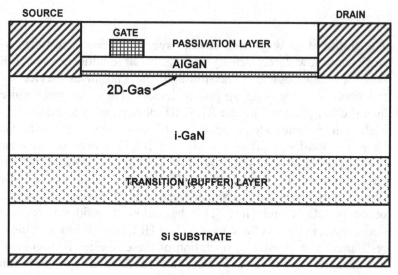

Fig. 14.1 Lateral GaN HFET structure.

The basic structure of the lateral GaN power HFET structure is illustrated in Fig. 14.1. The device structure is fabricated by the growth of a transition (or buffer) layer on (111) silicon substrates before the growth on an un-doped (i-GaN) gallium nitride layer. Some groups utilize an AlN transition layer[6,7] while other groups report using a stack of AlGaN layers[8] with different compositions of Al to relieve the strain due to lattice mismatch. The thickness of the transition and un-doped GaN layers has an impact on the maximum breakdown voltage of the lateral GaN HEMT transistors.

A thin (typically 10-20 nm thick) AlGaN layer is grown on top of the i-GaN layer. The Al composition of this layer is typically 25 percent. The piezoelectric properties of this material produces the two-dimensional electron gas in the GaN layer just below the AlGaN layer as discussed in chapter 2. The device fabrication process consists of formation of the source and drain ohmic contacts using Ti/Al/Ni/Au metal stack subjected to rapid thermal annealing at   825 °C in a nitrogen ambient[9]. The passivation layer consisting typically of silicon nitride is then deposited by plasma-enhanced chemical-vapor-deposition. A window is cut in the passivation layer followed by deposition of the Ni-Au Schottky gate contact metal. Isolation around the device can be performed by etching through the GaN layer by reactive-ion-etching or by multiple energy ion-implantation of nitrogen[9] or argon[3]. These ion-implants produce a high resistivity layer that is stable even after annealing at high temperatures[3].

The presence of the 2D-electron gas layer between the source and drain contacts allows current flow between them at zero gate bias. The specific on-resistance of the lateral GaN HFET structure is small due to the low sheet resistance of the 2D-electron gas. The device can support a high drain bias voltage if a negative gate bias is applied to remove the 2D-electron gas. The design of the breakdown voltage capability is discussed in the next section.

14.1.1 Blocking Voltage Capability

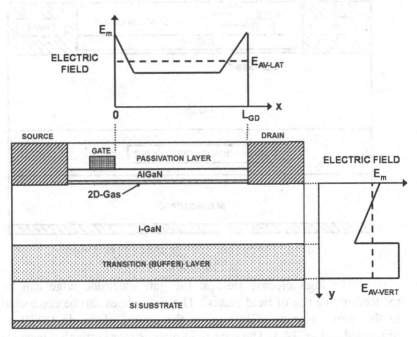

Fig. 14.2 Breakdown paths in the lateral GaN power HFET.

The maximum blocking voltage capability of the lateral GaN HFET structure can be limited by the two paths illustrated in Fig. 14.2. The first path is along the surface between the drain and the gate electrodes. The electric field along this path can be non-uniform with a large peak formed at the gate and drain electrode as shown in the figure. Impact ionization occurs when the maximum electric field (E_m) becomes equal to the critical electric field for breakdown in GaN which is about 3 MV/cm. It has been found that the breakdown voltage of actual devices varies linearly with increasing drain to gate spacing (L_{GD})[2]. Based up on this behavior, an average electric field (E_{AV-LAT}) for breakdown in lateral GaN HFET devices can be defined as illustrated by the dashed line. A typical value for this average lateral electric field for breakdown is 1 MV/cm which is considerably smaller that the critical electric field for GaN due to the high electric field peaks at the gate electrode.

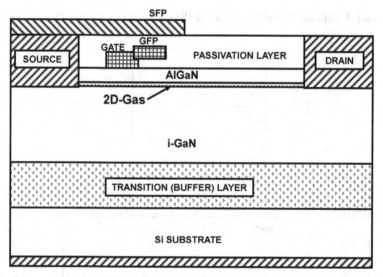

Fig. 14.3 Lateral GaN HFET structure with field plates.

The high electric field at the gate electrode edge can be reduced by the use of field plates[6]. The field plates can be connected to the gate electrode (GFP) or to the source electrode (SFP) as illustrated in Fig. 14.3. The gate field plate is more effective because it is located closer to the edge of the gate electrode with a thinner dielectric below it. Gate field plate structures with high breakdown voltages have been reported by Furukawa Electric Company[2,10]. However, the gate field plate increases the gate-to-drain capacitance which can degrade the switching performance. The source field plate is preferable because it does not impact switching performance.

Devices with source field plate structures have been reported by Toshiba Corporation[11] for devices made on silicon carbide substrates. According to this work, the electric field becomes distributed between the gate and drain electrodes with the source field plate while it is entirely confined near the gate electrode without it. By using a source field plate extending 5 microns beyond the gate edge, a breakdown voltage of 600 volts was achieved with a gate to drain spacing of 10 microns.

Many designers employ both gate and source field plates. HRL laboratories has reported a lateral GaN HFET with a gate field plate and two source field plates to achieve a breakdown voltage of 1200 volts[12]. Their measurements of the gate-to-drain capacitance shows three capacitance plateaus associated with the depletion of the 2D-electron gas under each of the field plates. Improved breakdown voltage has been demonstrated by using both a source field plate and a drain field plate[13] for the lateral GaN HFET. The double field plate structure had twice the breakdown voltage of the device with just the source field plate. A source field plate length of 4 microns was shown to be sufficient. A thickness of 0.4 microns was found to be optimum for the silicon nitride dielectric layer below the field plate.

Another method for reducing the electric field at the gate edge is to employ the silicon substrate as a field plate. This requires growth of the AlN and GaN layers on a conductive Si substrate[14]. The Si substrate is connected to the source electrode of the GaN HFET by making a via down through the GaN layer and the AlN buffer layer. It was shown in this paper by numerical simulations that the electric field at the gate edge is relaxed by the substrate field plate. A breakdown voltage of 350-V was reported for these devices with a specific on-resistance of 1.9 mΩ-cm^2.

The second path for breakdown in the lateral GaN HFET structure is between the drain and the silicon substrate because the substrate is usually connected to the source electrode as a reference point. This produces an electric field profile along the y-direction as illustrated in Fig. 14.2. It has been reported that the vertical breakdown voltage is proportional to the total epitaxial layer (i-GaN plus transition layer) thickness[2]. Based up on this, an average electric field for vertical breakdown can be defined as illustrated in Fig. 14.2. The value for this average electric field for vertical breakdown ($E_{AV-VERT}$) is 3 MV/cm which is close to the critical electric field for breakdown in GaN. This high value can be attributed to most of the voltage being supported across the AlN transition layer which has a larger critical electric field for breakdown than GaN due to its large energy bandgap.

Improvement in the breakdown voltage of lateral GaN HEMT devices has also been achieved by removal of the silicon substrate. In the first approach[15], the Si substrate was thinned to 100 microns by grinding. The wafer was the bonded to a sapphire wafer using WaferBOND HT-250. The Si substrate was then completely removed by wet etching. Vias were etched through the GaN and AlN layers to reach the source, gate and drain contacts. The breakdown voltage of devices was found to increase from 500-V with the Si substrate to over 1100-V after substrate removal without impacting the 2D-gas. In the second approach[16], the Si substrate was removed only in local regions below the gate-to-drain spacing. The breakdown voltage of devices was found to increase from 500-V with the Si substrate to over 2200-V after substrate removal without impacting the 2D-gas.

It is important to recognize that the lateral GaN HFET devices do not have any avalanche breakdown tolerance unlike silicon power devices. Operating the device close to the breakdown point leads to catastrophic failure due to dielectric breakdown[5]. Most manufacturers design the breakdown voltage of the devices to be far greater than the blocking voltage rating in their data sheets. The power electronics circuit designers for these devices must ensure that the circuit does not impose a voltage spike that could result in catastrophic failure. Various paths for the breakdown in lateral GaN HFETs have been analyzed using numerical simulations corroborated with electrical measurements[8].

14.1.2 On-Resistance

A 2D-electron gas is formed in the lateral GaN HFET structure with no bias applied to the gate electrode. Consequently, current can flow from the drain to the source via the shaded path indicated in Fig. 14.4 via the 2D-electron gas. The total on-resistance for the lateral GaN HFET structure is determined by the resistance of the components in the current path:

$$R_{on,sp} = R_{SC} + R_{GS} + R_G + R_{GD} + R_{DC} \qquad \text{[14.1]}$$

where R_{SC} is the source contact resistance, R_{GS} is the resistance between the gate and the source electrodes, R_G is the resistance under the gate electrode, R_{GD} is the resistance between the gate and the drain electrodes, and R_{DC} is the drain contact resistance. In addition, substantial contribution can arise from the current flow in the source and gate metal fingers orthogonal to the cross-section shown in the figure. This resistance can be reduced by using thick metal lines but they are difficult to fabricate with small widths for the source and drain fingers.

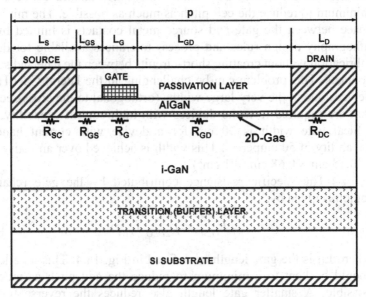

Fig. 14.4 On-resistance of the lateral GaN power HFET.

An analytical model for the specific on-resistance for the lateral GaN HFET structure can be generated assuming that the current conduction occurs via the 2D-electron gas layer. As previously discussed in chapter 2, the typical sheet resistance ($\rho_{2\text{-DEG}}$) for the 2D-electron gas layer reported[17] in the literature is 300 Ω/sq.

The source contact resistance is given by:

$$R_{SC,SP} = \frac{\rho_C \cdot p}{L_S}$$ [14.2]

where L_S is the length of the source contact, ρ_C is the specific contact resistance, and p is the cell pitch as defined in Fig. 14.4. Due to the symmetric boundary conditions on the edges of the unit cell shown in the figure, the actual physical width of the metal lines for the source contact is twice L_S.

The specific resistance contributed by the region between the gate and the source electrodes is given by:

$$R_{GS,SP} = \rho_{2DEG}.L_{GS}.p \qquad [14.3]$$

where L_{GS} is the length of the space between the gate and the source contacts as defined in Fig. 14.4. This space should be kept to a minimum to reduce the cell pitch as much as possible. The minimum space between the gate and source metal contacts is limited by the lithography design rules and process tolerances available for device fabrication without creating short-circuits between the gate and source contacts. This problem can be challenging in the lateral GaN HEFT devices due to the very large widths (orthogonal to the cross-section) that are required to make devices with high current capability. The typical gate width is 30 cm for a device with current handling capability of 50 amperes[2]. This width is achieved over an active area of 0.15 cm × 0.68 cm (0.1 cm²).

The specific resistance contributed by the gate (channel) region is given by:

$$R_{G,SP} = \rho_{2DEG}.L_G.p \qquad [14.4]$$

where L_G is the gate length as defined in Fig. 14.4. This gate length should be kept to a minimum to reduce the cell pitch as much as possible. A smaller gate length also reduces the reverse transfer capacitance (C_{GD}) which is beneficial for fast switching performance. The minimum gate width is limited by the lithography design rules and process tolerances available for device fabrication. It is typically 1 micron in size.

The specific resistance contributed by the region between the gate and the drain electrodes is given by:

$$R_{GD,SP} = \rho_{2DEG}.L_{GD}.p \qquad [14.5]$$

where L_{GD} is the length of the space between the gate and the drain contacts as defined in Fig. 14.4. As discussed in the previous section, it has been found that the breakdown voltage of lateral GaN HFETs increases linearly with increasing gate-drain spacing. The gate-drain spacing can be computed by using:

$$L_{GD} = \frac{BV}{E_{AV-LAT}} \qquad [14.6]$$

where BV is the breakdown voltage and E_{AV-LAT} is the average lateral electric field as discussed in the previous section. A typical value[2] for the average electric field in the lateral direction for GaN HFETs is 1 MV/cm. From this value, a typical gate-drain spacing for a lateral GaN HFET is found to be 6 microns in order to support 600 volts.

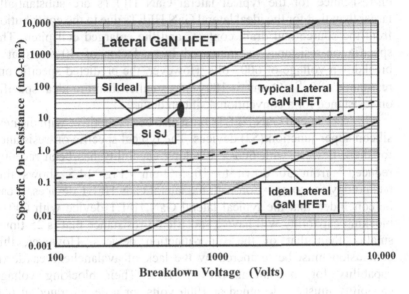

Fig. 14.5 Specific on-resistance for the lateral GaN HFET structure.

The drain contact resistance is given by:

$$R_{DC,SP} = \frac{\rho_C \cdot p}{L_D} \qquad [14.7]$$

where L_D is the length of the drain contact and p is the cell pitch as defined in Fig. 14.4. Due to the symmetric boundary conditions on the edges of the unit cell shown in the figure, the actual physical width of the metal lines for the source contact is twice L_D.

The specific on-resistances of the lateral GaN HFETs with various breakdown voltages can be modeled using the above analytical expressions. As an example, the source and drain contact lengths will be assumed to be 2 microns in size; the space between the gate and source contacts and the gate electrode length will be assumed to be

1 micron; and the space between the gate and drain will be determined using Eq. [14.6]. A sheet resistance for the 2D-electron gas layer will assumed to be 300 Ω/sq.

The specific on-resistance predicted by the analytical model is shown in Fig. 14.5 for breakdown voltages ranging from 100 to 10,000 volts. The values for the ideal silicon vertical devices and those for the ideal lateral GaN HFETs obtained using Eq. [4.19] are also included for comparison. It can be observed from this figure that the specific on-resistance for the typical lateral GaN HFETs are substantially larger than that for the ideal lateral GaN HFETs due to the contribution from the source and drain contacts and the enlarged cell pitch. The specific on-resistance is increased by a factor of 40-times at a breakdown voltage of 600 volts. However, the predicted specific on-resistance of 0.4 mΩ-cm^2 is 180-times smaller than the ideal specific on-resistance for a conventional silicon device.

The lateral GaN HFETs must also be competitive with the new silicon super-junction (SJ) devices. The typical specific on-resistance for these devices with 600-V blocking capability has been recently reduced[18] from 30 down to 10 mΩ-cm^2. Fig. 14.5 shows this technology for comparison with the lateral GaN HFET devices. It can be concluded that the typical lateral GaN HFET device with 600-V blocking capability will have a specific on-resistance that is 25-times smaller than that of the super-junction devices. However, this conclusion must be tempered by the lack of avalanche breakdown capability for the lateral GaN HFETs. Their blocking voltage capability must be designed at 1000 volts for a device rated at 600 volts. This increases the specific on-resistance to 0.74 mΩ-cm^2 resulting in a reduced advantage of 13-times over the super-junction devices.

The performance of various devices reported in the literature will be compared with the values indicated by analytical model for the typical devices in subsequent sections.

14.1.3 Gate Control

An analytical model for the change in the on-resistance of the lateral GaN HFET as a function of gate bias can be derived by analysis of the 2D-electron gas sheet charge density as a function of gate bias. In these devices, the output characteristics are commonly measured as a function of both positive and negative gate bias[2].

The sheet charge density in the 2D-electron gas has been obtained by deriving the Fermi level position in the quantum well that confines the electrons[19]. The 2D-electron gas sheet charge density (n_S) is given by:

$$n_S = \left(\frac{C_g V_{g0}}{q}\right)\left\{\frac{V_{g0} + \left(\frac{kT}{q}\right)\left[1 - ln\left(\beta V_{g0}\right)\right] - \frac{\gamma_0}{3}\left(\frac{C_{g0} V_{g0}}{q}\right)^{2/3}}{V_{g0} + \left(\frac{kT}{q}\right) + \frac{2\gamma_0}{3}\left(\frac{C_g V_{g0}}{q}\right)^{2/3}}\right\}$$ [14.8]

In this equation:

$$C_g = \frac{\varepsilon_S}{d}$$ [14.9]

$$V_{g0} = V_G - V_{OFF}$$ [14.10]

$$\beta = \frac{C_g}{kTD}$$ [14.11]

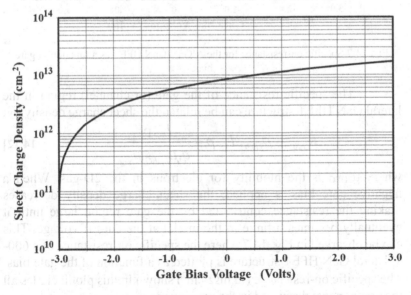

Fig. 14.6 Sheet charge density in the 2D-gas of the AlGaN/GaN HFET.

The parameter d is the thickness of the AlGaN layer (typically 20 nm); V_{OFF} is the cut-off voltage (or threshold voltage) at which the drain current goes to zero; D is the density of states (1×10^{14}/cm^2-V); and γ_0 has a value of 4.57×10^{-15} V-cm$^{4/3}$.

The sheet charge density computed by using the analytical model is provided in Fig. 14.6 for gate bias ranging from +3 to –3 volts. The sheet charge density has a value of about 1×10^{13} cm^{-3} at above zero volts. It reduces sharply near the cut-off voltage.

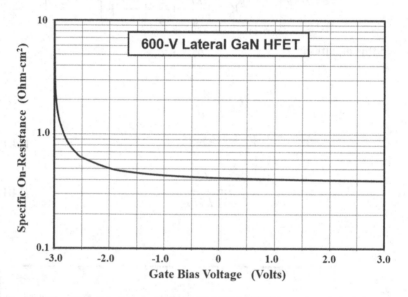

Fig. 14.7 Specific on-resistance for the AlGaN/GaN HFET as a function of gate bias.

The specific resistance of the gate (or channel) region in the lateral GaN HFET structure can be related the sheet charge density by:

$$R_{G,SP} = \rho_{2DEG} \cdot L_G \cdot p = \frac{L_G \cdot p}{q \cdot \mu_{2DEG} \cdot n_S} \qquad [14.12]$$

where μ_{2DEG} is the mobility for electrons in the 2D-gas. When a negative gate bias is applied, the sheet charge density decreases making the resistance contributed by the gate region large until it eventually becomes infinite in the model at the cut-off voltage. This can be observed in Fig. 14.7 where the specific on-resistance of a 600-V lateral GaN HFET structure is plotted as a function of the gate bias. The specific on-resistance (in mΩ-cm^2) shown in this plot includes all the components discussed in the previous section.

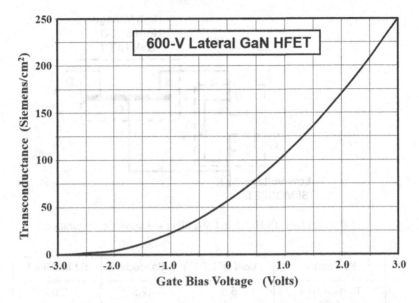

Fig. 14.8 Transconductance for the AlGaN/GaN HFET as a function of gate bias.

The transconductance of the lateral GaN HFET in its linear region of operation can be obtained from the rate of variation of the on-resistance with change in gate bias. The transconductance computed from the analytical model is provided in Fig. 14.8 for the case of a drain bias of 1 volt. As expected, the values obtained for the linear region are about one-order of magnitude smaller than the usually reported values in the current saturation region.

14.1.4 Baliga-Pair or Cascode Configuration

The lateral GaN HFET structure discussed above has normally-on characteristics that is not acceptable for power circuit applications. These devices are commercially used[20,21] in the Baliga-Pair (or Cascode) configuration as shown in Fig. 14.8. This is the same Baliga-Pair or Cascode configuration described in chapter 8 with the high voltage SiC JFET replaced by the high voltage normally-on GaN HFET. The operation of this device topology to create a normally-off power switch with an MOS-input gate drive interface compatible with gate drivers for silicon power MOSFETs and high voltage blocking capability with low specific on-resistance is described in detail in chapter 8.

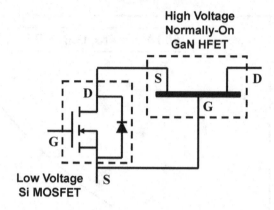

Fig. 14.9 GaN HFET in Baliga-Pair or Cascode configuration.

Parameter	GaN HFET	Cascode	SJ MOSFET
Turn-on Time (nS)	5.5	12.5	23.4
Turn-off Time (ns)	17.5	33.5	125
Turn-on Power Loss (μJ)	8.2	5.6	45.4
Turn-off Power Loss (μJ)	25.7	13.7	46.9

Table. 14.1 Comparison of switching performance.

The switching behavior of the GaN cascode circuit has been compared[22] with that for the GaN device and a silicon DMOSFET device with the same breakdown voltage of 600 volts. These lateral GaN HFETs were fabricated using a P^+ silicon substrate. The $[R_{ON}*Q_{GD}]$ figure-of-merit for the lateral GaN transistor with the silicon substrate connected to the source electrode was found to be 0.78 Ω*nC compared with 2.5 Ω*nC for the cascode configuration and 7.9 Ω*nC for the silicon DMOSFET. This demonstrates that the switching FOM for the lateral GaN HFET is superior to that for the conventional silicon DMOSFET but the performance of the cascode topology is degraded by the Q_{GD} of the silicon power MOSFET. In high frequency applications, the turn-off [dV/dt] of the switch is important because of generation of EMI noise that can disrupt circuit operation. The trade-off curve between the turn-off [dV/dt] and turn-off switching losses was found to be worse than the silicon DMOSFET with the cascode topology. The authors concluded that improved performance can be achieved by optimization of the silicon MOSFET and the stray inductance in the package.

The switching performance of the lateral GaN-on-Si HFET structure in a cascade configuration has been compared[23] with the silicon super-junction MOSFET. The data obtained by the authors with a resistive load and 400-V DC power supply are given in Table 14.1. The power losses for the cascode configuration are much lower than those observed with the silicon SJ MOSFET.

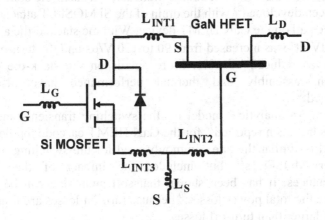

Fig. 14.10 Parasitic inductances in the GaN Baliga-Pair or Cascode configuration.

The GaN cascode topology enables increasing the switching frequency for the power circuits to the megahertz range. However, the switching performance becomes limited by the parasitic inductances[24] shown in Fig. 14.10. Several loops impact this performance during the turn-on and turn-off transients. The first loop is the Si MOSFET driving loop consisting of the gate inductance (L_G), the Si MOSFET input capacitance (C_{IN}), the internal package inductance (L_{INT3}), and the source inductance (L_S). The second loop is power loop consisting of the drain inductance (L_D), the GaN HEMT device, the internal package inductance (L_{INT1}), the Si MOSFET, the internal package inductance (L_{INT3}), and the source inductance (L_S). The third loop is the GaN driving loop consisting of the Si MOSFET, the internal package inductance (L_{INT3}), , the internal package inductance (L_{INT2}), the GaN HEMT gate-to-source, and the package inductance (L_{INT1}). Typical values for the parasitic inductances are: L_{INT1} = 0.26 nH; L_{INT2} = 0.20 nH; L_{INT3} = 0.30 nH; L_S = 0.70 nH; L_D = 1.70 nH; and L_G = 3.00

nH. It was found that these parasitic inductances produce ringing losses during the switching events which increases the power losses. The turn-on energy loss was 8 times larger than the turn-off loss at high current levels. The high turn-on losses are attributed to the discharge of the output capacitance through the channel of the GaN HEMT device and the Si MOSFET. The power losses in the Si MOSFET are about 3-times larger than in the GaN HEMT device. In addition, the common-source inductance (L_{INT3}) slows down the turn-on and turn-off transitions resulting in longer voltage and current cross-over times which enhance the switching power losses. In order to eliminate the common-source inductance, a stack-die configuration has been developed[25] with the drain of the Si MOSFET attached to the source pad on the GaN HEMT device. With the stacked-die approach, the [dV/dt] was increased from 90 to 130 V/µs and the turn-on power loss was reduced by 30%. More details on the stack-die package design, assembly and thermal performance have also been reported[26,27].

An analytical model for the switching transients and power losses has been reported[28] for the GaN HEMT cascode topology. The model is similar the one commonly used for the switching of silicon power MOSFETs[29] but includes the impact of the parasitic inductances. It has been shown that soft switching can be used to reduce the total power losses[30] because turn-on losses are found to be much larger than turn-off losses.

The interaction between the high voltage GaN transistor and the low voltage silicon transistor within the cascode topology can lead to avalanche breakdown of the silicon device during the turn-off transition[31,32]. In the cascode topology, the Si MOSFET is first turned-off by the gate bias. The voltage across the Si MOSFET increases charging the capacitances $C_{DS,Si}$ and $C_{GD,GaN}$ through the GaN transistor as illustrated in Fig. 14.11(a). After the GaN transistor source potential rises above its threshold voltage, the GaN transistor turns-off and begins to support voltage as shown in Fig. 14.11(b). Its output capacitance $C_{DS,GaN}$ is then charged in series with capacitances $C_{DS,Si}$ and $C_{GS,GaN}$ in parallel. In a properly matched Si and GaN transistor pair, the voltages across the GaN transistor reaches the drain supply voltage while the voltage across the Si transistor remains below its breakdown voltage.

If the transistors are not properly matched, the Si transistor undergoes avalanche breakdown as shown in Fig. 14.11(c) while the voltage across the GaN transistor rises to the drain supply voltage. The output capacitance $C_{DS,GaN}$ is charged during this time resulting in an undesirable power loss:

$$P_{AVAL} = BV_{Si} \cdot Q_{DS,GaN} \cdot f \qquad [14.13]$$

where BV_{Si} is the breakdown voltage of the Si transistor, $Q_{DS,GaN}$ is the stored charge in the output capacitance of the GaN transistor, and f is the switching frequency.

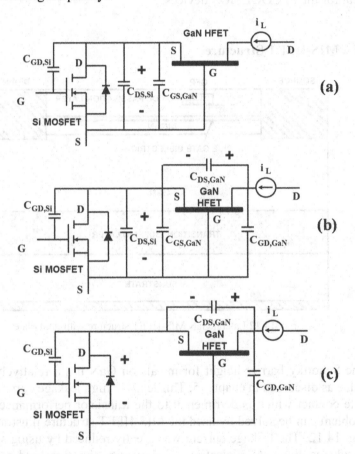

Fig. 14.11 Voltage distribution in the GaN Baliga-Pair or Cascode configuration.

A major advantage of GaN transistors is the ability to operate at higher frequencies to reduce the size of power circuits. However, this is accompanied by very high [dV/dt] and [dI/dt] transients that can disrupt proper circuit operation. The design of a gate driver for the GaN cascode topology has been reported[33] for controlling these parameters. In the case of a totem-pole circuit topology, the input-output capacitance between the high side gate driver and the common switching node was found to lead to shoot through if the bottom switch is operated as a synchronous rectifier. Any common source inductance in the bottom switch was found to lead to high voltage ringing. The [dV/dt] and [dI/dt] for the GaN cascode topology could be reduced by increasing the gate resistance but they were found to be much larger than for the Si COOLMOS devices.

14.2 MIS-HFET Structure

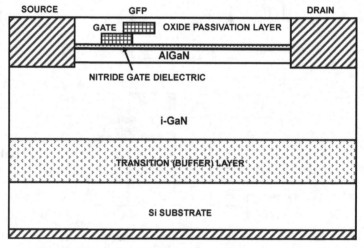

Fig. 14.12 Lateral GaN MIS-HFET structure with field plate.

The Schottky barrier height for metals on GaN has a relatively low value as discussed in chapter 5. This leads to high leakage currents for gate contact which is detrimental to the transistor performance. The problem can be solved by used the MIS-HFET structure illustrated in Fig. 14.12. The leakage current was greatly reduced by using silicon nitride as the gate dielectric[10]. However, the threshold voltage increased from –2.0 volts to –20 volts when the nitride thickness was increased from zero to 80 nm. A breakdown voltage of 1500-V was reported for a gate-drain spacing of 10 microns by using silicon dioxide as the passivation layer. A device with breakdown voltage of 1.7 kV was found to have a specific on-resistance of 5.9 mΩ-cm^2.

14.3 Normally-Off GaN HFET Structures

It is essential for the GaN transistors to have normally-off or enhancement-mode characteristics for use in power electronic circuits. Normally-on devices, such as the ones described in previous sections, create shoot through currents during powering up the circuits leading to catastrophic failures. There has been a concerted effort to develop lateral GaN HFET structures on Si substrates with normally-off behavior. These structures are discussed in this section.

14.3.1 Recessed-Gate Structure

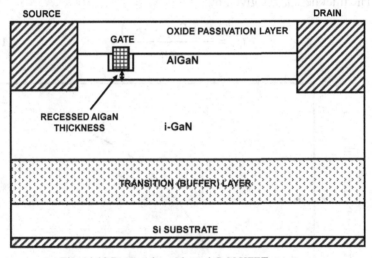

Fig. 14.13 Recessed-gate lateral GaN HFET structure.

The first approach to achieving a normally-off lateral GaN HFET is by reducing the thickness of the AlGaN layer under the gate electrode[34]. The reduction of AlGaN layer thickness alters the band structure leading to a smaller concentration of electron charge in the 2D-gas. This structure is illustrated in Fig. 14.13 with the recessed AlGaN thickness (t_{RAlGaN}) indicated below the gate electrode. The fabricated devices had an AlGaN layer thickness of 30 nm. The AlGaN layer thickness was selectively reduced in the gate region by using reactive-ion-etching. It was found that the threshold voltage without the recess was -4 volts leading to a normally-on device. The threshold voltage reduced at the rate of 0.184 volts per nm of AlGaN layer thickness (t_{RA}) reduction:

$$V_{TH} = 1.51 - \left(0.184 . t_{RA}\right) \qquad \text{[14.14]}$$

Consequently, normally-off characteristics could be achieved with a recessed AlGaN layer thickness below 8 nm.

An analytical model for the electron sheet charge in the 2D-gas layer has been developed[33]. The sheet charge density is given by:

$$n_{SR} = n_S \left(1 - \frac{t_{CR}}{t_{RA}}\right) \qquad \text{[14.15]}$$

The thickness t_{CR} is given by:

$$t_{CR} = \frac{\varepsilon_S \left(E_{DD} - \Delta E_C\right)}{q\, n_S} \qquad \text{[14.16]}$$

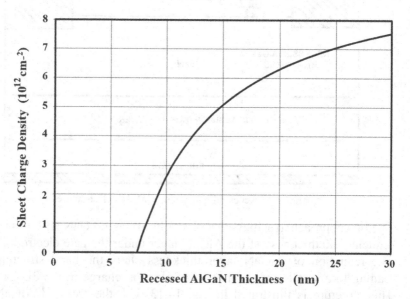

Fig. 14.14 Sheet charge density in the recessed-gate AlGaN/GaN HFET.

The band offset energies E_{DD} and ΔE_c are defined in Fig. 2.14. The value for the surface level E_{DD} is 1.65 eV for AlGaN and the value for ΔE_c is 0.37 eV for the case of $Al_{0.25}Ga_{0.75}N$. Using Eq. [14.16] in Eq. [4.15]:

$$n_{SR} = n_S - \frac{\varepsilon_S (E_{DD} - \Delta E_C)}{q \, t_{RA}}$$ [14.17]

The values for the electron sheet charge density computed by using this expression are provided in Fig.14.14. It can be observed that the sheet charge density goes to zero when the AlGaN thickness is reduced to 7.37 nm. This value corresponds to t_{CR} as given by Eq. [4.16].

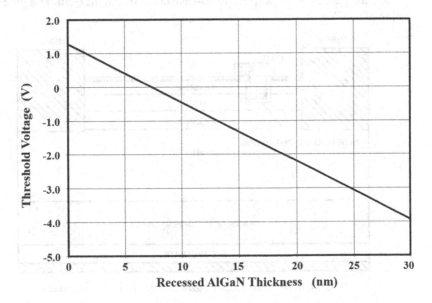

Fig. 14.15 Threshold voltage for the recessed-gate AlGaN/GaN HFET.

The threshold voltage is related to the recessed AlGaN layer thickness by:

$$V_{TH} = \frac{q \, n_S (t_{CR} - t_{RA})}{\varepsilon_S}$$ [14.18]

The threshold voltage is plotted in Fig. 14.15 as a function of the recessed AlGaN layer thickness. It can be observed that the threshold voltage is zero at the critical layer thickness and becomes positive below this value. It can be concluded that the threshold voltage is very sensitive to the AlGaN layer thickness making this method difficult to implement in high current power devices with large values for the gate width.

An alternate recessed-gate lateral GaN HFET structure[35] has been demonstrated by using the structure illustrated in Fig. 14.16. It makes use of additional GaN layer 2 and AlGaN layer 2. The recessed is etched down through the AlGaN-2 layer and partially through the GaN-2 layer. The AlGaN layer 1 is 5 nm thick with 15% Al content while AlGaN layer 2 is 10 nm thick with 25% Al content. A threshold voltage of 0.6 to 1.0 volts was obtained by decreasing the GaN-2 layer thickness from 15 to 2 nm. Devices had a breakdown voltage of 640-V at zero gate bias and a specific on-resistance of 3.8 mΩ-cm^2 at a gate bias of + 5-V.

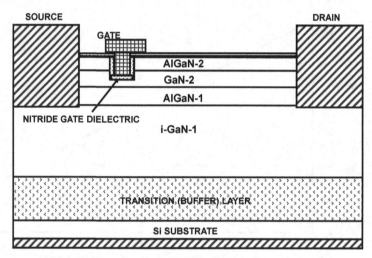

Fig. 14.16 Alternate recessed-gate lateral GaN HFET Structure.

A variation of this structure containing yet another GaN layer on top has been reported[36]. The epitaxial layers are: i-GaN of 1 micron thickness; $Al_{0.2}Ga_{0.8}N$ with 20 nm thickness; N-GaN with 2 nm thickness; i-AlN with 2 nm thickness; and n-GaN with 2 nm thickness. The gate dielectric was Al_2O_3 with a thickness of 20 nm. The threshold voltage for the structure was increased from −5.8V without the recess to +3V with an etch depth of 23 nm.

14.3.2 Fluoride-Plasma Treated Structure

The threshold voltage of the normally-on lateral GaN HFET structure has been successfully shifted from large negative values to positive values by exposing the AlGaN layer in the gate region to a CF$_4$ plasma for 150 seconds in a reactive ion etcher[37]. The plasma treatment produces incorporation of Flourine ions with negative charge into the

AlGaN layer. This has been verified by secondary ion mass spectrum (SIMS) measurements. The negatively charged Flourine ions deplete the 2D-electron gas at the AlGaN/GaN interface and shift the threshold voltage to above zero. A rapid-thermal-anneal at 400 °C for 10 minutes is required to remove damage produced by the plasma treatment without removal of the Flourine ions.

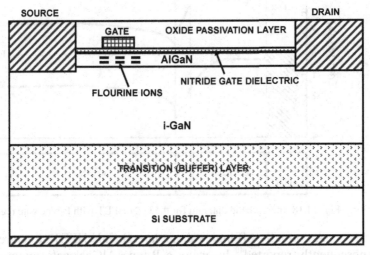

Fig. 14.17 Lateral GaN HFET Structure with Florine ions under the gate.

An improved version of this concept is shown in Fig. 14.17 where a silicon nitride gate dielectric is added[38] to reduce the gate leakage current. In this case, a CF_4 plasma treatment was performed for 250 seconds to achieve a threshold voltage of 3.6 V. The devices fabricated with a gate-drain spacing of 15 microns had a breakdown voltage of 600-V with a specific on-resistance of 2.1 mΩ-cm^2 at a gate bias of 14 V.

14.3.3 P-GaN Gate Structure

A positive threshold voltage can be achieved in the lateral GaN HFET device by making use of a P-type GaN gate region instead of the Schottky metal gate. This was first demonstrated[39] by selective epitaxial growth of a Magnesium-doped P-type GaN layer with silicon dioxide as a mask. The threshold voltage was shifted towards the positive side by about 2.5 V – the difference between the energy gap of GaN and the Schottky barrier height. The device can be turned-on by the application of a positive gate bias. The largest positive bias that can be applied becomes limited by the on-set of injection of holes across the gate P-N junction.

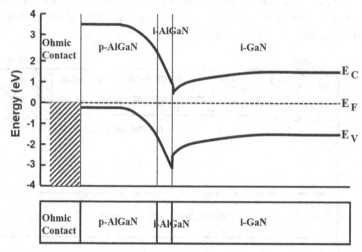

Fig. 14.18 Energy band diagram for the GaN HFET with P-type gate region.

A normally-off high voltage lateral GaN HFET structure was subsequently reported[40] by using a P-type AlGaN gate region. The energy band diagram for the gate region is illustrated in Fig. 14.18. The P-type AlGaN gate region raises the potential above the i-AlGaN layer preventing the formation of the 2D-electron gas at the i-AlGaN/GaN interface. Consequently, no drain current flows at zero gate bias due to the absence of a conductive channel under the gate region. Drain current flow can be induced by the application of a positive gate bias. For gate bias values below 3 volts, the gate P-N junction is not forward biased sufficiently to produce injection of holes. The device then operates like the conventional lateral GaN HFET structure with current flow in the channel via the 2D-electron gas. A larger drain current flow can be produced by forward biasing the gate junction above 3 volts to produce the injection of holes into the i-GaN layer below the gate. The presence of holes in the i-GaN layer attracts electrons from the source contact to preserve charge neutrality. These electrons aid the transport of current to the drain due to the high mobility of electrons in the 2D-gas between the gate and the drain. The holes remain in the gate region due to their much lower (~100-times) mobility when compared with electrons.

A 800-V normally-off lateral GaN HFET structure with P-type AlGaN gate region was successfully fabricated with a positive threshold voltage of 1 volt. This structure is illustrated in Fig. 14.19. A specific on-resistance of 2.6 mΩ-cm2 was observed for these devices. The current gain (beta) under the 'gate injection transistor' or GIT mode was found to range from 100 to 1000. The turn-off time for the devices under this mode was found to be 100 ns.

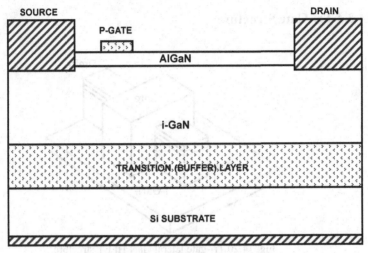

Fig. 14.19 Lateral GaN HFET Structure with P-type gate region.

Normally-off lateral GaN HFETs have been reported[41] with a P-type GaN gate region. In these devices, an $Al_{0.05}Ga_{0.95}N$ buffer layer was added below a 10 nm thick GaN layer to increase the threshold voltage to 1.25 volts. Devices with 600-V blocking capability were fabricated using a gate-drain spacing of 15 microns to yield a specific on-resistance of $2.5 \ m\Omega\text{-cm}^2$.

Normally-off lateral GaN HFETs were reported[42] by the growth of a 50 nm thick P-type GaN layer on top of the AlGaN layer. It was found that a spurious leakage current path forms between the drain and the source in the presence of this p-type GaN layer. This path can be suppressed by etching the P-type GaN layer in the portion between the drain and source to reduce its thickness to 5 nm. Devices with positive threshold voltages of up to 4 volts were achieved. A breakdown voltage of 800-V was observed for a gate-drain spacing of 10 microns after removal of the P-type GaN layer between the gate and drain.

The temperature dependence of the performance of the GIT normally-off GaN HFET devices has been reported[43] up to 150 °C. The threshold voltage was found to reduce by only 5% with increasing temperature. The on-resistance was found to increase by 50% with temperature. The turn-on and turn-off times were found to remain independent with temperature. These results demonstrate that the GIT structure offers stable performance over a broad temperature range.

14.3.4 Tri-Gate Structure

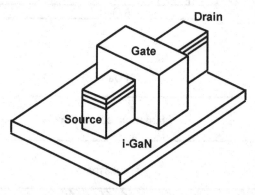

Fig. 14.20 Tri-gate lateral GaN HFET structure.

The tri-gate structure is similar to that used in silicon FINFETs to surround the channel region by the gate. The tri-gate structure for the lateral GaN HFET device is illustrated in Fig. 14.20 on the gate side[44]. A 3.3 micron thick AlGaN/GaN buffer layer is grown on a Si substrate followed by a 1.2 micron thick i-GaN layer, 18 nm thick $Al_{0.26}Ga_{0.74}N$ barrier layer, and a 2 nm GaN cap layer. A 30 nm deep recess with a length of 120 nm is etched through the upper layers. A gate dielectric of 9 nm of SiO_2/ 7 nm of Al_2O_3 is then deposited followed by Ni/Au gate metal. The devices had normally-off characteristics with a low threshold voltage of close to zero. Devices with a breakdown voltage of 565 V were reported to have a specific on-resistance of 2 mΩ-cm^2.

14.4 MOS-HFET Structure

Lateral GaN HFETs have been developed with conventional MOS-gate structures. This requires the growth of Magnesium doped P-type GaN layers on the i-GaN or the buffer layer. A trench is etched through upper AlGaN and GaN layers into the P-type GaN layer as shown in Fig. 14.21. The current between the source and drain must flow through a channel formed on the surface of the P-type GaN layer. Devices were fabricated using 60 nm thick SiO_2 deposited by PE-CVD as a gate dielectric. A threshold voltage of 2.8-V was obtained. The breakdown voltage for a device with gate-drain spacing of 15 microns was 500-V with a relatively large specific on-resistance of 16 $m\Omega$-cm^2. This indicates that the channel mobility must be poor.

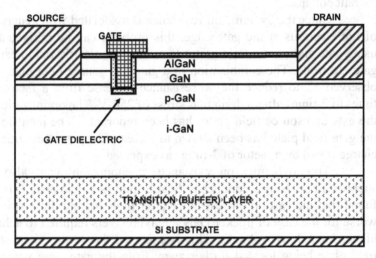

Fig. 14.21 Lateral GaN MOS-HFET structure.

14.5 Dynamic On-Resistance

One of the major problems associated with the lateral GaN HFET devices has been a phenomenon known as 'dynamic on-resistance' or 'current collapse'. This effect was first observed in microwave transistors[45]. Current collapse refers to a reduction of the drain current after application of a high drain bias voltage during switching. Dynamic on-resistance refers to an increased on-resistance in the devices after application of a high drain bias voltage during switching. In both cases, the devices recover slowly to their static characteristics leading to an increase in the switching losses. The origin of this phenomenon has been identified to be traps in the passivation dielectric between the gate and drain. During high drain bias stress, hot electrons are generated in the GaN which can be injected into the passivation layer and captured by the traps. The negative charge in the traps suppresses the 2D-electron gas near the gate electrode producing a larger on-resistance. The increase in on-resistance of commercial GaN transistors due to drain bias stress was documented in 2011 indicating that the problem had not been solved at that time[46]. The mechanisms responsible for the charge trapping have been elucidated[47].

The current collapse problems has been reduced by passivation of the gate-drain surface using silicon nitride[48]. A 100 nm thick Si_3N_4 layer deposited by PE-CVD was found to suppress the current collapse.

Since the dynamic on-resistance is associated with generation of hot electrons at the gate edge, this problem can be mitigated by using gate and source field plates to reduce the electric field near the gate electrode. The combination of a gate and source field plate was observed[12,49] to reduce the on-resistance increase from a factor 10-times to 2-times after a drain bias stress of 250-V. The optimization of the gate and source field plates has been reported[50]. The inclusion of the gate field plate has been shown to increase the gate-drain transfer charge (Q_{GD}) by a factor of 3-times as expected.

The dynamic on-resistance phenomenon can also be suppressed by using a conductive Si substrate as a field plate[51]. It was found that increase in on-resistance after high voltage drain stress is worse for the case of thicker GaN epitaxial layers required to achieve higher breakdown voltages. This is expected due to the Si substrate field plate being located further away from the gate edge when the GaN layer thickness is increased. The suppression of the increase in on-resistance was also shown to be greater by using a gate field plate in conjunction with the conductive Si substrate[10].

Four types of field plate structures have been compared with regard to suppressing the dynamic on-resistance phenomenon[52]. It was demonstrated that the gate field plate is critical to suppressing the dynamic on-resistance because it is most effective in reducing the electric field at the gate edge. Analysis of hot electron trapping at the gate edge has been performed and related to the electric field at the gate edge[53]. A 2 μm long gate field plate was found to reduce the increase in on-resistance to only 1.1-times the steady-state value.

The increase in on-resistance after stress has been correlated to the edge dislocation density[54]. On the other hand, no correlation was found with the screw dislocation density. In addition, the dynamic on-resistance has been correlated with the properties of the buffer layer[55]. The authors point out that Carbon-doped buffer layers exhibit high isolation voltages of 120 V/μm compared with Iron-doped buffer layers that exhibit isolation voltages of 50 V/μm. Unfortunately, GaN HFET devices fabricated with the Carbon-doped buffer layers exhibit very large dynamic on-resistance of 10-times the steady-state value. The on-resistance decays very slowly with a time constant of several microseconds making the switching losses high. C-doped buffer layers combined with AlGaN buffer layers provide both high isolation voltage and reduced dynamic on-resistance.

14.6 Device Passivation

The passivation of the surface between the gate and drain electrodes of the lateral GaN HFET structure has been found to impact the breakdown voltage and on-resistance. A comparison of passivation using SiO_2 and Si_3N_4 deposited using inductively coupled plasma (ICP) chemical vapor deposition (CVD) has been reported[56]. The ICP-CVD process has a remote plasma density to avoid damaging the GaN surface. It was found that silicon dioxide provided the best results with an increase in breakdown voltage from 238-V to 455-V due to suppressing the leakage current.

The optimization of silicon nitride passivation layers for lateral GaN HFETs has been reported[57]. It was found that the 2D-electron gas density increased from 8.0×10^{12} cm^{-2} to 9.2×10^{12} cm^{-2} as the Si_3N_4 thickness was increased from zero to 40 nm. In addition, annealing the passivation layer using RTA at 900 °C increased the 2D-electron gas density to 1.05×10^{13} cm^{-2}. These results are for the case of an $Al_{0.25}Ga_{0.75}N$ layer. The authors provide a model for analysis of changes in the AlGaN layer composition.

Commercialization of the lateral GaN HFET devices requires statistical data on devices fabricated in a manufacturing environment. An 'industrial process' for these devices has been reported with in-situ growth of a silicon nitride passivation and gate dielectric[58]. These devices have a 2D-elecron gas density of 9×10^{12} cm^{-2} and mobility of 1800 cm^2/V-s. A gate field plate together with two source field plates were used in the design. Measurements of the interface between the SiN layer and the AlGaN layer showed that a thin (2 monolayers) monocrystalline SiN is present which terminates the III-site dangling bonds. A very high lateral breakdown field strength of 1,5 MV/cm was observed with a floating Si substrate but the breakdown voltage was limited to 650-V at 150 °C by vertical breakdown from the drain to the

Si substrate. The on-resistance increased by a factor of 2-times for devices without field plates but was reduced to less than 20% with field plates. The output and reverse transfer capacitances decreased by 10-100 times when the drain bias was increased above 20 volts.

14.7 Chip Design

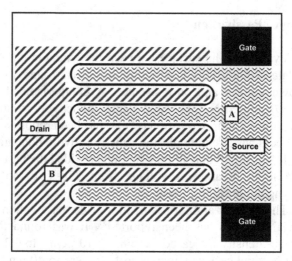

Fig. 14.22 Interdigitated finger design of lateral GaN HFETs.

The GaN HFET power devices require an interdigitated finger layout as shown in Fig. 14.22 due to the lateral configuration. The cross-sections for the devices shown in earlier sections of this chapter apply to all places along the fingers. The source and drain current is collected at each point along the fingers and must flow through the source and drain metallization to the source and drain pads. The current density in the metal has the highest value where the fingers terminate at the source (location A) and drain (location B) pads. Very high current density in the metal fingers can cause reliability problems due to electro-migration failures. The metal must be thick enough to not only prevent electro-migration but to avoid adding to the on-resistance of the device. Any significant resistance in the source metal will also de-bias the gate at the ends of the fingers producing non-uniform current flow within the lateral GaN device.

A fractal branch structure design approach[59] has been utilized to achieve homogeneous current flow in the active area of large GaN HFET devices. The drain and source metallization is designed with bus bars between the active area of the device. These bus bars are tapered to minimize the resistance of the metals and to acjieve uniform current distribution within the active area.

14.8 Summary

Significant progress has been achieved in the development of lateral GaN HFET structures with very low specific on-resistance. These devices utilize the 2D-electron gas created at an AlGaN/GaN interface with high sheet charge density and mobility. The best specific on-resistance has been obtained with normally-on device structures. These devices must be operated in the Baliga-Pair or cascode topology in power circuit applications. Normally-off lateral GaN HFETs have been achieved more recently by a variety of techniques described in this chapter. These devices are preferable for power applications but their specific on-resistance is not as good as the normally-on devices.

References

[1] B.J. Baliga, "Power Integrated Circuits – A Brief Overview", IEEE Transactions on Electron Devices, Vol. ED-33, pp. 1936-1939, 1986.

[2] N. Ikeda *et al.*, "GaN Power Transistors on Si Substrates for Switching Applications", Proceedings of the IEEE, Vol. 98, pp. 1151-1161, 2010.

[3] S. Tamura *et al.*, "Recent Advances in GaN Power Switching Devices", IEEE Compound Semiconductor Integrated Circuits Symposium, pp. 1-4, 2010.

[4] K. Boutros, R. Chu and B. Hughes, "Recent Advances in GaN Power Electronics", IEEE Custom Integrated Circuits Conference, pp. 1-4, 2013.

[5] E.A. Jones, F. Wang and B. Ozpineci, "Application-Based Review of GaN HFETs", IEEE Workshop on Wide bandgap Power Devices and Applications", pp. 24-29, 2014.

[6] W. Saito *et al.*, "600V AlGaN/GaN Power-HEMT: Design, Fabrication, and Demonstration on High Voltage DC-Dc Converter", IEEE International Electron Devices Meeting, pp. 23.7.1-23.7.4, 2003.

[7] M. Wosko *et al.*, "Different Buffer Approaches for AlGaN/GaN Heterostructures Epitaxy on Si(111) Substrates", IEEE International Conference on Advanced Semiconductor Devices and Microsystems, pp. 1-4, 2014.

[8] F.A. Marino *et al.*, "Breakdown Investigation in GaN-based MIS-HEMT Devices", European Solid State Device Research Conference, pp. 377-380, 2014.

[9] J.W. Johnson *et al.*, "12 W/mm AlGaN-GaN HFETs on Silicon Substrates", IEEE Electron Device Letters, Vol. 25, pp. 459-461, 2004.

[10] N. Ikeda *et al.*, "High-Power AlGaN/GaN MIS-HFET with Field-Plates on Si substrates", IEEE International Symposium on Power Semiconductor Devices and ICs, pp. 251-254, 2009.

[11] W. Saito *et al.*, "High Breakdown Voltage AlGaN-GaN Power-HEMT Design and High Current Density Switching Behavior", IEEE Transactions on Electron Devices, Vol. 50, pp. 2528-2531, 2003.

[12] R. Chu *et al.*, "1200-V Normally-Off GaN-on-Si Field-Effect Transistors with Low Dynamic On-Resistance", IEEE Electron Device Letters, Vol. 32, pp. 632-634, 2011.

[13] W. Saito *et al.*, "Design Optimization of High Breakdown Voltage AlGaN-GaN Power HEMT on an Insulating Substrate", IEEE Transactions on Electron Devices, Vol. 52, pp. 106-111, 2005.

[14] M. Hikita *et al.*, "AlGaN/GaN Power HFET on Silicon Substrate with Source-Via Grounding (SVG) Structure", IEEE Transactions on Electron Devices, Vol. 52, pp. 1963-1968, 2005.

[15] P. Srivastava *et al.*, "Silicon Substrate removal of GaN DHFETs for Enhanced (>1000 V) Breakdown Voltage", IEEE Electron Device Letters, Vol. 31, pp. 851-853, 2010.

[16] P. Srivastava *et al.*, "Record Breakdown Voltage (2200V) of GaN DHFETs on Si with 2 micron Buffer Thickness by Local Substrate Removal", IEEE Electron Device Letters, Vol. 32, pp. 30-32, 2011.

[17] O. Ambacher *et al.*, "Two-dimensional Electron Gas Induced by Spontaneous and Piezoelectric Polarization Charges in N- and Ga-face AlGaN/GaN Heterostructures", Journal of Applied Physics, Vol. 85, pp. 3222-3233, 1999.

[18] R. Rupp *et al.*, "Application Specific Trade-Offs for WBG SiC, GaN, and High-end Si Power Switch Technologies", IEEE International Electron Devices Meeting, pp. 2.3.1-2.3.4, 2014.

[19] S. Khandelwal, N. Goyal and T.A. Fjeldly, "A Physics based Analytical Model for 2DEG Charge Denisty in AlGaN/GaN HEMT

Devices", IEEE Transactions on Electron Devices, Vol. 58, pp. 3622-3625, 2011.

[20] Y. Wu, "600V-Class GaN Power Devices", Transphorm Exhibitor Presentation, IEEE Applied Power Electronics Conference, 2013.

[21] M.A. Briere, "Status of GaN-on-Si Development at IRF", International Rectifier Exhibitor Presentation, IEEE Applied Power Electronics Conference, 2013.

[22] W. Saito et al., "Switching Controllability of High Voltage GaN-HEMTs and the Cascode Connection", IEEE International Symposium on Power Semiconductor Devices and ICs, pp. 229-232, 2012.

[23] T. Hirose et al., "Dynamic Performances of GaN-HEMT on Si in Cascode Configuration", IEEE Applied Power Electronics Conference, pp. 174-181, 2014.

[24] A. Liu et al., "Package Parasitic Inductance Extraction and Simulation Model Development for the High-Voltage Cascode GaN HEMT", IEEE Transactions on Power Electronics, Vol. 29, pp. 1977-1985, 2014.

[25] Z. Liu et al., "Evaluation of High-Voltage Cascode GaN HEMT in Different Packages", IEEE Applied Power Electronics Conference, pp. 168-173, 2014.

[26] W. Zhang et al., "A New Package of High-Voltage Cascode Gallum Nitride Device for Megahertz Operation", IEEE Transactions on Power Electronics, Vol. 31, pp. 1344-1353, 2016.

[27] S. She et al., "Thermal Analysis and Improvement of Cascode GaN HEMT in Stack-Die Structure", IEEE Energy Conversion Congress and Exposition, pp. 57090-5715, 2014.

[28] X. Huang et al., "Analytical Loss Model of High Voltage GaN HEMT in Cascode Configuration", IEEE Transactions on Power Electronics, Vol. 29, pp. 2208-2219, 2014.

[29] B.J. Baliga, "Fundamentals of Power Semiconductor Devices", Chapter 6, Springer-Science, 2008.

[30] X. Huang et al., "Evaluation and Application of 600 V GaN HEMT in Cascode Structure", IEEE Transactions on Power Electronics, Vol. 29, pp. 2453-2461, 2014.

[31] X. Huang et al., "Characterization and Enhancement of High Voltage Cascode GaN Devices", IEEE Transactions on Electron Devices, Vol. 62, pp. 270-277, 2015.

[32] X. Huang et al., "Avoiding Si MOSFET Avalanche and achieving Zero-Voltage Switching for Cascode GaN Devices", IEEE Transactions on Power Electronics, Vol. 31, pp. 593-600, 2016.

[33] W. Zhang *et al.*, "Gate Drive Design considerations for High Voltage Cascode GaN HEMT", IEEE Applied Power Electronics Conference, pp. 1484-1489, 2014.

[34] W. Saito *et al.*, "Recessed-Gate Structure approach toward Normally-Off High-Voltage AlGaN/GaN HEMT for Power Electronics Applications", IEEE Transactions on Electron Devices, Vol. 53, pp. 356-362, 2006.

[35] M. Kuraguchi *et al.*, "Normally-off GaN-MISFET with Well-controlled Threshold Voltage", Physica Status Solidi, Vol. 204, pp. 2010-2013, 2007.

[36] M. Kanamura *et al.*, "Enhancement-Mode GaN MIS-HEMTs with n-GaN/i-AlN/n-GaN Tripple Cap Layer and High-k Gate Dielectrics", IEEE Electron Device Letters, Vol. 31, pp. 189-191, 2010.

[37] Y. Cai *et al.*, "High-Performance Enhancement-Mode AlGaN/GaN HEMTs using Flouride-based Plasma Treatment", IEEE Electron Device Letters, Vol. 26, pp. 435-437, 2005.

[38] Z. Tang *et al.*, "600-V Normally-Off SiNx/AlGaN/GaN MIS-HEMT with Large gate Swing and Low Current collapse", IEEE Electron Device Letters, Vol. 34, pp. 1373-1376, 2013.

[39] X. Hu *et al.*, "Enhancement Mode AlGaN/GaN HFET with Selectively Grown PN Junction Gate", Electronics Letters, Vol. 36, pp. 753-754, 2000.

[40] Y. Uemoto *et al.*, "Gate Injection Transistor (GIT) – A Normally-Off AlGaN/GaN Power Transistor using Conductivity Modulation", IEEE Transactions on Electron Devices, Vol. 54, pp. 3393-3399, 2007.

[41] O. Holt *et al.*, "Normally-Off AlGaN/GaN HFET with p-type Gan Gate and AlGaN Buffer", IEEE International Conference on Integrated Power Electronic Systems, pp. 1-4, 2010.

[42] L-Y. Su *et al.*, "Enhancement-Mode GaN-based High-Electron Mobility Transistors on the Si Substrate with a P-type GaN Cap Layer", IEEE Transactions on Electron Devices, Vol. 61, pp. 460-465, 2014.

[43] H. Li *et al.*, "Evaluation of 600 V GaN Based Gate Injection Transistors for High Temperature and High Efficiency Applications", IEEE Workshop on Wide bandgap Power Devices and Applications, Paper T5.8067, pp. 85-91, 2015.

[44] B. Lu, E. Matioli and T. Palacios, "Tri-Gate Normally-Off GaN Power MOSFET", IEEE Electron Device Letters, Vol. 33, pp. 360-362, 2012.

[45] S.C. Binari, B. Klein and T.E. Kazior," Trapping Effects in GaN and SiC Microwave FETs", Proceedings of the IEEE, Vol. 90, pp. 1048-1058, 2002.

[46] B. Lu *et al.*, "Extraction of Dynamic On-Resistance in GaN Transistors", IEEE Compound Semiconductor Integrated Circuits Symposium, pp. 1-4, 2011.

[47] D. Jin and J.A. del Alamo, "Mechanisms responsible for Dynamic On-Resistance in GaN High-Voltage HEMTs", IEEE International Symposium on Power Semiconductor Devices and ICs, pp. 333-336, 2012.

[48] T. Mizutani *et al.*, "A Study of Current Collapse in AlGaN/GaN HEMTs induced by Bias Stress", IEEE Transactions on Electron Devices, Vol. 50, pp. 2015-2020, 2003.

[49] W. Saito *et al.*, "On-Resistance Modulation of High Voltage GaN HEMT on Sapphire Substrate under High Applied Voltage", IEEE Electron Device Letters, Vol. 28, pp. 676-678, 2007.

[50] W. Saito *et al.*, "Suppression of Dynamic On-Resistance increase and Gate Charge Measurements in High-Voltage GaN-HEMTs with Optimized Field-Plate Structure", IEEE Transactions on Electron Devices, Vol. 54, pp. 1825-1830, 2007.

[51] N. Ikeda *et al.*, "High Power AlGaN/GaN HFET with a High Breakdown Volatge of over 1.8 kV on 4 inch Si Substrates and the Suppression of Current Collapse", IEEE International Symposium on Power Semiconductor Devices and ICs, pp. 287-290, 2008.

[52] W. Saito *et al.*, "Field-Plate Structure dependence of Current Collapse Phenomenon in High Voltage GaN HEMTs", IEEE Electron Device Letters, Vol. 31, pp. 659-661, 2010.

[53] H. Huang *et al.*, "Effects of Gate Field Plates on the Surface State related Current Collapse in AlGaN/GaN HEMTs", IEEE Transactions on Power Electronics, Vol. 29, pp. 2164-2173, 2014.

[54] W. Saito *et al.*, "Effect of Buffer layer Structure on Drain Leakage Current and Current Collapse Phenomena in High Voltage GaN-HEMTs", IEEE Transactions on Electron Devices, Vol. 56, pp. 1371-1376, 2009.

[55] J. Wurfl *et al.*, "Techniques towards GaN Power Transistors with improved High Voltage Dynamic Switching Properties", IEEE International Electron Devices Meeting, pp. 144-147, 2013.

[56] M-W. Ha *et al.*, "Silicon Dioxide Passivation of AlGaN/GaN HEMTs for High Breakdown Voltage", IEEE International Symposium on Power Semiconductor Devices and ICs, pp. 1-4, 2006.

[57] I.R. Gatabi *et al.*, "PECVD Silicon Nitride Passivation of AlGaN/GaN Heterostructures", IEEE Transactions on Electron Devices, Vol. 60, pp. 1082-1087, 2013.

[58] P. Moens *et al.*, "An Industrial Process for 650V rated GaN-on-Si Power Devices using in-situ SiN as a Gate Dielectric", IEEE International Symposium on Power Semiconductor Devices and ICs, pp. 374-377, 2014.

[59] R. Reiner *et al.*, "Fractal Structures for Low-Resistance Large Area AlGaN/GaN Power Transistors", IEEE International Symposium on Power Semiconductor Devices and ICs, pp. 341-344, 2012.

Chapter 15

SiC Bipolar Junction Transistors

The silicon bipolar transistor (BJT) was replaced by the silicon IGBT in the 1980s due its many shortcomings. These include its low current gain and poor safe-operating-area. Silicon BJTs required bulky discrete circuits to control the device and lossy snubbers to ensure fail-safe operation.

Silicon carbide BJTs have been investigated as an alternative to SiC power MOSFETs because of problems with inversion channel mobility. Recent progress with making SiC power MOSFETs with excellent specific on-resistance has made the SiC BJT less attractive from an applications stand point. SiC BJTs are discussed only briefly in this chapter for this reason but included in the book for the sake of completeness.

15.1 SiC BJT Structure and Operation

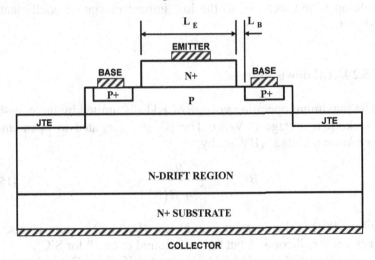

Fig. 15.1 SiC bipolar junction transistor structure.

The bipolar junction transistor (BJT) structure has been discussed in detail in the textbook[1]. This chapter will focus on highlighting the structural differences of the SiC BJT and unique aspects of device physics. The basic structure of the SiC BJT is shown in Fig. 15.1. It is fabricated by the growth of all the layers – the collector drift region, the P-base region, and the N^+ emitter region because ion-implantation results in poor minority carrier lifetime in SiC. As a consequence, the doping concentrations for all the layers can be assumed to be constant unlike in the case of silicon devices fabricated by deep diffusions. The P-base region is accessed by etching through the N^+ emitter layer. Another etching step is performed through the P-base region to make the edge terminations by the junction termination extension approach.

The devices support the blocking voltage across the drift layer. The specific resistance of the drift layer for SiC devices is much smaller than that for silicon devices as discussed in chapter 4. The relatively high doping level in the drift region for the SiC devices results in much less conductivity modulation than in Si BJTs during current flow. This eliminates the quasi-saturation region in SiC BJTs.

The SiC BJT carries current by injection of electrons into the base region in response to a base drive current. The current gain for the SiC BJT is degraded by additional factors than for silicon devices as discussed in this chapter. However, the SiC BJTs have excellent safe-operating-area due to the low impact ionization coefficients in 4H-SiC.

15.2 Breakdown Voltage

The maximum operating voltage of BJTs is limited by the open-base breakdown voltage (BV_{CEO}). The BV_{CEO} is related to open-emitter breakdown voltage (BV_{CBO}) by:

$$BV_{CEO} = \frac{BV_{CB0}}{\left[1 + \beta(0)\right]^{1/n}} \qquad [15.1a]$$

where $\beta(0)$ is the current gain at low (leakage) current levels. The factor n for silicon is 6 but has been found to be 10 for SiC[2].

The BV_{CEO} of Si BJTs is about half of its BV_{CBO} due to high values for the current gain. This requires designing the doping concentration and thickness of the N-drift region for a parallel-plane (or diode) breakdown voltage of twice the rated breakdown voltage of the BJT. In contrast, in the case of SiC BJT[3,4,5,6], the current gain at low current levels is very small leading to BV_{CE0} nearly equal to BV_{CB0}.

The drift region for SiC BJTs can therefore be designed using parallel plane breakdown consideration as discussed in chapter 3 taking into account edge terminations. This allows exploitation of the very low drift region specific on-resistance in SiC resulting in a low on-state voltage drop without conductivity modulation of the drift region.

15.3 Current Gain

The common-base current gain (alpha) for the power bipolar transistor is given by:

$$\alpha_{NPN} = \frac{\delta I_C}{\delta I_E} = \left(\frac{\delta I_{nE}}{\delta I_E}\right)\left(\frac{\delta I_{nC}}{\delta I_{nE}}\right)\left(\frac{\delta I_C}{\delta I_{nC}}\right)$$ [15.1b]

The first term is the emitter injection efficiency:

$$\gamma_E = \left(\frac{\delta I_{nE}}{\delta I_E}\right)$$ [15.2]

The second term is the base transport factor:

$$\alpha_T = \left(\frac{\delta I_{nC}}{\delta I_{nE}}\right)$$ [15.3]

The third term is the collector efficiency:

$$\gamma_C = \left(\frac{\delta I_C}{\delta I_{nC}}\right)$$ [15.4]

15.3.1 Emitter Injection Efficiency

When the base-emitter junction is forward biased, current flow across the junction occurs by the injection of electrons into the base region as well as the injection of holes into the emitter region. The emitter injection efficiency is a measure of the emitter current produced by the injection of electrons into the P-base region. The carrier distribution profiles in the base and emitter regions for the SiC BJT are illustrated in Fig. 15.2 using a linear scale for the concentrations.

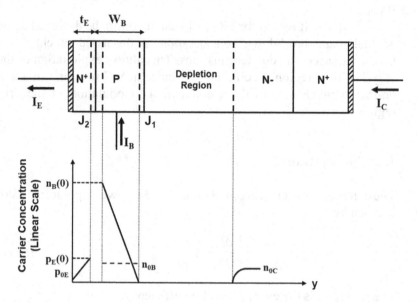

Fig. 15.2 Minority carrier distribution in the SiC N-P-N bipolar power transistor.

An equation for the injection efficiency has been derived in the textbook[1] for the case of Si BJTs. In these devices, the N^+ emitter thickness is made large enough so that the diffusion length for holes in it is much smaller than its thickness. This assumption cannot be made for the SiC BJTs because the emitter thickness is usually small (~ 1 μm)[7,8,9]. In this case, the hole concentration takes a linear distribution in the emitter as shown in Fig. 15.2 with a value of zero at the metal ohmic contact.

The injected carrier concentrations on both sides of the base-emitter junction are related to the corresponding minority carrier concentrations[1]:

$$n_B(0) = n_{0B} . e^{qV_{BE}/kT} \qquad\qquad [15.5]$$

$$p_E(0) = p_{0E} . e^{qV_{BE}/kT} \qquad\qquad [15.6]$$

where V_{BE} is the forward bias across the base-emitter junction. In these equations, n_{0B} and p_{0E} are the minority carrier concentrations in equilibrium within the P-base and N^+ emitter regions, respectively.

The hole current density flowing at the base-emitter junction for the case of SiC BJTs is given by:

$$J_p(0) = -qD_{pE}\left(\frac{dp}{dy}\right)_{y=0} = \frac{qD_{pE}P_E(0)}{t_E} = \frac{qD_{pE}P_{0E}}{t_E}.e^{qV_{BE}/kT} \qquad [15.7]$$

where t_E is the thickness of the emitter and D_{pE} is the diffusion coefficient for holes in the emitter. The electron current component for the case of no recombination in the P-base region is given by[1]:

$$J_n(0) = \frac{qD_{nB}}{W_B}n_B(0) = \frac{qD_{nB}}{W_B}n_{0B}.e^{qV_{BE}/kT} \qquad [15.8]$$

where W_B is the width of the P-base region and D_{nB} is the diffusion coefficient for electrons in the base.

The emitter injection efficiency can be obtained by using the electron and hole current components:

$$\gamma_E = \frac{J_n(0)}{J_n(0) + J_p(0)} \qquad [15.9]$$

Substituting Eq. [15.7] and Eq. [15.8]:

$$\gamma_E = \frac{D_{nB}t_E n_{0B}}{D_{nB}t_E n_{0B} + D_{pE}W_B P_{0E}} \qquad [15.10]$$

The common-emitter current gain (β_E), as determined purely by the emitter injection efficiency, can also be obtained by using the electron and hole current components of the total emitter current:

$$\beta_E = \frac{J_n(0)}{J_p(0)} \qquad [15.11]$$

Substituting Eq. [15.7] and Eq. [15.8]:

$$\beta_E = \frac{D_{nB}t_E n_{0B}}{D_{pE}W_B P_{0E}} \qquad [15.12]$$

In these equations, the diffusion coefficients must be calculated by taking into account the reduction of mobility in the highly doped base and emitter regions. More importantly, the minority carrier densities in the equation are strongly impacted by incomplete ionization of the dopants in SiC. Incomplete ionization of dopants in SiC was discussed in section 2.13.2. Consequently:

$$\gamma_E = \frac{D_{nB} t_E N_{DE}^+}{D_{nB} t_E N_{DE}^+ + D_{pE} W_B N_{AB}^-}$$

[15.13]

and

$$\beta_E = \frac{D_{nB} t_E N_{DE}^+}{D_{pE} W_B N_{AB}^-}$$

[15.14]

The ionized donor concentration in the emitter and ionized acceptor concentration in the base are used in these equations. The impact of incomplete ionization has been recognized to influence the change in current gain with temperature[10].

As an example, consider the case of a SiC BJT with typical emitter doping concentration of 1×10^{19} cm^{-3}, base doping concentration of 2.5×10^{17} cm^{-3} and P-base thickness of 1 micron[3,4,5,6]. The calculated ionized emitter donor concentration at 300 °K is 1.15×10^{17} cm^{-3} and the calculated ionized base acceptor concentration at 300 °K is 1.09×10^{16} cm^{-3}. The emitter injection efficiency related values for γ_E and β_E obtained using the above equations are 0.9957 and 232 for an emitter thickness of 1 micron. The values for γ_E and β_E can be increased to 0.9986 and 697 by increasing the emitter thickness to 3 microns. This conclusion was also suggested by Ivanov *et al.*[11]

15.3.2 Base Transport Factor

The base transport factor is a measure of the ability for the minority carriers injected from the base-emitter junction to reach the base-collector junction. It is the ratio of the electron current at the base-collector junction to the electron current at the base-emitter junction. In the case of the Si BJT, the base transport factor is determined by loss of electrons due to recombination in the P-base region[1]:

$$\alpha_T = \frac{J_{nC}}{J_{nE}} = \frac{1}{\cosh(W_B / L_{nB})}$$

[15.15]

where W_B is the width of the P-base region.

In the case of SiC BJTs, the base width is made much smaller[5,6,7] (~1 μm) than for Si BJTs. Consequently, the diffusion length for electrons is much larger than the base width leading to a base transport factor of close to unity for case of diffusion of electrons through the P-base region. However, several other recombination paths for the electrons degrade the base transport factor.

The formation of an ohmic contact to the P-base region requires an increased P-type doping concentration under the metal contact as illustrated in Fig. 15.1. One method for achieving this is by ion implantation of Aluminum. It has been found that the ion-implantation produces a low lifetime in the P^+ region despite annealing to activate the dopants. This allows some of the injected electrons from the emitter to be recombined at the P^+ contact region resulting in a degradation of the base transport factor. The current gain was found to improve by increasing the space (L_B in Fig. 15.1) between emitter and the P^+ contact region[5,12,13].

The problem of recombination at the P^+ contact region can be mitigated by growth of a P^+ contact layer instead of ion-implantation[14]. The P^+ layer is grown after etching away the emitter to expose the P-base region. The P^+ layer must be selectively removed from the top and sidewalls of the N^+ emitter region.

In addition, recombination of the injected electrons at the emitter-base junction can occur at the SiC surface of the P-base region, i.e. in the space (L_B in Fig. 15.1) between the emitter and the P^+ base contact region[15]. The surface recombination can be reduced by improved passivation of the surface between the emitter and the P^+ contact region. The passivation of the SiC surface[4] by thermally grown SiO_2 at 1240 °C in an N_2O ambient has been found to improve the current gain by a factor of 2 compared with oxide grown without the N_2O.

A subsequent study[16] of various passivation dielectrics was reported in 2011. A summary of the information from this study is provided in Table 15.1. It was found that the highest current gain was obtained by using a PECVD-N2O layer deposited at 300 °C and annealed in N_2O at 1100 °C for 3 hours to passivate the edge of the emitter and the surface of the P-base region. The high current gain was correlated with a low effective charge and interface state density as seen in Table 15.1. However, the open-base breakdown voltage for this case was reduced by 20% compared with the case of N_2O. The authors ascribe this to the influence of surface charge on the JTE edge termination that had the same passivation layer rather than the increased current gain (see Eq. [15a]).

Passivation Layer	Q_{EFF} (10^{11} cm^{-2})	D_{it} (10^{11} cm^{-2}eV^{-1})	Beta
N$_2$O	7.5	23.1	23
O$_2$	26.7	129.2	17
TEOS N$_2$O	9.1	28.7	30
PECVD N$_2$O	4.6	4.25	40

Table 15.1 Impact of passivation dielectrics on current gain of SiC N-P-N BJTs.

Once the surface recombination and recombination at the P$^+$ contact have been suppressed, the current gain can be improved by increasing the lifetime in the SiC. SiC BJTs with a maximum current gain of 257 were reported[17] by (a) first increasing the lifetime in the P-base layer by thermal oxidation in O$_2$ at 1150 °C for 5 hours before activation annealing of the ion-implants in the P$^+$ region and (b) repeating this process after annealing the ion implants in the P$^+$ region. These oxidation steps eliminate the $Z_{1/2}$ and EH$_{6/7}$ defects that are responsible for bulk recombination in 4H-SiC (see chapter 2).

15.3.3 Net Current Gain

The net current gain for the SiC BJT can be computed by taking the product of the emitter injection efficiency and the base transport factor. The collector efficiency is unity at typical collector bias voltages due to very little impact ionization. The base current in the SiC BJT supports recombination in the emitter, in the P-base region, and at the surface:

$$J_{p,T} = \frac{J_C}{\beta} = J_{p,E} + J_{p,B} + J_{p,SR} \qquad [15.16]$$

These currents can be obtained using:

$$J_{p,E} = \frac{J_C}{\beta_E} \qquad [15.17]$$

where β_E is given by Eq. [15.14].

$$J_{p,T} = \frac{J_C}{\beta_T}$$ [15.18]

β_T is given by:

$$\beta_T = \frac{\alpha_T}{1 - \alpha_T}$$ [15.19]

with α_T given by Eq. [15.15].

$$J_{p,SR} = K_{SR}.v_{SR}$$ [15.20]

where K_{SR} is the surface recombination coefficient and v_{SR} is the surface recombination velocity. Typical values for K_{SR} and v_{SR} are 5 × 10^{-4} Coulombs/cm^3 and 4000 cm/s.

Early work[3,6,8] on SiC BJTs produced devices with relatively low current gains with values below 10. With improved understanding of the role of surface recombination and the ability to enhance the bulk lifetime, SiC BJTs with high current gains of over 300 have been achieved[17].

15.4 Emitter Current Crowding

It is well known that the current crowds at the edge of the emitter fingers in silicon bipolar transistors[1]. This phenomenon is caused by the voltage drop in the P-base region produced by base current flow. Unfortunately, the effect is much worse in SiC BJTs due to high sheet resistance of the P-base region.

The sheet resistance of the P-base region is given by:

$$\rho_{SQ,PB} = \frac{\rho_{PB}}{W_B} = \frac{1}{q.\mu_p.N_A^-.W_B}$$ [15.21]

where ρ_{PB} is the resistivity of the P-base region. The mobility for holes (μ_p) in 4H-SiC (see section 2.8) is only 80 cm^2/V-s at the typical doping concentration of 2.5 × 10^{17} cm^{-3}. Furthermore, the ionized acceptor concentration (N_A^-) or majority hole carrier density in the P-base region is only 1.09 × 10^{16} cm^{-3} due to the large dopant ionization energy in 4H-SiC. Using these values together with a typical P-base thickness of 1 micron yields a very high sheet resistance of 7.17 × 10^4 Ohms/sq.

The collector current density distribution has the form[1]:

$$J_E(x) = \frac{J_E(0)}{(1 + x/x_{0LL})^2}$$ [15.22]

where x_{0LL} is a current crowding parameter under low-level injection conditions in the P-base region given by:

$$x_{0LL} = \sqrt{\frac{2kTW_B}{q(1-\alpha)\rho_{B0}J_E(0)}}$$ [15.23]

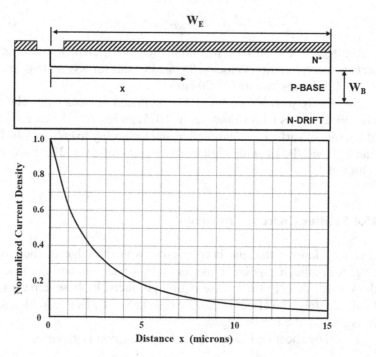

Fig. 15.3 Emitter current crowding in the SiC N-P-N bipolar power transistor.

The emitter current crowding in a typical 4H-SiC BJT is shown in Fig. 15.3 for the case of emitter fingers with a typical total width of 30 microns[3,10,11]. The current crowding parameter was found to be 3.8 microns for the case of a P-base doping concentration of 2.5 × 10^{17} cm^{-3} (ionized acceptor concentration of 1.09 × 10^{16} cm^{-3}) and thickness of 1 micron. The current gain was assumed to be 20. It can be seen from Fig. 15.3 that the current crowding is severe in 4H-SiC BJTs. They must be designed with narrow emitter fingers to avoid large amount of wasted space. However, a narrower emitter finger degrades the current gain due to recombination at the surface at the edges of the emitter[11,12,13]. The high current density at the emitter edges produces a rapid fall in current gain at high current levels in the devices[3,5,6,9].

15.5 On-State Voltage Drop

The specific resistance of the drift region in Si BJTs is large due to the low doping concentrations required to sustain high breakdown voltages. This problem is worsened by the large difference between the BV_{CE0} and BV_{CB0} values. The on-state voltage drop for the Si BJT is reduced by high level injection of electrons into the drift region. Conductivity modulation of the entire drift region results in operation in the saturation-mode[1]. If only a portion of the drift region has conductivity modulation, the device operates in the quasi-saturation region. The large drift region resistance in Si BJTs defines the boundary between the active region and the quasi-saturation region.

4H-SiC BJTs can be designed with drift regions with much larger drift region doping concentration to support the same breakdown voltage as Si BJTs. The doping concentration and thickness of the drift region for 4H-SiC was related to the breakdown voltage in chapter 3. If the impact of edge termination is neglected, the drift region specific resistance is given by:

$$R_{D,SP}(Ideal) = \frac{W_{PP}}{q.\mu_n.N_D} \qquad [15.24]$$

where the thickness of the drift region was assumed to be the parallel-plane breakdown value. The doping concentration of the drift region (N_D) can be used in this equation because all the donors are ionized at low doping levels. This ideal specific on-resistance for the drift region is plotted in Fig. 4.3.

In the case of actual fabricated SiC BJTs, the specific resistance of the drift region is given by:

$$R_{D,SP}(actual) = \frac{W_D}{q.\mu_n.N_D} \qquad [15.25]$$

where W_D is the actual thickness of the drift region. In many fabricated devices, the thickness used produces a punch-through of the electric field in the blocking mode.

The relatively high doping concentration of the drift region in 4H-SiC BJTs makes the drift region resistance very low even without conductivity modulation. In the case of SiC BJTs, the measured specific on-resistance in the on-state is found to be very close to the value obtained using Eq. [15.25] for unipolar operation[18]. Consequently, typical SiC BJT devices do not exhibit a quasi-saturation region[19,20,21].

15.6 Switching Speed

The turn-off speed for the Si BJT becomes limited by a large storage time that is required to extract the stored charge in the drift region[1]. As pointed out in section 15.5, SiC BJTs operate without conductivity modulation of the drift region which eliminates the storage phase during turn-off. The switching times for SiC BJTs have been reported[22,23,24] to be very fast (~10 ns) with a storage time of about 100 ns. In addition, there is no current tail like that observed in Si IGBTs making the energy loss small.

15.7 Safe-Operating-Area

The boundary of the locus for the current-voltage trajectory during the turn-off of Si BJTs becomes limited by the reverse-biased-safe-operating-area (RBSOA)[1]. When a reverse base current is applied to remove the stored charge in the base region of the BJT, the collector current constricts towards the middle of the emitter finger. This produces an increase in the local collector current density.

The electron concentration in the drift region due to the collector current flow is given by:

$$n_D = \frac{J_C}{q \cdot v_{sat}} \qquad [15.26]$$

where v_{sat} is the saturated drift velocity. The saturated drift velocity for electrons in 4H-SiC is 2.2×10^7 cm/s compared with 0.985×10^7 cm/s in Si as discussed in chapter 2.

The electric field in the drift region is enhanced by the presence of these electrons in the drift region leading to a reduced breakdown voltage. The breakdown voltage that defines the RBSOA boundary is given by[1]:

$$BV_{RBSOA} = \frac{\varepsilon_S . E_C^2}{2q.\left[N_D - \left(\dfrac{J_C}{q.v_{sat}} \right) \right]} \qquad [15.27]$$

where N_D is the doping concentration in the drift region.

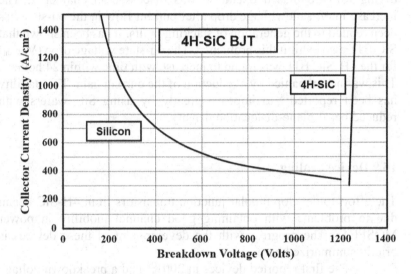

Fig. 15.4 Reverse-biased-safe-operating-area for 4H-SiC N-P-N BJT.

A comparison of RBSOA boundaries for the 4H-SiC and Si BJTs can be performed by considering devices with the same blocking voltage capability. In case of devices with blocking voltage of 1200-V, the doping concentration for the Si BJT is 1×10^{14} cm^{-3} while that for the 4H-SiC BJT is 1.5×10^{16} cm^{-3}. As an example, at a current density of 500 A/cm^2, the electron concentration in the drift region for the Si BJT is 3.2×10^{14} cm^{-3} compared with 1.4×10^{14} cm^{-3} for 4H-SiC. Consequently, the electron concentration in the drift region exceeds the doping concentration in the case of the Si BJT. This produces a reversal of the slope of the electric field profile with peak electric field shifting to the N-drift/N$^+$ substrate interface[1]. The peak electric field increases with collector current density limiting the RBSOA as shown in Fig. 15.4. In contrast, the electron concentration in the drift region is well below the doping concentration in the case of the SiC BJT. It compensates the donor charge making the peak electric field at the collector-base junction becoming smaller with increasing current density. This produces a vertical RBSOA boundary as shown in Fig. 15.4. This behavior has been discussed in more detail in the literature[25].

15.8 Bipolar Degradation

The degradation of the on-state characteristics of P-i-N rectifiers during forward biased operation was discussed in chapter 7. The increase in on-state voltage drop after current flow in the on-state has been related to the generation of stacking faults. It has been shown that SiC BJT also exhibit this problem[26]. The on-state voltage drop ($V_{CE,sat}$) for the 4H-SiC BJT was found to increase with time within 24 hours[27]. This was accompanied by a reduction of the current gain. The stability has been reported[28] to improve greatly by using SiC wafers with reduced basal-plane-dislocation density.

15.9 Device Scaling

The effort to develop bipolar junction transistors from 4H-SiC began due to problems with obtaining good channel mobility in power MOSFETs. The progress with the development of these devices is briefly summarized here.

The first reported devices in 2001[29] had a breakdown voltage of 1800-V and a peak current gain of 20. Subsequently, 4H-SiC BJTs with breakdown voltage of 1677-V were reported[30] in 2005 with specific on-resistance of 5.7 mΩ-cm^2 corresponding to a BV^2/R$_{ON,SP}$ Figure-of-Merit of 493 MW/cm^2. 4H-SiC BJTs with BV$_{CB0}$ of 4-kV were reported[31] in 2005 with specific on-resistance of 56 mΩ-cm^2. These devices suffered from a low current gain of about 9. 4H-SiC BJTs with current gain of 40 and breakdown voltage of 1 kV were also reported[32] at this time. Later, an ion implantation free process was implemented[33] for 4H-SiC BJTs to achieve a breakdown voltage of 2.7 kV with a current gain of 50.

In 2012, 4H-SiC BJTs with a breakdown voltage of 21-kV were achieved[34] with a current gain of 63 and specific on-resistance of 321 mΩ-cm^2. This required improvement of bulk lifetime by thermal oxidation and reduction of surface recombination by improved surface passivation. Optimization of the emitter cell geometry was reported[35] in 2015. It was demonstrated that the hexagonal emitter geometry produced the best maximum current gain at high collector current densities when compared with the linear cell geometry because of reduced peripheral recombination. A good review of the design of 4H-SiC BJTs has also been recently prepared[36].

15.10 Summary

Although the silicon bipolar junction transistor was rendered obsolete by the development of the silicon IGBT in the 1980s[37,38], there has been interest in the development of BJTs from 4H-SIC due to difficulties with making SiC power MOSFETs and IGBTs. The first SiC BJTs were reported in 2001 with a blocking voltage of 1800-V. Since then the blocking voltage has been enhanced to 21-kV. In addition, the initial current gains for the SiC BJTs were poor (<10) due to surface recombination at the edges of the emitter. Efforts to enhance the bulk lifetime and reduce the surface recombination with better passivation have allowed increasing the current gain to over 200.

References

[1] B.J. Baliga, "Fundamentals of Power Semiconductor Devices", Chapter 7, Springer-Science, New York, 2008.

[2] T.P. Chow, "High-Voltage SiC and GaN Power Devices", Microelectronics Engineering, Vol. 83, pp. 112-122, 2006.

[3] A.K. Agarwal et al., "Recent Progress in SiC Bipolar Junction Transistors", IEEE International Symposium on Power Semiconductor Devices and ICs, Abstract 5.2, pp. 361-364, 2004.

[4] S. Balachandran et al., "4 kV 4H-SiC Epitaxial Emitter Bipolar Junction Transistors", IEEE Electron Device Letters, Vol. 26, pp. 470-472, 2005.

[5] H-S. Lee et al., "1200-V 5.2 mW-cm2 4H-SiC BJTs with a High Common-Emitter Current Gain", IEEE Electron Device Letters, Vol. 28, pp. 1007-1009, 2007.

[6] M. Nawaz *et al.*, "Static and Dynamic Characterization of High Power Silicon Carbide BJT Modules", IEEE Energy Conversion Congress and Exposition, pp. 2824-2831, 2014.

[7] S-H. Ryu *et al.*, "1.8 kV, 3.8 A Bipolar Junction Transistors in 4H-SiC", IEEE International Symposium on Power Semiconductor Devices and ICs, Abstract 2.1, pp. 37-40, 2001.

[8] C-F. Huang and J.A. Cooper, "4H-SiC npn Bipolar Junction Transistors with $BV_{CEO} > 3,200$ V", IEEE International Symposium on Power Semiconductor Devices and ICs, pp. 57-60, 2002.

[9] Y. Lou *et al.*, "High Voltage (>1 kV) and High Current Gain (32) 4H-SiC Power BJTs using Al-Free Ohmic Contact to the Base", IEEE Electron Device Letters, Vol. 24, pp. 695-697, 2003.

[10] A.A. Agarwal *et al.*, "Large Area, 1.3 kV, 17 A, Bipolar Junction Transistors in 4H-SiC", IEEE International Symposium on Power Semiconductor Devices and ICs, pp. 135-138, 2003.

[11] P.A. Ivanov *et al.*, "Temperature Dependence of the Current Gain in Power 4H-SiC NPN BJTs", IEEE Transactions on Electron Devices, Vol. 53, pp. 1245-1249, 2010.

[12] M. Domeij *et al.*, "Geometrical Effects in High Current Gain 100-V 4H-SiC BJTs", IEEE Electron Device Letters, Vol. 26, pp. 743-746, 2005.

[13] B. Buono *et al.*, "Influence of Emitter Width and Emitter-Base Distance on the Current Gain in 4H-SiC Power BJTs", IEEE Transactions on Electron Devices, Vol. 57, pp. 2664-2670, 2010.

[14] H-S. Lee *et al.*, "High-Current-Gain SiC BJTs with Regrown Extrinsic Base and Etched JTE", IEEE Transactions on Electron Devices, Vol. 55, pp. 1894-1898, 2008.

[15] B. Buono *et al.*, "Modeling and Characterization of Current Gain versus Temperature in 4H-SiC Power BJTs", IEEE Transactions on Electron Devices, Vol. 57, pp. 704-711, 2010.

[16] R. Ghandhi *et al.*, "Surface-Passivation Effects on the Performance of 4H-SiC BJTs", IEEE Transactions on Electron Devices, Vol. 58, pp. 259-265, 2011.

[17] H. Miyake *et al.*, "4H-SiC BJTs with Record Current Gains of 257 on (0001)", IEEE Electron Device Letters, Vol. 32, pp. 841-843, 2011.

[18] B. Buono *et al.*, "Modeling and Characterization of the On-Resistance in 4H-SiC Power BJTs", IEEE Transactions on Electron Devices, Vol. 58, pp. 2081-2087, 2011.

[19] A. Lindgren and M. Domeij, "Degradation Free Fast Switching 1200 V 50 A Silicon Carbide BJTs", IEEE Applied Power Electronics Conference, pp. 1064-1070, 2011.

[20] R. Singh *et al.*, "1200 V SiC "Super" Junction Transistors operating at 250 °C with Extremely Low Energy Losses for Power Conversion Applications", IEEE Applied Power Electronics Conference, pp. 2516-2520, 2012.

[21] R. Singh *et al.*, "10 kV SiC BJTs – Static, Switching, and Reliability Characteristics", IEEE Workshop on Wide Bandgap Power Devices and Applications, Paper T7.8086, 2015.

[22] Y. Gao *et al.*, "Comparison of Static and Switching Characteristics of 1200 V 4H-SiC BJT and 1200 V Si-IGBT", IEEE Transactions on Industry Applications, Vol. 44, pp. 887-893, 2008.

[23] M. Ostling *et al.*, "SiC Bipolar Devices for High Power and Integrated Drivers", IEEE Bipolar/BiCMOS Circuits and Technology Meeting, pp. 227-234, 2011.

[24] H. Zhu *et al.*, "A Comparison of Switching Characteristics between SiC BJT and Si IGBT at Junction Temperature above 200 oC", IEEE International Conference on Power Electronics, In Telligent Motion, Renewable Energy, and Energy Management, pp. 283-290, 2015.

[25] Y. Gao *et al.*, "Analysis of SiC BJT RBSOA", IEEE International Symposium on Power Semiconductor Devices and ICs, pp. 135-138, 2003.

[26] S. Krishnaswami *et al.*, "4 kV, 10 A Bipolar Junction Transistors in 4H-SiC", IEEE International Symposium on Power Semiconductor Devices and ICs, pp. 1-4, 2006.

[27] J. Zhang *et al.*, "Fabrication and Characterization of High-Current-Gain 4H-SiC Bipolar Junction Transistors", IEEE Transactions on Electron Devices, Vol. 55, pp. 1899-1906, 20108.

[28] Q. Zhang *et al.*, "4H-SiC Bipolar Junction Transistors: From Research to Development", IEEE International Symposium on Power Semiconductor Devices and ICs, pp. 339-342, 2009.

[29] S-H. Ryu *et al.*, "1800 V NPN Bipolar Junction Transistors in 4H-SiC", IEEE Electron Device Letters, Vol. 22, pp. 124-126, 2001.

[30] J. Zhang *et al.*, "1677 V, 5.7 mΩ-cm^2 4H-SiC BJTs", IEEE Electron Device Letters, Vol. 26, pp. 188-190, 2005.

[31] S. Balachandran *et al.*, "4kV 4H-SiC Epitaxial Emitter Bipolar Junction Transistors", IEEE International Symposium on Power Semiconductor Devices and ICs, pp. 1-4, 2005.

[32] S. Krishnaswami *et al.*, "1000-V, 30-A 4H-SiC BJTs with High Current Gain", IEEE Electron Device Letters, Vol. 26, pp. 175-177, 2005.

[33] R. Ghandhi *et al.*, "Fabrication of 2700-V, 12 mΩ-cm^2 Non Ion-Implanted 4h-SiC BJTs with Common-Emitter Current Gain of 50", IEEE Electron Device Letters, Vol. 29, pp. 1135-1137, 2008.

[34] H. Miyake *et al.*, "21-kV SiC BJTs with Space-Modulated Junction Termination Extension", IEEE Electron Device Letters, Vol. 33, pp. 1598-1600, 2012.

[35] A. Salemi *et al.*, "Optimal Emitter Cell Geometry in High Power 4H-SiC BJTs", IEEE Electron Device Letters, Vol. 36, pp. 1069-1071, 2015.

[36] A. Salemi *et al.*, "Area and Efficiency Optimized Junction Termination for a 5.6 kV SiC BJT Process with Low On-Resistance", IEEE International Symposium on Power Semiconductor Devices and ICs, pp. 249-252, 2015.

[37] B.J. Baliga, "The IGBT: The GE Story", IEEE Power Electronics Magazine, Vol. 2, pp. 16-23, June 2015.

[38] B.J. Baliga, "The IGBT Device", Elsevier Press, Amsterdam, April 2015.

Chapter 16

SiC Gate Turn-Off Thyristors

Silicon thyristors and Gate Turn-Off Thyristors (GTOs) were developed in the 1960s and served many high power applications because their voltage and current handling capability could be scaled to high levels[1]. Silicon GTOs were utilized for driving the motors in traction (electric trains and locomotives) applications for urban and long distance transportation. The devices required snubbers for fail safe operation increasing the power losses and making the systems bulky. More importantly, a very large gate current was required to turn-off the anode (load) current in the GTO. In some application, the turn-off was performed using a unity current gain making the gate control circuit large and expensive. These drawbacks of silicon GTOs resulted in their replacement with silicon IGBTs in the 1990s due to availability of modules with sufficiently high power ratings[2].

Interest in silicon carbide thyristors and GTOs was motivated by military applications such as electric guns[3], compact army vehicle drives[4], and pulse power[5]. With this support, the voltage blocking capability of 4H-SiC GTOs was increased from 700-V in 1997[6] to 20-kV in 2013[5]. In this chapter, the attributes of SiC GTOs are described. The characteristics of devices that have been successfully fabricated over the years are reviewed.

16.1 SiC GTO Structure and Operation

The gate turn-off thyristor structure has been discussed in detail in the textbook[1]. This chapter will focus on highlighting the benefits of the SiC GTO and any unique aspects of its design and operation. The basic structure of the SiC GTO is shown in Fig. 16.1. It is fabricated by the epitaxial growth of all the layers – the N^+ buffer layer, the P^+ buffer layer, the P-drift region, the N-base region, and the P^+ anode region because ion-implantation results in poor minority carrier lifetime in SiC. As a consequence, the doping concentrations for all layers can be assumed to be constant unlike in the case of silicon devices fabricated by deep diffusions. The N-base region is accessed by etching through the P+ anode layer. Another etching step is performed through the N-base region to make the edge terminations by the junction termination extension approach. The N+ contacts to the N-base region are formed by ion-implantation of nitrogen followed by the annealing process.

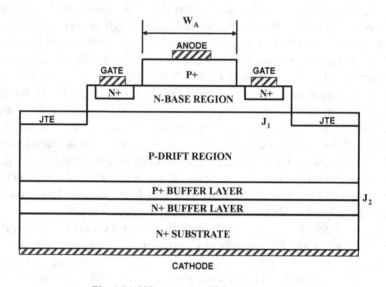

Fig. 16.1 SiC gate turn-off thyristor structure.

The asymmetric GTO device is designed to support voltage only in one direction. Since high quality P+ substrates are not available in 4H-SiC material, it is customary to make the SiC GTOs by using N+ substrates as shown in the Fig. 16.1. This is a complementary structure when compared with traditional silicon devices. The device can support a large negative voltage applied to the cathode electrode across the junction (J_1) between the N-base region and the P-drift region. The P+ buffer layer serves to prevent reach-through to the cathode junction (J_2) if its doping and thickness are sufficient. The charge in the P+ buffer layer for 4H-SiC devices must be made 10-times larger than that in the buffer layers for Si devices due to the larger electric field in this material. The device cannot support a high positive voltage applied to the cathode due to the P+ buffer layer leading to an asymmetrical blocking structure. This is similar to Si GTOs.

The forward blocking voltage capability for the asymmetric GTO structure is determined by the thickness of the drift region with a punch-through electric-field profile[1]. The thickness of the drift region for a 4H-SiC GTO is about 10 times smaller than that for the Si device due to the 10-times larger critical breakdown field strength for 4H-SiC material. This feature has been utilized for constructing devices with extremely high blocking voltages[5].

16.2 Forward Blocking Capability

The forward blocking voltage for the 4H-SiC GTO structure can be analyzed by using the procedure described for the asymmetric IGBTs in the textbook[7]. It is necessary to take open-base transistor breakdown into consideration while also accounting for the trapezoidal electric field distribution.

The first (1997) 4H-SiC GTOs with 700-V blocking capability were fabricated[6] using a P-drift region thickness of 6 microns and doping concentration of 5×10^{14} cm^{-3}. Devices with blocking voltage capability of 100-V were reported[8] in 1998 by using a P-drift region thickness of 14 microns. The blocking voltage was increased[9,10] to 3-kV by 2001 by using a P-drift region thickness of 50 microns and doping concentration of 8×10^{14} cm^{-3}.

In 2004, a 4H-SiC GTO with forward blocking capability of 12.7 kV was reported[11] for utility applications. The authors utilized P-drift layers with a doping concentration of $1-2 \times 10^{14}$ cm^{-3}. The forward blocking voltage scaled with P-drift layer thickness as expected with values of 6.2, 8.1, and 12.7 kV observed for thicknesses of 75, 90, and 120 microns, respectively. The peak electric field at junction J_1 is given by[1]:

$$E_M = \frac{\left[BV_{PT} + \left(\dfrac{q N_A W_P^2}{2\varepsilon_S} \right) \right]}{W_P}$$

[16.1]

where BV_{PT} is the punch-through breakdown voltage, q is the electron charge, N_A is the doping concentration of the P-base region, W_P is the thickness of the P-base region, and ε_S is the dielectric constant for 4H-SiC. Using the values for the 12.7 kV device, the maximum electric field is found to be 1.2 MV/cm which is slightly below the critical electric for breakdown of 1.47 MV/cm for 4H-SiC with a doping concentration of 1.5×10^{14} cm^{-3}. This is as expected from the open-base breakdown conditions in the GTO structure.

In the case of silicon thyristors, it is well known that the increase in leakage current at elevated temperatures can lead to triggering the device into the on-state[1]. This problem is suppressed by using cathode-shorts in silicon thyristors. In the case of silicon GTOs, the cathode fingers have a small width to allow device turn-off under gate control. This suppresses the turn-on by the leakage current. Similar performance has been reported for 4H-SiC GTOs with 9-kV blocking capability[12] with low leakage currents at 175 °C. The growth in die size for the 4H-SiC GTO from 0.2 cm × 0.2 cm in 1999 to 1 cm × 1 cm in 2011 has been reported[13] with improvement in blocking voltage from 3-kV to 12-kV. The 12-kV devices were achieved by using a novel 600 micron wide negative bevel edge termination formed with multiple steps etched into the N-base region. This edge termination improved the blocking voltage by 3.5-4.0 kV when compared with the 15 zone JTE termination.

16.3 On-State Characteristics

The on-state characteristics for the GTO resemble that of the P-i-N rectifier due to high level injection into the drift region from the anode and cathode regions[1]. The on-state characteristics for the 4H-SiC P-i-N rectifier was discussed in chapter 7. It was pointed out (see Fig. 7.5) that the on-state voltage drop for the 4H-SiC devices has a minimum value of about 3.2 volts which is much larger than that for typical Si devices due to the larger energy band gap of 4H-SiC. For this reason, the on-state voltage drop for unipolar devices, such as power MOSFETs, is much lower than that for the 4H-SiC GTO for breakdown voltages below 6-kV. The development of SiC GTOs with blocking voltages above 6-kV is of greatest interest from an applications standpoint.

16.3.1 On-State Voltage Drop

The on-state characteristics of 9-kV 4H-SiC GTOs with a size of 1 cm x 1 cm has been reported[12]. The devices exhibited a knee voltage of 3-V at 25 °C as expected with an on-state voltage drop of 3.7-V at an on-state current density of 100 A/cm^2. Subsequently, the on-state voltage drop for the 12-kV 4H-SiC GTOs was reported[13] to be 4.0-V at an on-state current density of 100 A/cm^2.

In the case of 4H-SiC GTOs with a blocking voltage of 20-kV, the on-state voltage drop was reported[5] as 7.0-V at an on-state current density of 25 A/cm^2. This indicates that the lifetime is not sufficiently large for these devices with a drift region thickness of 160 microns. The ambipolar diffusion coefficient for 4H-SiC is low (5.6 cm^2/s) resulting in a high-level diffusion length of 58 microns for a lifetime of 5 microseconds. This is the typical lifetime reported for the drift region after lifetime enhancement by oxidation at 1300 °C for 5 hours[14]. This results in a (d/L$_a$) ratio of 1.5 which should be reduced closer to unity to improve the on-state voltage drop.

As expected[1], the on-state voltage drop for the 4H-SiC GTO decreases with increasing lifetime for minority carriers in the drift region. This has been shown[11] using numerical simulations for a 12.7-kV 4H-SiC GTO structure. The on-state voltage drop reduced from 9-V to 3.2-V by increasing the lifetime by a factor of 4-times.

16.3.2 Surge Current Capability

An important application for the 4H-SiC GTOs has been as pulse power switches[3,5]. The performance of 4H-SiC GTO with 7-kV blocking capability was evaluated[15] under sinusoidal current pulses with a duration of 1 ms. The devices survived after a pulse with a peak current density of 4.4 kA/cm^2. The voltage drop across the device reached 10-V at the peak of the pulse. The on-state voltage drop at surge current levels increased after multiple current pulses due to heating according to the authors. It is worth pointing out that the on-state voltage drop decreases due to heating at normal on-state current density values[14].

16.3.3 On-State Voltage Instability

The on-state voltage drop of P-i-N rectifiers has been found to increase after current conduction as discussed in chapter 7 (see section 7.9). This bipolar degradation phenomenon has also been reported[12] for the 4H-SiC GTOs. It was found that the on-state voltage drop of the 6-kV 4H-SiC GTO increased from 4.1-V to 4.55-V after current flow for 10 hours at room temperature. This degradation has been related to the generation of stacking faults at basal plane dislocations. The degradation (increase) in on-state voltage drop has been eliminated by improved epitaxial growth processes[12].

16.4 Switching Characteristics

The physics of switching current on and off in thyristors and GTOs has been described in the textbook[1] for silicon devices. The same analysis is applicable to the silicon carbide devices. The switching behavior observed in actual devices fabricated from 4H-SiC are discussed in this section.

16.4.1 Turn-on

The 4H-SiC GTO can be turned-on by the application of a pulse of current at the gate electrode to forward bias the anode-base junction and initiate the injection of holes from the anode region. Early devices designed with low blocking voltages (< 3-kV) and small active area exhibited sub-microsecond turn-on delay times[16,17]. A very high gate trigger current, the same as the cathode current, was required at room temperature[17,18]. This was attributed to the poor ionization of dopants in the anode region by the authors. An involute gate design was compared with the linear cell design in this work. Similar observations were also reported by another group[19]. The improved impurity activation at elevated temperature reduces the turn-on time in these 4H-SiC GTOs. Current spreading issues that influence the turn-on of large area silicon thyristors are not relevant to the 4H-SiC GTOs due to their small area and highly inter-digitated gate geometry.

16.4.2 Turn-off

The turn-off behavior of 3-kV 4H-SiC GTOs was first reported[19] in 2002. The turn-off waveform at room temperature is identical to that for silicon devices[1], consisting of a rapid fall in current followed by a current tail. A maximum turn-off current gain of 12.5 was observed. The turn-off gain was found to decrease rapidly with increasing temperature to a value of about unity at 175 °C. The devices exhibited a turn-off dI/dt of 4.7 A/μs and dV/dt of 2220 V/μs when operated at a current gain of 1.4[20].

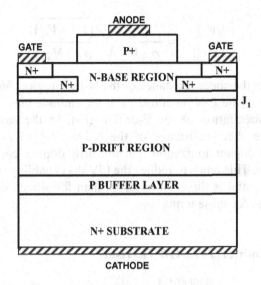

Fig. 16.2 Novel SiC gate turn-off thyristor structure.

Improvement in the turn-off capability of the SiC GTO was attempted by incorporation of an N^+ burred layer inside the N-base region[11] as shown in Fig. 16.2. The N^+ buried region was formed by ion-implantation. This allowed turning-off an anode current density of 240 A/cm^2 a current gain of 0.6. The turn-off time was reported as 2.68 μs consisting of a storage time of 0.55 μs and a fall time of   2.13 μs.

16.4.3 [dV/dt] Capability

The SiC GTO is subjected to a high rate of rise of voltage during operation in pulse power circuits. A thyrisior can be triggered into the on-state by the displacement current during a high [dV/dt] transient[1]. The [dV/dt] is improved by using cathode-shorts. Silicon GTOs tend to have a high [dV/dt] capability because of the narrow cathode fingers[1] eliminating the need for cathode shorts.

This should hold true for SiC devices as well. The [dV/dt] capability of 9-kV SiC GTOs was evaluated[21] for chips with 1 cm^2 area. The devices has an anode width (W_A in Fig. 16.1) of 22 μm with a cell pitch of 48 μm. The [dV/dt] capability was measured using R-C snubber between the gate and the anode to divert the displacement current. A maximum [dV/dt] of 18 kV/μs was observed.

The [dV/dt] is given by[1]:

$$\left[\frac{dV}{dt}\right]_{Max} = \frac{8V_{bi}}{\rho_{SB} W_A^2} \sqrt{\frac{2(V_K + V_{bi})}{q \varepsilon_S N_A}} \qquad [16.2]$$

where ρ_{SB} is the sheet resistance of the N-base region, V_{bi} is the built-in potential of the P-N junction, V_K is the cathode bias, and N_A is the doping concentration of the P-drift region. In the case of 4H-SiC devices, the sheet resistance of the N-base region is high due to incomplete dopant ionization and the drift doping concentration is much larger. This tends to reduce the [dV/dt] capability. However, the built-in potential is three times larger than for silicon devices which compensates for these terms.

16.5 The Emitter Turn-Off Thyristor

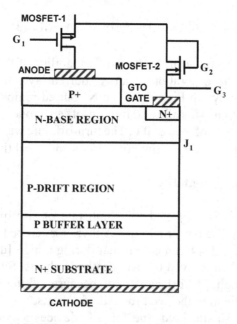

Fig. 16.3 MOS-gated emitter turn-off thyristor structure.

One of the major drawbacks of the GTO structure is the very large gate current required to turn-off the device. This problem can be addressed by integration of a MOSFET in series with the thyristor structure to form the emitter switched thyristor (EST)[22]. Experimental demonstration of this concept was reported[23,24] in 1991 with excellent on-state and MOS-gate controlled turn-off for Si devices. It is difficult to scale up the current handling capability of these devices due to loss of yield in MOS-gate structures for power devices.

An equivalent high current composite device can be created by using discrete MOSFETs and a GTO as illustrated in Fig. 16.3. The composite device operates in the on-state by turning-on p-channel MOSFET-1 and supplying a gate drive current via terminal G_3 to turn-on the GTO. The device can be turned off by turning-off MOSFET-1. This was first demonstrated in 1990 using silicon MOSFETs and a silicon GTO[22]. It was shown that the turn-off occurs with the anode current of the GTO diverted to the gate of the GTO with current flow via MOSFET-2. The unity current-gain turn-off produces a short storage time and fast fall time for the current. The concept was later demonstrated with high power handling capability under the name emitter turn-off thyristor (ETO)[25].

The ETO concept has been implemented using 4H-SiC GTOs[26]. A 4.5 kV SiC GTO with area of 0.36 cm^2 was used together with 100-V Si power MOSFETs. The composite device had an on-state voltage drop of 4.2-V when the SiC GTO was operating at a current density of 25 A/cm^2. During inductive load turn-off, the device was found to have a storage time of 504 ns followed by a voltage rise-time of 980 ns. The current fall time was 240 ns. An optically triggered version of the SiC ETO has been proposed and demonstrated with 15-kV blocking voltage capability[27].

16.6 Summary

Interest in the development of the SiC GTO structure was mainly driven by military and pulse-power applications. The chip size for the SiC-GTO was increased from 2 mm × 2 mm in 1999 to 3 mm × 3mm in 2003 to 4 mm × 4mm in 2006 to 7 mm × 7mm in 2008 and 10 mm × 10 mm in 2009[13]. At the same time the blocking voltage was increased from 700-V to 12-kV. The minority carrier lifetime in the thick drift layers for the high blocking voltage devices had to be enhanced to several microseconds in order to achieve a low on-state voltage drop.

A half-bridge inverter was demonstrated[28] using the SiC-GTO with pulse-width modulation at a switching frequency of 2 kHz and DC bus voltage of 600-V. The benefits of the SiC GTO were enumerated as[29]: *2-3 times reduction in series connected components when compared with 4-6 kV Si GTOs; higher switching frequency of 10 kHz compared with 1 kHz for the Si GTO; immunity from thermal runaway and latch-up at elevated temperatures.*" A 22-kV 4H-SiC GTO was achieved[30] in 2014 by using a p-type drift layer with thickness of 160 μm and doping concentration of 2×10^{14} cm^{-3}. A 2.5 mm thick layer with doping concentration of $1\text{-}5 \times 10^{17}$ cm^{-3} was used as the buffer region. An on-state voltage drop of 5 volts was observed for an on-state current density of 25 A/cm^2. Although these impressive results have been obtained, recent success with making high voltage IGBTs has shift interest away from these devices.

References

[1] B.J. Baliga, "Fundamentals of Power Semiconductor Devices", Chapter 8, Springer-Science, New York, 2008.

[2] B.J. Baliga, "The IGBT Device", Elsevier Press, Amsterdam, April 2015.

[3] T. Burke *et al.*, "Silicon Carbide Thyristors for Electric Guns", IEEE Transactions on Magnetics, Vol. 33, pp. 432-437, 1997.

[4] P.B. Shah *et al.*, "4H-SiC Gate Turn-Off Thyristor Designs for Very High Power Control", Material Science Forum, Vol. 338-342, pp. 1395-1398, 2000.

[5] L. Cheng *et al.*, "20 kV, 2 cm^2, 4H-SiC Gate Turn-Off Thyristors for Advanced Pulsed Power Applications", IEEE Pulse Power Conference, pp. 1-4, 2013.

[6] A.K. Agarwal *et al.*, "700-V Asymmetrical 4H-SiC Gate Turn-Off Thyristors (GTOs)", IEEE Electron Device Letters, Vol. 18, pp. 518-520, 1997.

[7] B.J. Baliga, "Fundamentals of Power Semiconductor Devices", Chapter 9, Section 9.4.4, Springer-Science, New York, 2008.

[8] R.R. Sergiej *et al.*, "1000V 4H-SiC Gate Turn Off (GTO) Thyristor", IEEE International Symposium on Compound Semiconductors, pp.363-366, 1998.

[9] A. Agarwal *et al.*, "2500 V, 12 A, 4H-SiC, Asymmetrical Gate Turn Off (GTO) Thyristor Development", Material Science Forum, Vol. 338-342, pp. 1387-1390, 2000.

[10] S-H. Ryu *et al.*, "3100 V, Asymmetrical, Gate Turn-Off (GTO) Thyristor in 4H-SiC", IEEE Electron Device Letters, Vol. 22, pp. 127-129, 2001.

[11] Y. Sugawara *et al.*, "12.7kV Ultra High Voltage SiC Commutated Gate Turn-Off Thyristor: SICGT", IEEE International Symposium on Power Semiconductor Devices and ICs, Paper 5.3, pp. 365-368, 2004.

[12] A. Agarwal *et al.*, "9 kV, 1 cm × 1 cm SiC Super GTO Technology development for Pulse Power", IEEE Pulse Power Conference, pp. 264-269, 2009.

[13] A. Agarwal *et al.*, "SiC Super GTO Thyristor Technology development: Present Status and Future Perspective", IEEE Pulse Power Conference, pp. 1530-1535, 2011.

[14] L. Cheng *et al.*, "Advanced Silicon Carbide Gate Turn-Off Thyristor for Energy Conversion and Power Grid Applications", IEEE Energy Conversion Congress and Exposition, pp. 2249-2252, 2012.

[15] H. O-Brien *et al.*, "Wide-Pulse Evaluation of 0.5 cm2 Silicon Carbide SGTO", IEEE Pulse Power Conference, pp. 260-263, 2009.

[16] J. Mooken *et al.*, "Switching Characteristics of an Asymmetrical Complementary 4H-SiC Gate Turn-Off (GTO) Thyristors", IEEE Industrial Applications Society Meeting, Vol. 2, pp. 1000-1005, 1997.

[17] J.B. Fedison *et al.*, "Dependence of Turn-on and Turn-off Characteristics on Anode/Gate Geometry of High-Voltage 4H-SiC GTO Thyristors", IEEE International Symposium on Power Semiconductor Devices and ICs, Paper 9.1, pp. 175-178, 2001.

[18] A. Elasser *et al.*, 'Silicon Carbide GTO: Static and Dynamic Characterization", IEEE Industrial Applications Society Meeting, Vol. 1, pp. 359-364, 2001.

[19] A.K. Agarwal *et al.*, "Dynamic Performance of 3.1-kV 4H-SiC Asymmetric GTO Thyristors", Material Science Forum, Vol. 389-393, pp. 1349-1352, 2002.

[20] S.V. Campen *et al.*, "100 A and 3.1-kV 4H-SiC GTO Thyristors", IEEE Lester Eastman Conference on High Performance Devices, pp. 58-64, 2002.

[21] A. Oguniyi *et al.*, "dV/dt Immunity and Recovery Time Capability of 1.0 cm^2 Silicon Carbide SGTO", IEEE Power Modulator and High Voltage Conference, pp. 354-357, 2012.

[22] B.J. Baliga, "The MOS-Gated Emitter Switched Thyristor", IEEE Electron Device Letters, Vol. 11, pp. 75-77, 1990.

[23] M.S. Shekar *et al.*, "Experimental Demonstration of the Emitter Switched Thyristor", IEEE International Symposium on Power Semiconductor Devices and ICs, pp. 128-130, 1991.

[24] M.S. Shekar *et al.*, "Characteristics of the Emitter Switched Thyristor", IEEE Transactions on Electron Devices, Vol. 38, pp. 1619-1623, 1991.

[25] Y. Li, A.Q. Huang and F.C. Lee, "Introducing the Emitter Turn-Off Thyristor (ETO)", IEEE Industrial Applications Society Meeting, pp. 860-864, 1998.

[26] J. Wang and A.Q. Huang, "Design and Characterization of High-Voltage Silicon Carbide Emitter Turn-Off Thyristor", IEEE Transactions on Power Electronics, Vol. 24, pp. 1189-1197, 2009.

[27] A. Mojab *et al.*, "15-kV Single-Bias All-Optical ETO Thyristor", IEEE International Symposium on Power Semiconductor Devices and ICs, pp. 313-316, 2014.

[28] C.W. Tipton *et al.*, "Half-Bridge Inverter using 4H-SiC Gate Turn-Off Thyristors", IEEE Electron Device Letters, Vol. 23, pp. 194-197, 2002.

[29] J.W. Palmour *et al.*, "SiC Power Devices for Smart Grid Systems", IEEE International Power Electronics Conference, pp. 1006-1013, 2010.

[30] J.W. Palmour, "Silicon Carbide Power Device Development for Industrial Markets", IEEE International Electron Devices Meeting, Paper 1.1.1, pp. 1-8, 2014.

Chapter 17

SiC IGBTs

The replacement of bipolar-mode silicon devices with unipolar-mode SiC devices was proposed to reduce the on-state voltage drop and switching losses in power electronic circuits[1]. This has been successfully achieved by the demonstration of high performance 4H-SiC power MOSFETs as described in chapter 11. Once the cost of the SiC wafers and fabrication technology decrease sufficiently, these devices will replace Si IGBTs for blocking voltages up to 6-kV. However, for even greater blocking voltages, the specific on-resistance for even 4H-SiC unipolar transistors becomes large creating high on-state power losses. This has motivated the development of IGBTs using 4H-SiC material.

    As in the case of silicon IGBTs, it is possible to make both n-channel and p-channel IGBTs using 4H-SiC. The development of p-channel IGBT took precedence over n-channel IGBT because the quality of N^+ substrates required for the p-channel devices was much superior to that of P^+ substrates required for the n-channel devices. The first p-channel silicon carbide IGBT structures with 10-kV blocking voltage capability were reported in 2005 with the trench-gate process[2]. In the case of all MOS-gated power devices formed using silicon carbide, it is necessary to shield the gate oxide from the high electric fields developed in the semiconductor[3]. This can be accomplished for the trench-gate MOSFET or IGBT structures by the incorporation of a junction at the bottom of the trenches[3,4].

    Subsequently, planar p-channel silicon carbide IGBT structures with 9-kV blocking voltage capability were reported in 2007[5] and 12-kV devices were reported in 2008[6]. Although this work demonstrated that conductivity modulation of the drift region can be achieved in the silicon carbide IGBT structure similar to that observed in silicon devices, the on-state voltage drop obtained for the silicon carbide IGBT was significantly worse that than shown in chapter 11 for the 10-kV silicon carbide power MOSFET structure. More recent interest has therefore shifted to even higher blocking voltage ratings from 15-kV to 20-kV. The on-state characteristics of asymmetric 4H-SiC p-channel IGBTs having a 175 micron thick P-base region with doping concentration of 2×10^{14} cm^{-3} were reported for the planar device architecture in 2007[7]. The devices had on-state voltage drop of 8.5 volts at an on-state current density of 25 A/cm^2. The high-level lifetime in the P-base region was measured at 0.46 microseconds. The measured blocking voltage capability of the devices was only 5-kV although the authors assumed that the parameters for the drift region are suitable for a 20-kV device. The optimization of p-channel asymmetric IGBT structures in 4H-SiC for blocking voltages between 15 and 20 kV has also been reported[8]. A comparison of the high frequency performance of 15-kV asymmetric and symmetric blocking p-channel IGBTs has been reported recently[9]. This study concludes that the lifetime in the buffer layer has to be optimized to obtain the best performance in the asymmetric p-channel 4H-SiC IGBT structure. However, current technology for silicon carbide does not allow selective adjustment of the lifetime in the buffer layer.

Progress with the development of n-channel 4H-SiC IGBT has accelerated since 2007 when the successful fabrication of 4H-SiC n-channel IGBT structures with 13-kV blocking voltage capability was reported[10]. The modulation of the conductivity of the drift region was verified in these devices from the on-state characteristics. Despite a low inversion layer mobility of 18 cm^2/V-s observed in the high-voltage structures, an on-state voltage drop of 3.8 volts was obtained an on-state current density of 25 A/cm^2 with a gate bias of 20 volts. The devices exhibited a voltage rise-time of 0.25 microseconds and a current fall-time of 0.1 microseconds when operated from a 7-kV power supply at a current density of 175 A/cm^2. Since that time, the blocking voltage of the n-channel IGBT has been steadily scaled until 10 kV devices were reported[11] in 2010, 12.5 kV devices were reported[12] in 2012, 21 kV devices were reported[13] in 2013, and 22-kV device were reported[14] in 2014.

The progress in development of high voltage silicon carbide power devices was reviewed in 2008[15]. The report concludes that the most appropriate applications for silicon carbide MOSFET structures are in the range of 4 to 10 kV blocking voltages while that for silicon carbide IGBTs are in the range of 15 to 30 kV blocking voltages. Based up on this, the discussion of silicon carbide IGBTs will be focused on 20-kV rated devices in this chapter. The characteristics of the p-channel and n-channel planar inversion-mode silicon carbide IGBT structures are compared here.

The basic operating principles and characteristics for the Si IGBT have been described in detail in the textbook[16]. The operating principles of the asymmetric 4H-SiC IGBT structure can be expected to be similar to those for the asymmetric Si IGBT structure. However, the doping concentration in the drift region for the 4H-SiC structure is much larger than that for the Si device structure. Consequently, although the minority carrier concentration in the space-charge region during the switching of the silicon carbide and silicon devices is nearly the same, it is much smaller than the doping concentration in the silicon carbide devices and greater than the doping concentration in the silicon devices. In the case of the asymmetric Si IGBT structure, the space-charge region does not reach-through the drift region during the voltage transient leaving stored charge in the drift region at the end of the voltage transient. In contrast, the space-charge region reaches-though the drift region when the collector voltage is much less than the collector supply voltage in the case of the typical asymmetric 4H-SiC IGBT structure. This alters the shape of the voltage and current transients as discussed in this chapter.

17.1 n-Channel Asymmetric Structure

The asymmetric n-channel 4H-SiC IGBT structure with the planar gate architecture is illustrated in Fig. 17.1 with its doping profile. Since the asymmetric IGBT structure is intended for use in DC circuits, its reverse blocking capability does not have to match the forward blocking capability allowing the use of an N-buffer layer adjacent to the P^+ collector region. The N-buffer layer has a much larger doping concentration than the lightly doped portion of the N-base region. The electric field in the asymmetric IGBT takes a trapezoidal shape allowing supporting the forward blocking voltage with a thinner N-base region. This allows achieving a lower on-state voltage drop and superior turn-off characteristics.

As in the case of silicon devices, the doping concentration of the buffer layer and the lifetime in the N-base region must be optimized to perform a trade-off between on-state voltage drop and turn-off switching losses[16]. Unlike the silicon device, the silicon carbide structure has uniform doping concentration for the various layers produced by using either epitaxial growth or by using multiple ion-implantation energies to form a box profile.

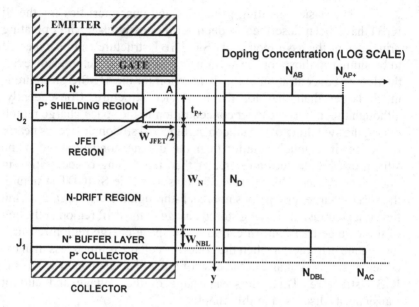

Fig. 17.1 The asymmetric n-channel SiC IGBT structure and its doping profile.

17.1.1 Blocking Characteristics

The design of the 20-kV asymmetric 4H-SiC n-channel IGBT structure is discussed in this section. The physics for blocking voltages in the first and third quadrants by the IGBT structure is discussed in detail in the textbook[16]. When a positive bias is applied to the collector terminal of the asymmetric 4H-SiC IGBT structure, the junction (J_2) between the P^+ shielding region and the N-base (drift) region becomes reverse biased while the junction (J_1) between the P^+ collector region and the N-buffer layer becomes forward biased. The forward blocking voltage is supported across the junction (J_2) between the P^+ shielding region and the N-base (drift) region with a depletion layer extending mostly within the N-base region. The doping concentration and width of the JFET region must be designed to allow suppression of the electric field at point A as discussed in chapter 11 for the 4H-SiC power MOSFET structure.

The forward blocking capability of the asymmetric n-channel 4H-SiC IGBT structure is determined by the open-base transistor breakdown phenomenon. The maximum blocking voltage occurs when the common base current gain of the PNP transistor of the n-channel structure becomes equal to unity. For the asymmetric IGBT structure, the emitter injection efficiency is smaller than unity due to the high doping concentration of the N-buffer layer. The emitter injection efficiency for the P$^+$ collector/N-buffer junction (J$_1$) can be obtained by using an analysis similar to that described in the textbook for the bipolar power transistor[16]:

$$\gamma_E = \frac{D_{pNBL} L_{nC} N_{AC}}{D_{pNBL} L_{nC} N_{AC} + D_{nC} W_{NBL} N_{DBL}}$$ [17.1]

where D$_{pNBL}$ and D$_{nC}$ are the diffusion coefficients for minority carriers in the N-buffer and P$^+$ collector regions; N$_{AC}$ and L$_{nC}$ are the acceptor concentration and diffusion length for minority carriers in the P$^+$ collector region; N$_{DBL}$ and W$_{NBL}$ are the donor concentration and width of the N-buffer layer. In determining the diffusion coefficients and the diffusion length, it is necessary to account for impact of the high doping concentrations in the P$^+$ collector region and N-buffer layer on the mobility. In addition, the lifetime within the highly doped P$^+$ collector region is reduced due to heavy doping effects, which shortens the diffusion length.

The open-base transistor breakdown condition for the asymmetric n-channel silicon carbide IGBT structure is given by:

$$\alpha_{PNP} = \left(\gamma_E . \alpha_T\right)_{PNP} M = 1$$ [17.2]

Based up on this expression, it can be concluded that the breakdown voltage for the asymmetric 4H-SiC IGBT structure will occur when the multiplication co-efficient is only slightly above unity. Using the avalanche breakdown criteria when the multiplication co-efficient becomes equal to infinity will lead to significant error in the design of the drift region for the IGBT structure.

When the collector bias exceeds the reach-through voltage (V$_{RT}$), the electric field is truncated by the high doping concentration of the N-buffer layer making the un-depleted width of the PNP transistor base region equal to the width of the N-buffer layer. The base transport factor is then given by:

$$\alpha_T = \frac{1}{\cosh\left(W_{NBL} / L_{pNB}\right)}$$ [17.3]

which is independent of the collector bias. Here, L$_{p,NB}$ is the diffusion length for holes in the N-buffer layer. This analysis neglects the depletion region extension within the N-buffer layer. The diffusion length for holes ($L_{p,NB}$) in the N-buffer layer depends upon the diffusion coefficient and the minority carrier lifetime in the N-buffer layer. The diffusion coefficient varies with the doping concentration in the N-buffer layer based upon the concentration dependence of the mobility. In addition, the minority carrier lifetime has been found to be dependent upon the doping concentration[17] in the case of silicon devices. Although this phenomenon has not been verified for silicon carbide, it is commonly used when performing numerical analysis of silicon carbide devices. The effect can be modeled by using the relationship:

$$\frac{\tau_{LL}}{\tau_{p0}} = \frac{1}{1 + \left(N_D / N_{REF}\right)}$$ [17.4]

where N_{REF} is a reference doping concentration whose value will be assumed to be 5×10^{16} cm^{-3} for 4H-SiC.

The multiplication factor for a P-N junction is given by:

$$M = \frac{1}{1 - \left(V_A / BV_{PP}\right)^n}$$ [17.5]

with the avalanche breakdown voltage of the P-base/N-base junction (BV_{PP}) *without the punch-through phenomenon*. The value for n of 6 for Si is also assumed for 4H-SiC. In order to apply this formulation to the punch-through case relevant to the asymmetric 4H-SiC IGBT structure, it is necessary to relate the maximum electric field at the junction for the punch-through case to the non-punch-through case. The electric field at the interface between the lightly doped portion of the N-base region and the N-buffer layer is given by:

$$E_1 = E_m - \frac{q N_D W_N}{\varepsilon_S}$$ [17.6]

The voltage supported by the device is given by:

$$V_C = \left(\frac{E_m + E_1}{2}\right) W_N = E_m W_N - \frac{q N_D}{2\varepsilon_S} W_N^2$$ [17.7]

From this expression, the maximum electric field is given by:

$$E_m = \frac{V_C}{W_N} + \frac{qN_DW_N}{2\varepsilon_S} \qquad [17.8]$$

The corresponding equation for the non-punch-through case is:

$$E_m = \sqrt{\frac{2qN_DV_{NPT}}{\varepsilon_S}} \qquad [17.9]$$

Consequently, the non-punch-through voltage that determines the multiplication coefficient 'M' corresponding to the applied collector bias 'V_C' for the punch-through case is given by:

$$V_{NPT} = \frac{\varepsilon_S E_m^2}{2qN_D} = \frac{\varepsilon_S}{2qN_D}\left(\frac{V_C}{W_N} + \frac{qN_DW_N}{2\varepsilon_S}\right)^2 \qquad [17.10]$$

The multiplication coefficient for the asymmetric 4H-SiC IGBT structure can be computed by using this non-punch-through voltage:

$$M = \frac{1}{1-\left(V_{NPT}/BV_{PP}\right)^n} \qquad [17.11]$$

The multiplication coefficient increases with increasing collector bias. The open-base transistor breakdown voltage (and the forward blocking capability of the asymmetric IGBT structure) is determined by the collector voltage at which the multiplication factor becomes equal to the reciprocal of the product of the base transport factor and the emitter injection efficiency.

The n-channel asymmetric 4H-SiC IGBT structure must have a forward blocking voltage of 22,000 volts for a 20-kV rated device. In the case of avalanche breakdown, there is a unique value for the doping concentration of 3.2×10^{14} cm^{-3} for the drift region with a width of 272 microns to obtain this blocking voltage. In the case of the asymmetric 4H-SiC IGBT structure, it is advantageous to use a much lower doping concentration for the lightly doped portion of the N-base region in order to reduce its width. The strong conductivity modulation of the N-base region during on-state operation favors a smaller thickness for the N-base region independent of its original doping concentration. A doping concentration of 2.0×10^{14} cm^{-3} for the N-base region will be assumed for the n-channel asymmetric

4H-SiC IGBT structure analyzed in this section. This typical for very high voltage 4H-SiC IGBTs reported in the literature[14].

The doping concentration of the N-buffer layer must be sufficiently large to prevent reach-through of the electric field to the P^+ collector region. Although the electric field at the interface between the N-base region and the N-buffer layer is slightly smaller than at the blocking junction (J_2), a worse case analysis can be done by assuming that the electric field at this interface is close to the critical electric field for breakdown in the drift region. The minimum charge in the N-buffer layer to prevent reach-through can be then obtained using:

$$N_{DBL}W_{NBL} = \frac{\varepsilon_S E_C}{q} \qquad [17.12]$$

Using a critical electric for breakdown in silicon carbide of 1.47×10^6 V/cm for a doping concentration of 2.0×10^{14} cm^{-3} in the N-base region, the minimum charge in the N-buffer layer to prevent reach-through for a silicon carbide asymmetric IGBT structure is found to be 8×10^{12} cm^{-2}. An N-buffer layer with doping concentration of 5×10^{16} cm^{-3} and thickness of 5 microns has a charge of 2.5×10^{13} cm^{-2} that satisfies this requirement.

The asymmetric n-channel 4H-SiC IGBT structure will be assumed to have a P^+ collector region with doping concentration of 1×10^{19} cm^{-3}. It will be assumed that all the acceptors are ionized even at room temperature although the relatively deep acceptor level in silicon carbide may lead to incomplete dopant ionization. In this case, the emitter injection efficiency computed using Eq. [17.1] is 0.971. When the device is close to breakdown, the entire N-base region is depleted and the base transport factor computed by using Eq. [17.3] in this case is 0.903. In computing these values, a lifetime of 1 microsecond was assumed for the N-base region resulting in a lifetime of 0.5 microseconds in the N-buffer layer due to the scaling according to Eq. [17.4]. Based up on Eq. [17.2], open-base transistor breakdown will then occur when the multiplication coefficient becomes equal to 1.14 for the above values for the injection efficiency and base transport factor.

The forward blocking capability for the silicon carbide n-channel asymmetric IGBT structure can be computed by using Eq. [7.2] for various widths for the N-base region. The analysis requires determination of the voltage V_{NPT} by using Eq. [17.10] for each width of the N-base region. The resulting values for the forward blocking voltage are plotted in Fig. 17.2. From this graph, the N-base region width required to obtain a forward blocking voltage of 23,000-V is 165 microns.

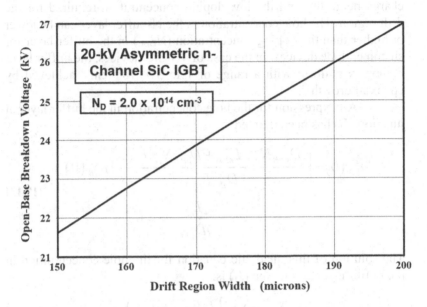

Fig. 17.2 Optimization of drift region width for the 20-kV asymmetric n-channel 4H-SiC IGBT structure.

A forward blocking voltage of 16 kV was reported[18] for an n-channel 4H-SiC IGBT fabricated using a drift region with doping concentration of 4×10^{14} cm^{-3} and thickness of 150 μm. The forward blocking voltage of 22.6 kV has been reported[14] for an n-channel asymmetric 4H-SiC IGBT by using a drift region thickness of 180 μm in the case of an N-base doping concentration of 2×10^{14} cm^{-3}. This larger drift region thickness is due to the smaller N-buffer layer thickness of 2 microns used by the authors. A blocking voltage of 27.5 kV was achieved[19] by using a drift region with doping concentration of 1×10^{14} cm^{-3} and thickness of 210 μm. These experimental results are consistent with the solutions obtained using the above analytical model.

17.1.2 On-State Voltage Drop

A generally applicable analytical model has been developed for the silicon asymmetric IGBT structure which is valid for any injection level in the buffer layer[20]. This analytical model can also be applied to silicon carbide devices. The carrier distribution profiles in the on-state for the asymmetric IGBT structure are shown in the textbook[16]. The hole and electron concentrations in the N-base region are equal due to charge neutrality and the low doping concentration required for the drift region. The hole concentration in the N-buffer layer can be lower or higher than the doping concentration (N_{DBL}) in the buffer layer for silicon carbide devices. In the case of 4H-SiC devices, the buffer layer doping is uniform with a range of 1 to 5×10^{17} cm^{-3} achieved by epitaxial growth[12].

An expression for the hole concentration in the buffer layer at junction (J_1) has been derived:

$$p_{NB}^2(0) + \left(\frac{D_{pNB} N_{AP+} L_{nP+} + D_{nP+} N_{DB} L_{pNB}}{D_{nP+} L_{pNB}} \right) p_{NB}(0)$$
$$- \frac{N_{AP+} L_{nP+} J_C}{q D_{nP+}} = 0 \quad \text{[17.13]}$$

The solution of this quadratic equation for the hole concentration in the buffer layer at junction (J_1) is:

$$p_{NB}(0) = \frac{1}{2} \left(\sqrt{b^2 - 4c} - b \right) \quad \text{[17.14]}$$

where

$$b = \frac{D_{pNB} N_{AP+} L_{nP+} + D_{nP+} N_{DB} L_{pNB}}{D_{nP+} L_{pNB}} \quad \text{[17.15]}$$

and

$$c = - \frac{N_{AP+} L_{nP+} J_C}{q D_{nP+}} \quad \text{[17.16]}$$

Since the unified analytical model presented is valid for all injection levels in the N-buffer layer, it can be used to predict the variation of the injected hole concentration with lifetime in the N-base region and the doping concentration in the N-buffer layer for the asymmetric 4H-SiC IGBT structure.

The holes diffuse through the buffer layer producing a concentration [$p(W_{NB-})$] inside the buffer layer at the boundary between the N-buffer layer and the N-base region:

$$p\left(W_{NB-}\right)= p_{NB}\left(0\right)e^{-\left(W_{NBL}/L_{pNB}\right)} \qquad [17.17]$$

where W_{NBL} is the thickness of the buffer layer. The hole concentration [$p(W_{NB-})$] in the N-base region at the boundary between the N-buffer layer and the N-base region can be obtained by equating the hole current density on the two sides of this boundary[16]:

$$p\left(W_{NB+}\right)=\frac{L_a\tanh\left[\left(W_N+W_{NBL}\right)/L_a\right]}{2qD_p}J_p\left(W_{NB-}\right) \qquad [17.18]$$

with

$$J_p\left(W_{NB-}\right)= J_p\left(0\right)e^{-\left(W_{NBL}/L_{pNB}\right)} \qquad [17.19]$$

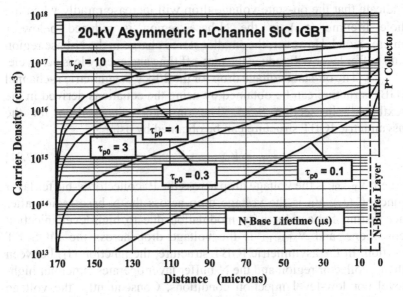

Fig. 17.3 Injected carrier profiles in the 20-kV n-channel asymmetric 4H-SiC IGBT structure: lifetime dependence.

The hole concentration profile in the N-base region as dictated by high-level injection conditions is given by[16]:

$$p\left(y\right)= p\left(W_{NB+}\right)\frac{\sinh\left[\left(W_N+W_{NBL}-y\right)/L_a\right]}{\sinh\left[\left(W_N+W_{NBL}\right)/L_a\right]} \qquad [17.20]$$

which is valid for y > W_{NBL}.

The free carrier distribution obtained by using the above equations is provided in Fig. 17.3 for the case of the 20-kV asymmetric n-channel IGBT structure with an N-base region thickness of 165 microns and a buffer layer thickness of 5 microns. The hole lifetime (τ_{p0}) in the N-base region was varied for these plots from 0.1 to 10 microseconds. Note that the high-level lifetime (τ_{HL}) in these cases is two-times the hole lifetime (τ_{p0}). It can be observed that the hole concentration [$p_{NB}(0)$] decreases at the collector side of the N-buffer layer (at y = 0) from 1.8×10^{17} cm^{-3} to 2.25×10^{16} cm^{-3}. In addition, the hole concentration is significantly reduced at the emitter side when the lifetime in the N-base region decreases. The carrier density falls below 1×10^{15} cm^{-3} over a significant portion of the N-base region when the lifetime becomes smaller than 1 microsecond. These results indicate that the on-state voltage drop will increase rapidly when the hole lifetime (τ_{p0}) in the N-base region is reduced below 1 microsecond. However, the smaller stored charge in the N-base region and buffer layer will reduce the turn-off time and energy loss per cycle.

The on-state voltage drop for the 20-kV asymmetric n-channel IGBT structure can be obtained by using the equations derived in the textbook[16] in section 9.5.5. The on-state voltage drop for the asymmetric IGBT structure can be obtained by using:

$$V_{ON} = V_{P+NBL} + V_B + V_{MOSFET} \qquad [17.21]$$

where V_{P+NBL} is the voltage drop across the P$^+$ collector/N-buffer layer junction (J_1), V_B is the voltage drop across the N-base region after accounting for conductivity modulation due to high-level injection conditions, and V_{MOSFET} is the voltage drop across the MOSFET portion. In the asymmetric IGBT structure, the junction (J_1) between the P$^+$ collector region and the N-buffer layer operates at neither high-level nor low-level injection conditions. Consequently, the voltage drop across the junction (J_1) must be obtained using:

$$V_{P+NB} = \frac{kT}{q} \ln \left(\frac{p_{NB}(0) N_{BL}}{n_i^2} \right) \qquad [17.22]$$

The voltage drop across the N-base region can be obtained by integrating the electric field inside the N-base region. The voltage drop is obtained by taking the sum of two parts. The first part is given by:

$$V_{B1} = \frac{2L_a J_C \sinh\left(W_N / L_a\right)}{qp\left(W_{NB+}\right)\left(\mu_n + \mu_p\right)} \left\{ \tanh^{-1}\left[e^{-(W_{ON}/L_a)} \right] - \tanh^{-1}\left[e^{-(W_N/L_a)} \right] \right\}$$

[17.23]

The depletion width (W_{ON}) across the P-base/N-base junction (J_2) in the on-state depends on the on-state voltage drop. The voltage drop associated with the second part is given by:

$$V_{B2} = \frac{kT}{q}\left(\frac{\mu_n - \mu_p}{\mu_n + \mu_p} \right) \ln\left[\frac{\tanh\left(W_{ON}/L_a\right)\cosh\left(W_{ON}/L_a\right)}{\tanh\left(W_N/L_a\right)\cosh\left(W_N/L_a\right)} \right]$$

[17.24]

For the planar gate IGBT structure considered here, the voltage drop across the MOSFET portion includes the channel, accumulation, and JFET regions[16]. The contribution from channel is given by:

$$V_{CH} = \frac{J_C L_{CH} W_{CELL}}{2\mu_{ni} C_{OX}\left(V_G - V_{TH}\right)}$$

[17.25]

The contribution from JFET region is given by:

$$V_{JFET} = \frac{J_C \rho_{JFET}\left(t_{P+} + W_0\right)W_{CELL}}{W_{JFET} - 2W_0}$$

[17.26]

The contribution from accumulation layer is given by:

$$V_{ACC} = \frac{J_C K_A W_{JFET} W_{CELL}}{4\mu_{nA} C_{OX}\left(V_G - V_{TH}\right)}$$

[17.27]

The on-state voltage drop (at an on-state current density of 25 A/cm^2) computed for the 20-kV asymmetric n-channel silicon IGBT structure by using the above equations is provided in Fig. 17.4 as a function of the high-level lifetime in the N-base region. This asymmetric 4H-SiC IGBT structure had the optimized N-base region width of 165 microns and N-buffer layer width of 5 microns. The N-buffer layer doping concentration was kept at 5×10^{16} cm^{-3} for all the cases. From the figure, it can be observed that the on-state voltage drop is close that of the collector/N-buffer layer junction (~3.2 V) when the high-level lifetime is greater than 2 microseconds. The on-state voltage drop increases rapidly when the high-level lifetime is reduced below 0.6 microseconds due an increase in the voltage drop across the N-base region. This is consistent with the lack of conductivity modulation of the drift region when the lifetime (τ_{p0}) becomes less than 0.3 microseconds as shown in Fig. 17.3. The on-state voltage drop is predicted as 3.65 volts for a high-level lifetime of 1 µs in the N-base region using the analytical model. It increases rapidly to 7.5 volts for the case of a high-level lifetime of 0.4 µs.

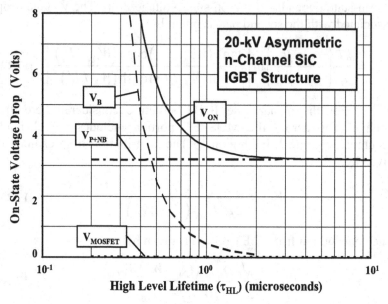

Fig. 17.4 On-state voltage drop for the 20-kV asymmetric 4H-SiC n-channel IGBT structure: N-base lifetime dependence.

The on-state voltage drop for a 16-kV n-channel 4H-SiC IGBT has been reported[18] to be 5-V at an on-state current density of 100 A/cm². The on-state voltage drop for a 22 kV, 1 cm² 4H-SiC n-channel IGBT has been reported[14] as 7.5 V at 20-A. The on-state voltage drop for an n-channel 4H-SiC IGBT has been reported[21] to increase when the buffer layer thickness is increased from 2 to 5 microns. This is consistent with the above analytical model because the injected excess carrier density in the N-base region is reduced when the buffer layer thickness is increased. The on-state voltage drop for an n-channel 4H-SiC IGBT capable of supporting 27.5-kV was reported[19] as 11.7 V due to its large (210 µm) thick drift layer.

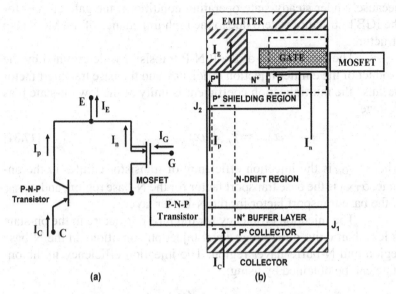

Fig. 17.5 Equivalent circuit for the 20-kV asymmetric 4H-SiC n-Channel
IGBT structure in the on-state.

The equivalent circuit for an n-channel IGBT structure[16] based
upon the P-N-P Transistor/MOSFET model consists of an n-channel
MOSFET providing the base drive current to a P-N-P transistor as
shown in Fig. 17.5(a). The P-N-P transistor and MOSFET portions are
identified by the dashed boxes in cross-section shown in Fig. 17.5(b).
The emitter current for the IGBT structure consists of the hole current
flow via the P-N-P transistor and the electron current via the MOSFET
portion:

$$I_E = I_p + I_n \qquad [17.28]$$

The electron current serves as the base drive current for the P-N-P
transistor. Consequently, these currents are inter-related by the
common base current gain of the P-N-P transistor:

$$I_p = \alpha_{PNP} I_E = \alpha_{PNP} I_C \qquad [17.29]$$

and:

$$I_n = (1 - \alpha_{PNP}) I_E = (1 - \alpha_{PNP}) I_C \qquad [17.30]$$

because, under steady-state operating conditions, the gate current for the IGBT structure is zero due to the high impedance of the MOS gate structure.

The current gain of the P-N-P transistor is determined by the product of the emitter injection efficiency and the base transport factor because the multiplication coefficient is unity at the low on-state bias voltages:

$$\alpha_{PNP} = \gamma_{E,ON} . \alpha_{T,NB} . \alpha_{T,NBL} \qquad [17.31]$$

where $\gamma_{E,ON}$ is the injection efficiency of transistor emitter in the on-state, $\alpha_{T,NB}$ is the base transport factor for the N-base region, and $\alpha_{T,NBL}$ is the base transport factor for the N-buffer layer.

The injection efficiency for the IGBT structure in the on-state is less than unity due to high-level injection conditions in the N-base region and N-buffer layer region. The injection efficiency in the on-state can be obtained by using:

$$\gamma_{E,ON} = \frac{J_p(J_1)}{J_{C,ON}} \qquad [17.32]$$

where $J_p(J_1)$ is the hole current density at junction J_1, which can computed using:

$$J_P(J_1) = \frac{qD_{pNB}p_{NB}(0)}{L_{pNB}} \qquad [17.33]$$

The base transport factor for the N-base region in the on-state can be obtained by using[16]:

$$\alpha_{T,N-Base,0} = \frac{J_p(W_N)}{J_p(W_{NB}+)} \qquad [17.34]$$

where $J_p(W_N)$ is the hole current density at junction J_2 and $J_p(W_{NB}+)$ is the hole current density at interface between the N-base region and the N-buffer layer. These current densities can be obtained by using[16]:

$$J_p(W_{NB}+) = \left[\left(\frac{\mu_p}{\mu_p + \mu_n}\right) + \left(\frac{\mu_n}{\mu_p + \mu_n}\right)K_{AS}\right]J_C \qquad [17.35]$$

and

$$J_p\left(W_N\right) = \left\{ \frac{\left[\left(\dfrac{\mu_p}{\mu_p+\mu_n}\right)-\left(\dfrac{\mu_n K_{AS}}{\mu_p+\mu_n}\right)\right]}{\left[\sinh\left(\dfrac{W_N}{L_a}\right)\tanh\left(\dfrac{W_N}{L_a}\right)-\cosh\left(\dfrac{W_N}{L_a}\right)\right]} \right\} J_C \qquad [17.36]$$

The base transport factor in the conductivity modulated lightly doped portion of the N-base region is enhanced by the combination of drift and diffusion due to the high-level injection conditions.

The base-transport factor associated with the N-buffer layer can be obtained from the decay of the hole current within the N-buffer layer as given by low-level injection theory[16]:

$$\alpha_{T,N-Buffer} = \frac{J_p\left(W_{NB}-\right)}{J_p\left(y_N\right)} = e^{-W_{NBL}/L_{pNB}} \qquad [17.37]$$

The case of the 20-kV asymmetric 4H-SiC n-channel IGBT structure with a N-base region width of 165 microns and a low-level lifetime of 1 microsecond in the N-base region can be analyzed using the above analytical solutions. The device will be assumed to have an N-buffer layer doping concentration of 5×10^{16} cm^{-3} and thickness of 5 microns; and P$^+$ collector region (emitter region of the internal PNP transistor) doping concentration of 1×10^{19} cm^{-3}. With these device parameters, the injected concentration of holes [$p_{NB}(0)$] in the N-buffer layer at the junction (J_1) is found to be 6.9×10^{16} cm^{-3} by using Eq. [17.14]. Using this value in Eq. [17.32], the injection efficiency in the on-state is found to be 0.959. With these parameters, the base transport factor for the N-base region is found to be 0.285 by using Eq. [17.34] and the base transport factor for the N-buffer layer is found to be 0.631 by using Eq. [17.37]. Combining these values, the common base current gain of the PNP transistor (α_{PNP}) in the on-state is found to be 0.172. Based up on this analysis, it can be concluded that only 17 percent of the emitter current of the 20-kV asymmetric 4H-SiC n-channel IGBT structure is due to the hole current component (I_p) and most of the current flow consists of the electron current (I_n) at the emitter side. This has been confirmed by numerical simulations[20].

17.1.3 Turn-Off Characteristics

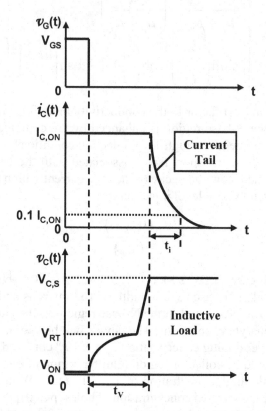

Fig. 17.6 Turn-off waveforms for the asymmetric n-channel 4H-SiC IGBT structure.

The turn-off behavior for the asymmetric 4H-SiC IGBT structure can be expected to be similar to that for the silicon asymmetric IGBT structure. However, the doping concentration of the drift region in the silicon carbide devices is two orders to magnitude larger than that for the silicon devices. This has a significant impact on the turn-off waveforms.

In the case of the silicon device, the hole concentration in the space charge region during the voltage rise-time becomes larger than the doping concentration in the drift region. This additional charge in the space charge region reduces its width to less than the width of the N-base region during the entire voltage transient. Consequently, the space-charge-region does not reach-though to the N-buffer layer at the end of the voltage rise-time when the collector voltage becomes equal to the supply voltage. During the current fall-time, the stored charge remaining in the N-base region must be first removed until the space-charge-region punches through to the N-buffer layer. This is followed by the recombination of the stored charge in the buffer layer. For the asymmetric silicon IGBT structure, a single phase is observed for the voltage transient while the current decays in two phases[20].

In contrast, the hole concentration in the space charge region during the voltage rise-time for the asymmetric 4H-SiC IGBT structure is much smaller than the doping concentration of the drift region. Consequently, the space-charge-region reaches-through to the buffer layer during the voltage rise-time at a collector bias that is well below the collector supply voltage. The electric field in the N-base region takes a trapezoidal shape after the space-charge-region reaches-through to the buffer layer, allowing the collector voltage to rise at a much more rapid rate until it reaches the supply voltage. All the stored charge in the N-base region is therefore removed during the voltage rise-time. During the current fall-time, the stored charge in the buffer layer is removed by recombination. This occurs with a single current decay transient.

In order to turn-off the IGBT structure, the gate voltage must simply be reduced from the on-state value (nominally 15 volts) to zero as illustrated in Fig. 17.6. The magnitude of the gate current can be limited by using a resistance in series with the gate voltage source. The waveform for the gate voltage shown in the figure is for the case of zero gate resistance. Once the gate voltage falls below the threshold voltage, the electron current from the channel ceases. In the case of an inductive load, the collector current for the IGBT structure is then sustained by the hole current flow due to the presence of stored charge in the N-base region. The collector voltage begins to increase in the IGBT structure immediately after the gate voltage reduces below threshold voltage.

17.1.3.1 Voltage Rise-Time:

The analysis of the turn-off waveform for the collector voltage transient for the asymmetric 4H-SiC IGBT structure can be performed by using the charge control principle. The concentration p_{WNB+} in the on-state at the interface between the lightly doped portion of the N-base region and the N-buffer layer was previously derived for the silicon carbide asymmetric IGBT structure in section 17.1.2. To develop the analysis of the collector voltage transient, it will be assumed that the hole concentration profile in the N-base region does not change due to recombination. In this case, the electric field profile in the asymmetric 4H-SiC IGBT structure during the collector voltage transient is illustrated in Fig. 17.7. As the space charge region expands towards the collector side, holes are removed from the stored chare region at its boundary. The holes then flow through the space charge region at their saturated drift velocity due to the high electric field in the space charge region.

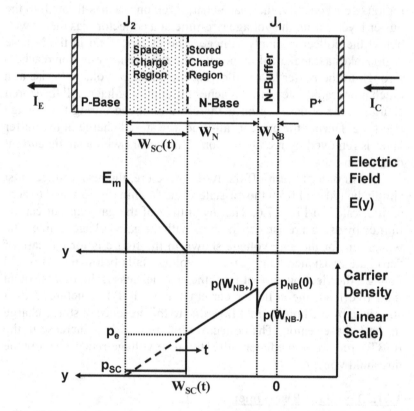

Fig. 17.7 Electric field and carrier concentration in the asymmetric n-channel 4H-SiC IGBT structure during the voltage transient.

The concentration of holes at the edge of the space-charge region (p_e) increases during the turn-off process as the space-charge width increases:

$$p_e(t) = p(W_{NB+}) \frac{\sinh\left[W_{SC}(t)/L_a\right]}{\sinh\left[(W_N + W_{NB})/L_a\right]} \qquad [17.38]$$

According to the charge-control principle, the charge removed by the expansion of the space-charge layer must equal the charge removed due to collector current flow:

$$J_{C,ON} = qp_e(t)\frac{dW_{SC}(t)}{dt} = qp(W_{NB+})\frac{\sinh\left[W_{SC}(t)/L_a\right]}{\sinh\left[(W_N + W_{NB})/L_a\right]}\frac{dW_{SC}(t)}{dt}$$

[17.39]

by using Eq. [17.38]. Integrating this equation on both sides and applying the boundary condition of width $W_{SC}(0)$ for the space-charge layer at time zero provides the solution for the evolution of the space-charge region width with time:

$$W_{SC}(t) = L_a\ a\cosh\left\{\begin{array}{l}\dfrac{J_{C,ON}\sinh\left[(W_N + W_{NBL})/L_a\right]}{qL_a p(W_{NB+})}t \\ +\cosh\left[W_{SC}(0)/L_a\right]\end{array}\right\}$$

[17.40]

The space-charge layer expands towards the right-hand-side as indicated by the horizontal time arrow in Fig. 17.7 with the hole concentration profile in the stored charge region remaining unchanged.

The collector voltage supported by the asymmetric silicon carbide IGBT structure is related to the space charge layer width by:

$$V_C(t) = \frac{q(N_D + p_{SC})W_{SC}^2(t)}{2\varepsilon_S}$$

[17.41]

The hole concentration in the space-charge layer can be related to the collector current density under the assumption that the carriers are moving at the saturated drift velocity in the space-charge layer:

$$p_{SC} = \frac{J_{C,ON}}{qv_{sat,p}}$$

[17.42]

The hole concentration in the space-charge region remains constant during the voltage rise-time because the collector current density is constant. Consequently, the slope of the electric field profile in the space-charge region also becomes independent of time. This analytical model for turn-off of the asymmetric 4H-SiC IGBT structure under inductive load conditions predicts a non-linear increase in the collector voltage with time.

The collector voltage increases in accordance with the above model until the space charge region reaches-through the N-base region. The reach-through voltage during the turn-off of the asymmetric 4H-SiC IGBT must be computed with inclusion of the positive charge due to the presence of holes in the space-charge-region associated with the collector current flow:

$$V_{RT}\left(J_{C,ON}\right) = \frac{q\left(N_D + p_{SC}\right)W_N^2}{2\varepsilon_S} \qquad [17.43]$$

For the asymmetric 4H-SiC n-channel IGBT structure, the hole concentration in the space-charge-region (p_{SC}) computed using Eq. [17.42] is 1.8×10^{13} cm^{-3} at an on-state current density of 25 A/cm^2 based up on a saturated velocity of 8.6×10^6 cm/s for holes. This is smaller than the doping concentration of the N-base region. For the case of the 20-kV asymmetric silicon carbide IGBT structure with an N-base width of 165 microns and doping concentration of 1.5×10^{14} cm^{-3}, the reach-through voltage is found to be 4262 volts. In contrast, the reach-through voltage under forward blocking operation is only 3150 volts. The time at which reach-through occurs can be derived from Eq. [17.40] by setting the space-charge-region width equal to the width of the N-base region:

$$t_{RT} = \frac{qL_a p\left(W_{NB}+\right)}{J_{C,ON}} \left\{ \frac{\cosh\left[W_N / L_a\right] - \cosh\left[W_{SC}\left(0\right)/L_a\right]}{\sinh\left[\left(W_N + W_{NB}\right)/L_a\right]} \right\} \qquad [17.44]$$

Once the space-charge-region reaches-through the N-base region, all the stored charge in the N-base region has been removed by the voltage transient. However, there is still substantial stored charge in the N-buffer layer. The expansion of the space-charge-region is now curtailed by the high doping concentration of the N-buffer layer.

The end of the first phase of the turn-off process occurs when the collector voltage rises to the reach-through voltage (V_{RT}). At this time, the space-charge-region has reached the edge of the N-buffer layer. This forces the hole concentration at the edge of the space-charge-region (at y = W_{NB} in Fig. 17.8) to the hole concentration (p_{SC}) inside the space-charge-region, which is close to zero when compared with the injected hole concentration at the junction (J_1). The hole concentration in the N-buffer layer at junction (J_1) changes abruptly when reach-through occurs in order to maintain the same current density because the collector current density is held fixed during the voltage rise-time.

The hole concentration [p(0,t$_{RT}$)] in the N-buffer layer at junction (J$_1$) during the second phase of the voltage rise-time can be obtained by analysis of current transport in the N-buffer layer. The hole concentration distribution in the N-buffer layer has the same boundary conditions as the base region of a bipolar transistor operating in its active region with finite recombination in the base region[16]:

$$p(y) = p(0, t_{RT}) \left\{ \frac{\sinh\left[(W_{NBL} - y) / L_{pNB} \right]}{\sinh\left(W_{NBL} / L_{pNB} \right)} \right\} \quad [17.45]$$

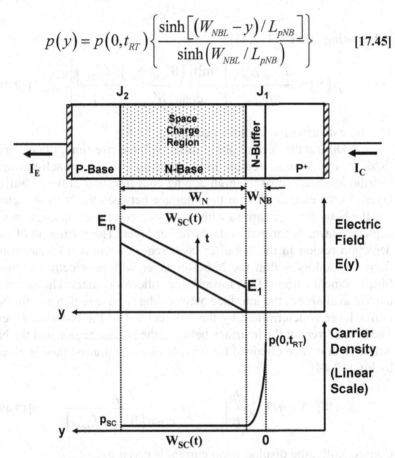

Fig. 17.8 Electric field and carrier distribution in the asymmetric 4H-SiC IGBT structure during the second phase of the voltage rise-time.

The hole current in the buffer layer at the junction (J$_1$) is equal to the collector current density:

$$J_p(0) = qD_{pNB} \frac{dp}{dy}\bigg|_{y=0} = J_{C,ON} \qquad \text{[17.46]}$$

Using Eq. [17.45] for the hole carrier distribution:

$$p(0, t_{RT}) = \frac{J_{C,ON} L_{pNB} \tanh\left(W_{NBL} / L_{pNB}\right)}{qD_{pNB}} \qquad \text{[17.47]}$$

Substituting into Eq. [17.45]:

$$p(y) = \frac{J_{C,ON} L_{pNB}}{qD_{pNB}} \left\{ \frac{\sinh\left[\left(W_{NBL} - y\right) / L_{pNB}\right]}{\cosh\left(W_{NBL} / L_{pNB}\right)} \right\} \qquad \text{[17.48]}$$

This hole distribution is illustrated in Fig. 17.8.

During the second phase of the voltage rise-time, the electric field in the N-base region must increase with a punch-through distribution because of the high doping concentration of the N-buffer layer. As the electric field at the interface between the N-base region and the N-buffer layer grows with increasing collector voltage, a small depletion layer is formed in the N-buffer layer. The formation of the depletion region in the N-buffer layer requires removal of electrons from the donors within the N-buffer layer with an electron current (displacement current) flow towards the collector contact. The electron current available at the interface between the N-base region and the N-buffer layer is determined by the hole current in the N-buffer layer. The hole current at the interface between the N-base region and the N-buffer layer can be obtained from the hole concentration profile given by Eq. [17.48]:

$$J_p(W_{NB}) = qD_{pNB} \frac{dp}{dy}\bigg|_{y=W_{NBL}} = \frac{J_{C,ON}}{\cosh\left(W_{NBL} / L_{pNB}\right)} \qquad \text{[17.49]}$$

Consequently, the displacement current is given by:

$$J_D = J_n(W_{NBL}) = J_{C,ON} - J_p(W_{NBL}) = J_{C,ON} \left[1 - \frac{1}{\cosh\left(W_{NBL} / L_{pNB}\right)}\right]$$

$$\text{[17.50]}$$

The capacitance of the space-charge-region during the second phase of the voltage rise-time is independent of the collector voltage because the space-charge-region width is essentially equal to the width of the N-base region because the depletion width in the N-buffer layer is very small due to its high doping concentration. The (specific) capacitance of the space-charge-region can be obtained using:

$$C_{SCR} = \frac{\varepsilon_S}{W_N}$$ [17.51]

The rate of rise of the collector voltage based up on charging the space-charge-region capacitance is given by:

$$\frac{dV_C}{dt} = \frac{J_D}{C_{SCR}}$$ [17.52]

Using Eq. [17.50]:

$$\frac{dV_C}{dt} = \frac{J_{C,ON}}{C_{SCR}} \left[1 - \frac{1}{\cosh\left(W_{NBL} / L_{pNB}\right)} \right]$$ [17.53]

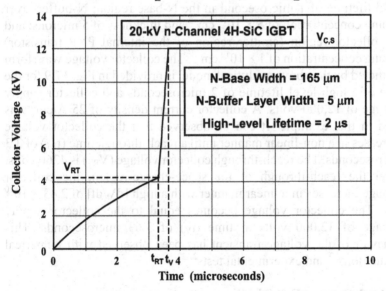

Fig. 17.9 Collector voltage waveform for the asymmetric n-channel 4H-SiC IGBT structure during inductive load turn-off.

The collector voltage waveform after reach-through is then given by:

$$V_C(t) = V_{RT}(J_{C,ON}) + \frac{J_{C,ON}}{C_{SCR}}\left[1 - \frac{1}{\cosh(W_{NBL}/L_{pNB})}\right]t \quad [17.54]$$

According to this analytical model, the collector voltage should increase linearly with time after the space-charge-region reaches-through the N-base region. The end of the voltage rise-time occurs when the collector voltage becomes equal to the collector supply voltage ($V_{C,S}$). Using this criterion in Eq. [17.54], the collector voltage rise-time interval is obtained:

$$t_V = t_{RT} + \frac{\varepsilon_S}{J_{C,ON}W_N}\left[\frac{\cosh(W_{NBL}/L_{pNB})}{\cosh(W_{NBL}/L_{pNB})-1}\right]\left[V_{C,S} - V_{RT}(J_{C,ON})\right]$$

$$[17.55]$$

Consider the case of the 20-kV asymmetric 4H-SiC n-channel IGBT structure with a N-base region width of 165 microns and a low-level lifetime of 1 microsecond in the N-base region; N-buffer layer doping concentration of 5×10^{16} cm^{-3} and thickness of 5 microns; and P$^+$ collector region (emitter region of the internal PNP transistor) doping concentration of 1×10^{19} cm^{-3}. The collector voltage waveform predicted by the above analytical model is provided in Fig. 17.9 for the case of a high-level lifetime of 2 microseconds and collector supply voltage of 12,000 volts. A collector current density of 25 A/cm^2 was used in this example. It can be observed that the collector voltage increases in a non-linear manner until a reach-through time (t_{RT}) of 3.3 microseconds. The reach-through collector voltage (V_{RT}) is 4260 volts. After the reach-through of the space-charge-region, the collector voltage increases in a linear manner with a high [dV/dt] of 2.43×10^{10} V/s. The collector voltage becomes equal to the collector supply voltage of 12,000 volts at time (t_V) of 3.63 microseconds. This behavior of the voltage transient has been observed with numerical simulations[20] and experimental tests[21,22].

17.1.3.2 Current Fall-Time:

At the end of the collector voltage transient in the asymmetric 4H-SiC IGBT, the space-charge-region has extended through the entire N-base region leaving stored charge only in the N-buffer layer. The collector current decays due to the recombination of this stored charge under low-level injection conditions. Unlike in the case of the silicon IGBT, the collector current transient occurs in a single phase as described by:

$$J_C(t) = J_{C,ON} e^{-t/\tau_{BL}}$$ [17.56]

where t_{BL} is the low-level lifetime in the N-buffer layer.

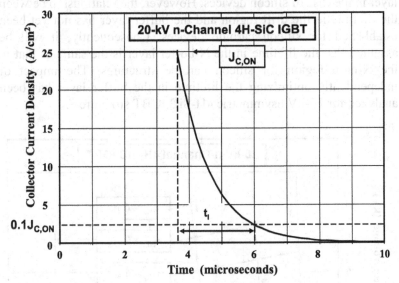

Fig. 17.10 Collector current waveform for the asymmetric n-Channel 4H-SiC IGBT structure during inductive load turn-off.

The collector current waveform for the 20-kV n-channel 4H-SiC asymmetric IGBT structure obtained by using the above model is provided in Fig. 17.10. The current fall-time is defined as the time taken for the current to reduce to 10 percent of the on-state value. In this case, the current fall time obtained by using Eq. [17.56] is:

$$t_I = 2.303 \tau_{BL}$$ [17.57]

For the above example, the current-fall time is found to be 2.3 microseconds if no scaling of the lifetime with buffer layer doping is taken into account. The results of numerical simulations support the analytical models derived here[20]. In addition, the turn-off waveforms for the 22 kV asymmetric 4H-SiC n-channel IGBT are very similar to those described by the analytical model[16].

17.1.4 Lifetime Dependence

From an applications perspective, the optimization of the power losses for the IGBT structure requires performing a trade-off between the on-state voltage drop and the switching losses. One approach to achieve this is by adjusting the lifetime in the drift (N-base) region. A reduction of the lifetime in the drift region also alters the lifetime in the N-buffer layer in the case of silicon devices. However, the relationship between the lifetime in the drift region and the buffer layer has not yet been established for silicon carbide devices. Consequently, it will be assumed that the lifetime in the N-buffer layer is the same as that in the N-base region for silicon carbide structures. The impact of independently optimizing the lifetime in the buffer layer has been analyzed for 15-kV asymmetric 4H-SiC IGBT structures[9].

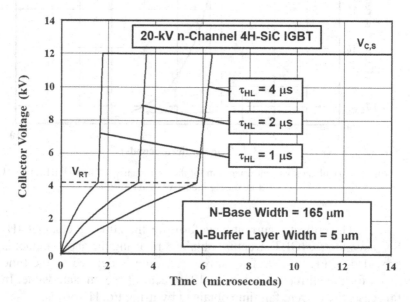

Fig. 17.11 Collector voltage transients during turn-off for the 20-kV asymmetric n-channel 4H-SiC IGBT structure: lifetime dependence.

The impact of reducing the lifetime in the drift region on the on-state voltage drop for the 20-kV asymmetric n-channel 4H-SiC IGBT structure was previously discussed in section 17.1.2. As in the case of silicon devices, the on-state voltage drop increases when the lifetime is reduced. The analytical model developed for turn-off of the asymmetric 4H-SiC IGBT structure can be used to analyze the impact of changes to the lifetime in the drift region on the turn-off characteristics. The collector voltage transients predicted by the analytical model are shown in Fig. 17.11 for the case of the 20-kV asymmetric 4H-SiC IGBT structure operating with an on-state current density of 25 A/cm^2. The voltage rise-time increases when the lifetime is increased because of the larger concentration for the holes in the N-base region that are being removed during the collector voltage transient. The voltage rise-times obtained by using the analytical model are 1.77, 3.63, and 6.38 microseconds for high-level lifetime values of 1, 2, and 4 microseconds, respectively. In all cases, the reach-through voltage has the same value as predicted by Eq. [17.42]. However, the rate of increase in the collector voltage [dV/dt] during the second phase of the voltage transient becomes larger when the lifetime is reduced. The collector voltage [dV/dt] increases from 1.24 × 10^{10} V/s to 4.66 × 10^{10} V/s when the high-level lifetime is reduced from 4 to 1 microsecond.

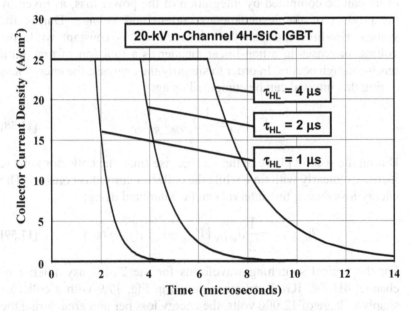

Fig. 17.12 Collector current transients during turn-off for the 20-kV asymmetric n-channel 4H-SiC IGBT structure: lifetime dependence.

The collector current transients predicted by the analytical model are shown in Fig. 17.12. It can be observed that the current transient becomes longer when the lifetime in the N-base region increases. The current fall-time increases when the lifetime is increased because of the reduced recombination rate in the N-buffer layer during the current transient. According to the analytical model, the current fall-times obtained by using the analytical model are 1.15, 2.30, and 4.61 microseconds for high-level lifetime values of 1, 2, and 4 microseconds, respectively.

17.1.5 Switching Energy Loss

The power loss incurred during the turn-off switching transient limits the maximum operating frequency for the IGBT structure. Power losses during the turn-on of the IGBT structure are also significant but strongly dependent on the reverse recovery behavior of the fly-back rectifiers in circuits. Consequently, it is common practice to use only the turn-off energy loss per cycle during characterization of IGBT devices. The turn-off losses are associated with the voltage rise-time interval and the current fall-time interval. The energy loss for each event can be computed by integration of the power loss, as given by the product of the instantaneous current and voltage. During the voltage rise-time interval, the anode current is constant while the voltage increases in a non-linear manner as a function of time until reach-through occurs. In order to simplify the analysis, the energy loss during this interval can be computed using:

$$E_{OFF,V1} = \frac{1}{2} J_{C,ON} V_{RT} t_{RT}$$

[17.58]

During the second phase of the voltage rise-time, the collector voltage increases linearly with time while the collector current is constant. The energy loss during this interval can be computed using:

$$E_{OFF,V2} = \frac{1}{2} J_{C,ON} \left(V_{C,S} - V_{RT} \right) \left(t_V - t_{RT} \right)$$

[17.59]

For the typical switching waveforms for the 20-kV asymmetric n-channel 4H-SiC IGBT structure shown in Fig. 17.9 with a collector supply voltage of 12,000 volts, the energy loss per unit area during the collector voltage rise-time is found to be 0.24 Joules/cm^2 if the on-state current density is 25 A/cm^2.

During the collector current fall-time interval, the collector voltage is constant while the current decreases exponentially with time. The energy loss during the collector current fall-time interval can be computed using:

$$E_{OFF,I} = J_{C,ON} V_{C,S} \tau_{BL} \qquad [17.60]$$

For the typical switching waveform for the 20-kV asymmetric n-channel 4H-SiC IGBT structure shown in Fig. 17.10 with a collector supply voltage of 12,000 volts, the energy loss per unit area during the collector current fall-time is found to be 0.30 Joules/cm^2 if the on-state current density is 25 A/cm^2. The total energy loss per unit area ($E_{OFF,V}$ + $E_{OFF,I}$) during the turn-off process for the 20-kV asymmetric n-channel 4H-SiC IGBT structure is then found to be 0.54 Joules/cm^2.

Using the results obtained from the numerical simulations[20], the on-state voltage drop and the total energy loss per cycle can be computed. These values are plotted in Fig. 17.13 to create a trade-off curve to optimize the performance of the 20-kV asymmetric n-channel 4H-SiC IGBT structure by varying the lifetime in the N-base region. Devices used in lower frequency circuits would be chosen from the left-hand-side of the trade-off curve while devices used in higher frequency circuits would be chosen from the right-hand-side of the trade-off curve.

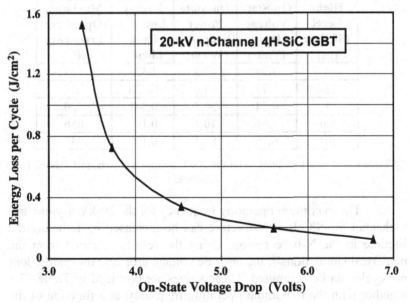

Fig. 17.13 Trade-off curve for the 20-kV asymmetric n-channel 4H-SiC IGBT structure: lifetime in N-base region.

17.1.6 Maximum Operating Frequency

The maximum operating frequency for operation of the 20-kV asymmetric n-channel 4H-SiC IGBT structure can be obtained by combining the on-state and switching power losses:

$$P_{D,TOTAL} = \delta\, P_{D,ON} + E_{OFF}\, f \qquad [17.61]$$

where δ is the duty cycle and f is the switching frequency. In the case of the baseline 20-kV asymmetric n-channel 4H-SiC IGBT device structure with a high-level lifetime of 2 microseconds in the N-base region, the on-state voltage drop is 3.728 volts at an on-state current density of 25 A/cm². For the case of a 50% duty cycle, the on-state power dissipation contributes 47 W/cm² to the total power loss. For this lifetime value, the energy loss per cycle during the voltage rise-time obtained from the numerical simulations is 0.425 Joules/cm² and the energy loss per cycle during the current fall-time obtained from the numerical simulations is 0.300 Joules/cm². Using a total turn-off energy loss per cycle of 0.725 Joules/cm² in Eq. [17.61] yields a maximum operating frequency of about 210 Hz.

High-Level Lifetime (μs)	On-State Voltage Drop (Volts)	On-State Power Dissipation (W/cm²)	Energy Loss per Cycle (J/cm²)	Maximum Operating Frequency (Hz)
4	3.39	42.3	1.523	104
2	3.73	46.6	0.725	212
1	4.54	56.7	0.341	420
0.6	5.61	70.1	0.198	656
0.4	6.76	84.5	0.122	947

Table 17.1 Power loss analysis for the 20-kV asymmetric n-channel 4H-SiC IGBT structure.

The maximum operating frequency for the 20-kV asymmetric n-channel 4H-SiC IGBT structure can be increased by reducing the lifetime in the N-base region. Using the results obtained from the numerical simulations[18], the on-state voltage drop and the energy loss per cycle can be computed. These values are provided in Table 17.1 together with the maximum operating frequency as a function of the high level lifetime in the N-base region under the assumption of a 50% duty cycle and a total power dissipation limit of 200 W/cm². The maximum operating frequency is plotted in Fig. 17.14 as a function of the high-level lifetime in the N-base region. It can be observed that the maximum operating frequency can be increased up to 950 Hz by reducing the high-level lifetime to 0.4 microseconds. The IGBT is often operated with pulse-width-modulation to synthesize variable frequency output power for motor control. In these applications, the duty cycle can be much shorter than 50 percent. In this case, the maximum operating frequency for the 20-kV asymmetric n-channel 4H-SiC IGBT structure can be increased. As an example, the maximum operating frequency for the 20-kV asymmetric n-channel 4H-SiC IGBT structure operated at a 10 percent duty cycle is included in Fig. 17.14. It can be seen that the maximum operating frequency can now exceed 1500 Hz.

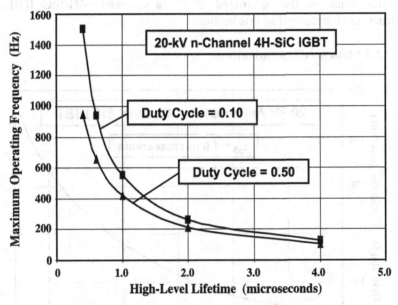

Fig. 17.14 Maximum operating frequency for the 20-kV asymmetric n-channel 4H-SiC IGBT structure.

The hard switching frequency limit of operation has been studied[23] for 15-kV n-channel 4H-SiC IGBTs. The authors concluded that operation at 6-kHz was feasible with liquid cooling and a junction temperature of 150 °C. The higher values for the switching frequency for these devices when compared with the analytical values plotted in Fig. 17.14 is due to the lower blocking voltage rating.

17.2 Optimized n-Channel Asymmetric Structure

In the previous section, it was found that the 20-kV asymmetric n-channel 4H-SiC IGBT structure exhibits a collector voltage turn-off waveform consisting of two phases. In the first phase, the collector voltage increases gradually up to a reach-through voltage and then during the second phase the collector voltage increases very rapidly with time until it reaches the collector supply voltage. This produces a high [dV/dt] which is not desirable during circuit operation. The second phase can be prevented from occurring by optimization of the doping concentration and width of the N-base region so that the reach-through of the space-charge-region occurs when the collector voltage becomes equal to the collector supply voltage. The design and performance of the optimized 20-kV asymmetric 4H-SiC IGBT structure is discussed in this section.

17.2.1 Structure Optimization

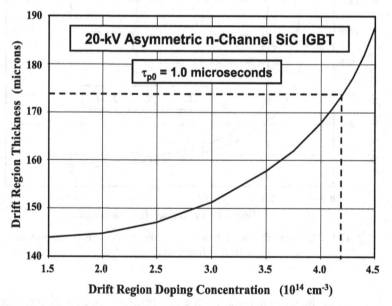

Fig. 17.15 Optimization of drift region width for the 20-kV asymmetric 4H-SiC n-channel IGBT structure.

An expression (see Eq. [17.43]) for the reach-through voltage was derived in the previous section. It can be concluded from this equation that the reach-through voltage is a function of the width and the doping concentration of the drift region, as well as the on-state current density. During optimization, it is necessary to choose these values after obtaining the hole concentration in the space-charge-region using the on-state current density. Although Eq. [17.43] indicates that the reach-through voltage can be increased by solely increasing the doping concentration of the N-base region, this approach results in a reduction of the blocking voltage. Consequently, the width and the doping concentration of the N-base region must be optimized together to simultaneously obtain the desired open-base breakdown voltage of 21-kV and a reach-through voltage equal to a collector supply voltage of 12-kV (as an example).

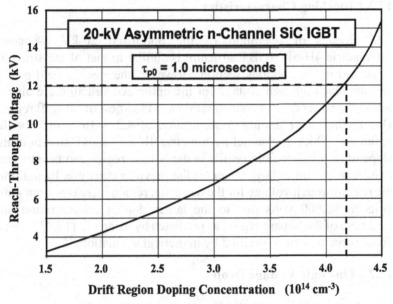

Fig. 17.16 Reach-through voltage for the 20-kV asymmetric 4H-SiC n-channel IGBT structure.

The width of the N-base region required to achieve a blocking voltage of 21-kV is shown in Fig. 17.15 based up on using open base transistor breakdown physics. A low-level lifetime of 1 microsecond was assumed in the drift region for the analysis. An N-buffer layer doping concentration of 5×10^{16} cm^{-3} was assumed with a thickness of 5 microns for this baseline device structure. For each value of the N-base doping concentration, its width was varied until the common-base current gain became equal to unity.

Using the optimum width for the N-base region corresponding to each doping concentration, the reach-through voltage can be computed by using Eq. [17.43]. The resulting values for the reach-through voltage are plotted in Fig. 17.16 as a function of the drift region doping concentration. From this plot, it can be observed that a reach-through voltage of 12-kV is obtained when the drift region doping concentration is 4.2×10^{14} cm^{-3}. For this drift region doping concentration, an open-base breakdown voltage of 21-kV is obtained if a drift region width of 175 microns is used according to Fig. 17.15. These values must be chosen for the optimized 20-kV n-channel asymmetric 4H-SiC IGBT structure.

17.2.2 Blocking Characteristics

The physics of operation of the optimized 20-kV n-channel asymmetric 4H-SiC IGBT structure is similar to that of the structure discussed in the previous section. However, the electric field profile and the reach-through voltage for the optimized structure are altered due to the larger doping concentration and thickness of the drift region. Due to the larger doping concentration of 4.2×10^{14} cm^{-3} for the optimized 20-kV n-channel asymmetric 4H-SiC IGBT structure, the slope of the electric field profile in the N-base region can be expected to be nearly 3-times larger than for the previous structure. In addition, the reach-through voltage for the depletion region increases from 4000 volts to 12,000 volts due to the larger doping concentration and thickness of the N-base region as predicted by using Eq. [17.43]. These conclusion have been verified by numerical simulations[20].

17.2.3 On-State Voltage Drop

The physics of operation in the on-state for the optimized 20-kV n-channel asymmetric 4H-SiC IGBT structure is identical to that for the structure discussed in the previous section. Based up on the high-level injection model for the IGBT structure, the on-state voltage drop for the optimized structure can be expected to be slightly greater than that for the previous structure for the same lifetime in the N-base region due to its larger width. This has been verified by numerical simulations[20]. The elimination of the high [dV/dt] during the turn-off transient in the 4H-SiC IGBT must be traded-off against the higher on-state power loss.

17.2.4 Turn-Off Characteristics

The turn-off behavior for the optimized 20-kV n-channel asymmetric 4H-SiC IGBT structure can be expected to be quite different from that for the 20-kV n-channel asymmetric 4H-SiC IGBT structure discussed in the previous section. In the optimized structure, the collector voltage should increase during a single phase to the collector supply voltage and the collector current should then decay by the recombination of holes in the N-buffer layer. The rise of the collector voltage is described by the physics developed for the previous structure during the first phase (see Eq. [17.41]). The current fall occurs with the same physics that governs the recombination of holes in the buffer layer for both structures (see Eq. [17.56]).

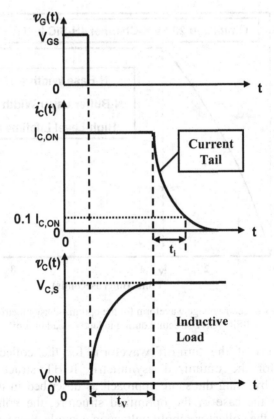

Fig. 17.17 Turn-off waveforms for the optimized asymmetric 4H-SiC IGBT structure.

In order to turn-off the IGBT structure, the gate voltage must simply be reduced from the on-state value (nominally 15 volts) to zero as illustrated in Fig. 17.17. The magnitude of the gate current can be limited by using a resistance in series with the gate voltage source. The waveform for the gate voltage shown in the figure is for the case of zero gate resistance. Once the gate voltage falls below the threshold voltage, the electron current from the channel ceases. In the case of an inductive load, the collector current for the IGBT structure is then sustained by the hole current flow due to the presence of stored charge in the N-base region. The collector voltage begins to increase in the IGBT structure immediately after the gate voltage reduces below the threshold voltage.

17.2.4.1 Voltage Rise-Time:

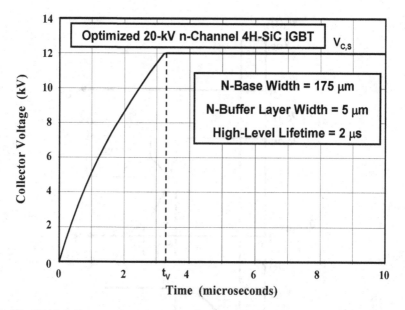

Fig. 17.18 Collector voltage waveform for the optimized asymmetric n-channel 4H-SiC IGBT structure during inductive load turn-off.

The analysis of the turn-off waveform for the collector voltage transient for the optimized asymmetric IGBT structure can be performed by using the same approach as described in the previous section. In the case of the optimized structure, the voltage should increase to the collector supply voltage in a single phase as described by Eq. [17.40]. The collector voltage supported by the optimized asymmetric 4H-SiC IGBT structure is related to the space charge layer width by Eq. [17.41]. The collector voltage increases in accordance with the above model until the space charge region reaches-through the N-base region when the collector voltage becomes equal to the collector supply voltage.

The time at which the collector voltage transient is completed is given by Eq. [17.44]. Once the space-charge-region reaches-through the N-base region, all the stored charge in the N-base region has been removed by the voltage transient. However, there is still substantial stored charge in the N-buffer layer. The expansion of the space-charge-region is now curtailed by the high doping concentration of the N-buffer layer.

Consider the case of the optimized 20-kV asymmetric silicon carbide n-channel IGBT structure with a N-base region width of 175 microns and a low-level lifetime of 1 microsecond in the N-base region; N-buffer layer doping concentration of 5×10^{16} cm^{-3} and thickness of 5 microns; and P$^+$ collector region (emitter region of the internal PNP transistor) doping concentration of 1×10^{19} cm^{-3}. The collector voltage waveform predicted by the above analytical model is provided in Fig. 17.18 for this structure. A collector current density of 25 A/cm^2 was used in this example. It can be observed that the collector voltage increases monotonically until a reach-through time (t_{RT}) of 3.3 microseconds when it becomes equal to the collector supply voltage of 12,000 volts. This time is identical to the first phase for the IGBT device in the previous section (see Fig. 17.9).

17.2.4.2 Current Fall-Time:

At the end of the collector voltage transient, the space-charge-region has extended through the entire N-base region leaving stored charge only in the N-buffer layer in the case of the optimized 20-kV asymmetric silicon carbide n-channel IGBT structure. The collector current therefore decays due to the recombination of this stored charge under low-level injection conditions. As in the case of the 20-kV asymmetric silicon carbide n-channel IGBT structure discussed in the previous chapter, the collector current transient occurs in a single phase as described by Eq. [17.56] with the current fall time obtained by using Eq. [17.57]. For the above example, the current-fall time is found to be 2.3 microseconds if no scaling of the lifetime with buffer layer doping is taken into account. This switching behavior has been verified by numerical simulations[20].

17.2.5 Lifetime Dependence

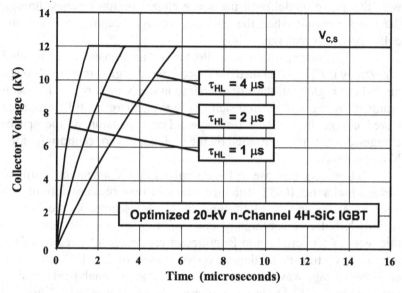

Fig. 17.19 Collector voltage transients during turn-off for the optimized 20-kV asymmetric n-channel 4H-SiC IGBT structure: lifetime dependence.

As in the case of the structure discussed in the previous section, it will be assumed that the lifetime in the N-buffer layer is the same as that in the N-base region for the optimized silicon carbide structure. The on-state voltage drop increases when the lifetime is reduced as discussed in section 17.2.3. The analytical model developed for turn-off of the optimized asymmetric IGBT structure presented in section 17.2.4 can be used to analyze the impact of changes to the lifetime in the drift region on the turn-off characteristics.

The collector voltage transients predicted by the analytical model are shown in Fig. 17.19 for the case of the optimized 20-kV asymmetric 4H-SiC IGBT structure operating with an on-state current density of 25 A/cm^2. The voltage rise-time increases when the lifetime is increased because of the larger concentration for the holes in the N-base region that are being removed during the collector voltage transient. The voltage rise-times obtained by using the analytical model are 1.5, 3.3, and 5.8 microseconds for high-level lifetime values of 1, 2, and 4 microseconds, respectively. In all cases, the collector voltage increases monotonically to the collector supply voltage of 12,000 volts as expected.

The collector current transients predicted by the analytical model are identical to that for the previous structure (see Fig. 17.12). This switching behavior has been verified by numerical simulations[20].

17.2.6 Switching Energy Loss

The turn-off loss for the optimized asymmetric 4H-SiC IGBT structure during the voltage rise-time interval is different from that for the device structure discussed in the previous section. Since the collector voltage transient for the optimized structure occurs in a single phase until it reaches the collector supply voltage, the energy loss during this interval can be computed using:

$$E_{OFF,V} = \frac{1}{2} J_{C,ON} V_{C,S} t_V \qquad [17.62]$$

For the typical switching waveforms for the optimized 20-kV asymmetric n-channel 4H-SiC IGBT structure shown in Fig. 17.18 with a collector supply voltage of 12,000 volts, the energy loss per unit area during the collector voltage rise-time is found to be 0.50 Joules/cm^2 if the on-state current density is 25 A/cm^2 which is more than that for the previous structure.

During the collector current fall-time interval, the collector voltage is constant while the current decreases exponentially with time. The energy loss during the collector current fall-time interval can be computed using:

$$E_{OFF,I} = J_{C,ON} V_S \tau_{BL} \qquad [17.63]$$

For the typical switching waveform for the optimized 20-kV asymmetric n-channel 4H-SiC IGBT structure with a collector supply voltage of 12,000 volts, the energy loss per unit area during the collector current fall-time is found to be 0.30 Joules/cm^2 if the on-state current density is 25 A/cm^2. This is the same as that for the structure discussed in the previous section. The total energy loss per unit area ($E_{OFF,V} + E_{OFF,I}$) during the turn-off process for the optimized 20-kV asymmetric n-channel 4H-SiC IGBT structure is found to be 0.80 Joules/cm^2. It can be concluded that, in order to eliminate an abrupt change in the collector voltage for the asymmetric silicon carbide IGBT structure, it is necessary to tolerate an increase in the switching power loss.

Using the results obtained from the numerical simulations, the on-state voltage drop and the total energy loss per cycle can be computed. These values are plotted in Fig. 17.20 to create a trade-off curve to optimize the performance of the optimized 20-kV asymmetric n-channel 4H-SiC IGBT structure by varying the lifetime in the N-base region. Devices used in lower frequency circuits would be chosen from the left-hand-side of the trade-off curve while devices used in higher frequency circuits would be chosen from the right-hand-side of the trade-off curve. The optimized structure has a superior trade-off curve when compared to the previous structure due to its reduced on-state voltage drop in spite of larger switching losses.

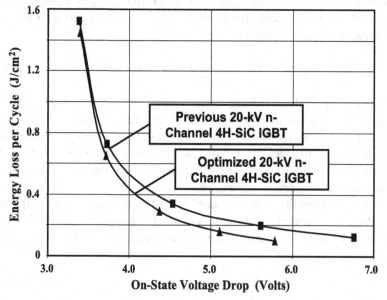

Fig. 17.20 Trade-off curve for the optimized 20-kV asymmetric n-channel 4H-SiC IGBT structure: lifetime dependence.

17.2.7 Maximum Operating Frequency

The maximum operating frequency for operation of the optimized 20-kV asymmetric n-channel 4H-SiC IGBT structure can be obtained by combining the on-state and switching power losses (see Eq. [17.61]. In the case of the baseline optimized 20-kV asymmetric n-channel 4H-SiC IGBT device structure with a high-level lifetime of 2 microseconds in the N-base region, the on-state voltage drop is 3.714 volts at an on-state current density of 25 A/cm². For the case of a 50% duty cycle, the on-state power dissipation contributes 46 W/cm² to the total power loss. For this lifetime value, the energy loss per cycle during the voltage rise-time obtained from the numerical simulations[20] is 0.348 Joules/cm^2 and the energy loss per cycle during the current fall-time obtained from the numerical simulations[20] is 0.300 Joules/cm^2. Using a total turn-off energy loss per cycle of 0.648 Joules/cm^2 in Eq. [17.61] yields a maximum operating frequency of about 240 Hz.

High-Level Lifetime (μs)	On-State Voltage Drop (Volts)	On-State Power Dissipation (W/cm²)	Energy Loss per Cycle (J/cm²)	Maximum Operating Frequency (Hz)
4	3.397	42.5	1.45	109
2	3.714	46.4	0.648	237
1	4.376	54.7	0.290	502
0.6	5.106	63.8	0.160	854
0.4	5.781	72.3	0.099	1291

Table. 17.2 Power loss analysis for the optimized 20-kV asymmetric n-channel 4H-SiC IGBT structure.

The maximum operating frequency for the optimized 20-kV asymmetric n-channel 4H-SiC IGBT structure can be increased by reducing the lifetime in the N-base region. Using the results obtained from the numerical simulations[20], the on-state voltage drop and the energy loss per cycle can be computed. These values are provided in Table 17.2 together with the maximum operating frequency as a function of the high level lifetime in the N-base region under the assumption of a 50% duty cycle and a total power dissipation limit of 200 W/cm^2. The maximum operating frequency is plotted in   Fig. 17.21 as a function of the high-level lifetime in the N-base region. It can be observed that the maximum operating frequency can be increased up to 1300 Hz by reducing the high-level lifetime to 0.4 microseconds.

The IGBT is often operated with pulse-width-modulation to synthesize variable frequency output power for motor control. In these applications, the duty cycle can be much shorter than 50 percent. In this case, the maximum operating frequency for the optimized 20-kV asymmetric n-channel 4H-SiC IGBT structure can be increased. As an example, the maximum operating frequency for the optimized 20-kV asymmetric n-channel 4H-SiC IGBT structure operated at a 10 percent duty cycle is included in Fig. 17.21. It can be seen that the maximum operating frequency can now approach   2000-Hz.

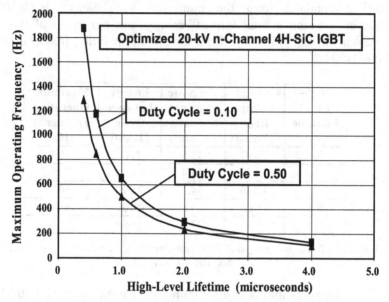

Fig. 17.21 Maximum operating frequency for the optimized 20-kV asymmetric n-channel 4H-SiC IGBT structure.

17.3 p-Channel Asymmetric Structure

The asymmetric p-channel 4H-SiC IGBT structure with the planar gate architecture is illustrated in Fig. 17.22 with its doping profile. As mentioned at the beginning of this chapter, the asymmetric p-channel silicon carbide IGBT structure received more attention initially than the n-channel structure because of concerns with the high resistance of available P^+ silicon carbide substrates. Since the asymmetric IGBT structure is intended for use in DC circuits, its reverse blocking capability does not have to match the forward blocking capability allowing the use of a P-buffer layer adjacent to the N^+ collector region. The P-buffer layer has a much larger doping concentration than the lightly doped portion of the P-base region. The electric field in the asymmetric IGBT takes a trapezoidal shape allowing supporting the forward blocking voltage with a thinner P-base region. This allows achieving a lower on-state voltage drop and superior turn-off characteristics. As in the case of silicon devices[16], the doping concentration of the buffer layer and the lifetime in the P-base (drift-region) region must be optimized to perform a trade-off between on-state voltage drop and turn-off switching losses. Like the asymmetric n-channel 4H-SiC IGBT structure, the asymmetric p-channel 4H-SiC IGBT structure has uniform doping concentration for the various layers produced by using either epitaxial growth or by using multiple ion-implantation energies to form a box profile.

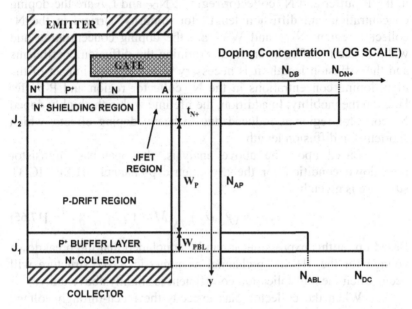

Fig. 17.22 Asymmetric p-channel 4H-SiC IGBT structure and its doping profile.

17.3.1 Blocking Characteristics

The design of the 20-kV asymmetric 4H-SiC p-channel IGBT structure is discussed in this section. The physics for blocking voltages is similar to the n-channel structure discussed in section 17.1. The forward blocking capability of the asymmetric 4H-SiC p-channel IGBT structure is determined by the open-base transistor breakdown phenomenon. In the case of the asymmetric 4H-SiC p-channel IGBT structure, the maximum blocking voltage occurs when the common base current gain of the NPN transistor becomes equal to unity. For the asymmetric p-channel IGBT structure, the emitter injection efficiency is smaller than unity due to the high doping concentration of the P-buffer layer. The emitter injection efficiency for the N^+ collector/P-buffer junction (J_1) can be obtained by using an analysis similar to that described in the textbook for the bipolar power transistor[16]:

$$\gamma_E = \frac{D_{nPBL} L_{pC} N_{DC}}{D_{nPBL} L_{pC} N_{DC} + D_{pC} W_{PBL} N_{ABL}} \qquad [17.64]$$

where D_{nPBL} and D_{pC} are the diffusion coefficients for minority carriers in the P-buffer and N^+ collector regions; N_{DC} and L_{pC} are the doping concentration and diffusion length for minority carriers in the N^+ collector region; N_{ABL} and W_{PBL} are the doping concentration and width of the P-buffer layer. In determining the diffusion coefficients and the diffusion length, it is necessary to account for impact of the high doping concentrations in the N^+ collector region and P-buffer layer on the mobility. In addition, the lifetime within the highly doped N^+ collector region is reduced due to heavy doping effects, which shortens the diffusion length.

Based upon the above analysis, the open-base transistor breakdown condition for the asymmetric p-channel 4H-SiC IGBT structure is given by:

$$\alpha_{NPN} = \left(\gamma_E . \alpha_T \right)_{NPN} M = 1 \qquad [17.65]$$

Based up on this expression, it can be concluded that the breakdown voltage for the 4H-SiC p-channel asymmetric IGBT structure will occur when the multiplication co-efficient is slightly above unity.

When the collector bias exceeds the reach-through voltage (V_{RT}), the electric field is truncated by the high doping concentration of the N-buffer layer making the un-depleted width of the NPN transistor base region equal to the width of the N-buffer layer. The base transport factor is then given by:

$$\alpha_T = \frac{1}{\cosh\left(W_{PBL} / L_{nPB} \right)} \qquad [17.66]$$

which is independent of the collector bias. Here, $L_{n,PB}$ is the diffusion length for electrons in the P-buffer layer. This analysis neglects the depletion region extension within the P-buffer layer. The diffusion length for electrons (L_{pPB}) in the P-buffer layer depends upon the diffusion coefficient and the minority carrier lifetime in the P-buffer layer. The diffusion coefficient varies with the doping concentration in the P-buffer layer based upon the concentration dependence of the mobility. In addition, the minority carrier lifetime has been found to be dependent upon the doping concentration[17] in the case of silicon devices. Although this phenomenon has not been verified for silicon carbide, it is commonly used when performing numerical analysis of silicon carbide devices. The effect can be modeled by using the relationship:

$$\frac{\tau_{LL}}{\tau_{n0}} = \frac{1}{1+\left(N_A / N_{REF}\right)} \qquad [17.67]$$

where N_{REF} is a reference doping concentration whose value will be assumed to be 5×10^{16} cm^{-3}.

The multiplication factor for a P-N junction is given by:

$$M = \frac{1}{1-\left(V_C / BV_{PP}\right)^n} \qquad [17.68]$$

with a value of n = 6 and the avalanche breakdown voltage of the N-base/P-base junction (BV_{PP}) *without the punch-through phenomenon*. In order to apply this formulation to the punch-through case relevant to the asymmetric p-channel silicon carbide IGBT structure, it is necessary to relate the maximum electric field at the junction for the two cases. The electric field at the interface between the lightly doped portion of the P-base region and the P-buffer layer is given by:

$$E_1 = E_m - \frac{qN_AW_P}{\varepsilon_S} \qquad [17.69]$$

The applied collector voltage supported by the device is given by:

$$V_C = \left(\frac{E_m + E_1}{2}\right)W_P = E_mW_P - \frac{qN_A}{2\varepsilon_S}W_P^2 \qquad [17.70]$$

From this expression, the maximum electric field is given by:

$$E_m = \frac{V_C}{W_P} + \frac{qN_AW_P}{2\varepsilon_S} \qquad [17.71]$$

The corresponding equation for the non-punch-through case is:

$$E_m = \sqrt{\frac{2qN_A V_{NPT}}{\varepsilon_S}} \qquad\qquad [17.72]$$

Consequently, the non-punch-through voltage that determines the multiplication coefficient 'M' corresponding to the applied collector bias 'V_A' for the punch-through case is given by:

$$V_{NPT} = \frac{\varepsilon_S E_m^2}{2qN_A} = \frac{\varepsilon_S}{2qN_A}\left(\frac{V_C}{W_P} + \frac{qN_A W_P}{2\varepsilon_S}\right)^2 \qquad [17.73]$$

The multiplication coefficient for the asymmetric silicon carbide IGBT structure can be computed by using this non-punch-through voltage:

$$M = \frac{1}{1 - \left(V_{NPT}/BV_{PP}\right)^n} \qquad\qquad [17.74]$$

The multiplication coefficient increases with increasing collector bias. The open-base transistor breakdown voltage (and the forward blocking capability of the asymmetric IGBT structure) is determined by the collector voltage at which the multiplication factor becomes equal to the reciprocal of the product of the base transport factor and the emitter injection efficiency.

The silicon carbide p-channel asymmetric IGBT structure must have a forward blocking voltage of 22,000 volts for a 20-kV rated device. In the case of avalanche breakdown, there is a unique value for the doping concentration of 3.2×10^{14} cm^{-3} for the drift region with a width of 272 microns to obtain this blocking voltage. In the case of the asymmetric silicon carbide IGBT structure, it is advantageous to use a much lower doping concentration for the lightly doped portion of the P-base region in order to reduce its width. The strong conductivity modulation of the P-base region during on-state operation favors a smaller thickness for the P-base region independent of its original doping concentration. A doping concentration of 1.5×10^{14} cm^{-3} will be assumed for the P-base region.

The doping concentration of the P-buffer layer must be sufficiently large to prevent reach-through of the electric field to the N$^+$ collector region. Although the electric field at the interface between the P-base region and the P-buffer layer is slightly smaller than at the blocking junction (J$_2$), a worse case analysis can be done by assuming that the electric field at this interface is close to the critical electric field for breakdown in the drift region. The minimum charge in the P-buffer layer to prevent reach-through can be then obtained using:

$$N_{ABL}W_{PBL} = \frac{\varepsilon_S E_C}{q} \qquad [17.75]$$

Using a critical electric for breakdown in silicon carbide of 2 × 10^6 V/cm for a doping concentration of 1.5 × 10^{14} cm^{-3} in the buffer layer, the minimum charge in the P-buffer layer to prevent reach-through for a silicon carbide p-channel asymmetric IGBT structure is found to be 1.07 × 10^{13} cm^{-2}. A P-buffer layer with doping concentration of 5 × 10^{16} cm^{-3} and thickness of 5 microns has a charge of 2.5 × 10^{13} cm^{-2} in which satisfies this requirement.

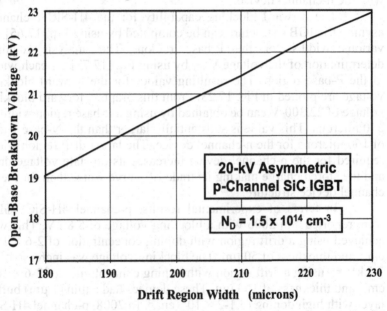

Fig. 17.23 Optimization of drift region width for the 20-kV asymmetric p-channel 4H-SiC IGBT structure.

The asymmetric p-channel 4H-SiC IGBT structure will be assumed to have an N$^+$ collector region with doping concentration of 1 × 10^{19} cm^{-3}. It will be assumed that all the donors are ionized even at room temperature although the relatively deep donor level in silicon carbide may lead to incomplete dopant ionization. In this case, the emitter injection efficiency computed using Eq. [17.64] is 0.997. When the device is close to breakdown, the entire P-base region is depleted and the base transport factor computed by using Eq. [17.66] in this case is 0.988. In computing these values, a lifetime of 1 microsecond was assumed for the P-base region resulting in a lifetime of 0.5 microseconds in the P-buffer layer due to the scaling according to Eq. [17.67]. Based up on Eq. [17.65], open-base transistor breakdown will then occur when the multiplication coefficient becomes equal to 1.02 for the above values for the injection efficiency and base transport factor. In comparison with the n-channel asymmetric 4H-SiC IGBT structure, the multiplication factor corresponding to open-base transistor breakdown has a much smaller value for the p-channel device. Consequently, it becomes necessary to utilize a thicker drift region for the p-channel device when compared with the n-channel device.

The forward blocking capability for the 4H-SiC p-channel asymmetric IGBT structure can be computed by using Eq. [17.65] for various widths for the P-base region. The analysis requires determination of the voltage V_{NPT} by using Eq. [17.73] for each width of the P-base region. The resulting values for the forward blocking voltage are plotted in Fig. 17.23. From this graph, a forward blocking voltage of 22,300-V can be obtained by using a P-base region width of 220 microns. This value is substantially larger than the N-base width of 160 microns for the n-channel device. The larger drift region width required for the p-channel device increases its on-state voltage drop and the stored charge making its trade-off curve worse than for the n-channel device structure.

In terms of experimental results, p-channel 4H-SiC IGBT were reported[24] in 2006 with a blocking voltage of 5.8 kV. This was achieved using a drift region with doping concentration of 2-6 × 10^{14} cm^{-3} and thickness of 50 μm. The blocking voltage was increased[25] to 7.5 kV by using a drift region with doping concentration of 2-6 × 10^{14} cm^{-3} and thickness of 100 μm. These devices had a thin (1 μm) buffer layer with high doping of 1-2 × 10^{17} cm^{-3}. In 2008, p-channel 4H-SiC IGBTs with the P-drift region doping concentration of 2-3 × 10^{14} cm^{-3} and thickness of 100 μm were reported[26] with blocking voltage of 11.5 kV.

In terms of 4H-SiC IGBTs with ultra-high (~15 kV) blocking voltage capability, the blocking voltage capability of n-channel and p-channel 4H-SiC IGBTs with same drift region doping concentration of 2 × 10^{14} cm^{-3} and thickness of 140 μm has been compared[27]. It was found that the n-channel 4H-SiC IGBT had superior blocking voltage capability consistent with the above analytical model. In 2014, p-channel 4H-SiC IGBTs were reported[28] with blocking voltage capability of 14.7 kV achieved using a drift region doping concentration of 5.2×10^{14} cm^{-3} and thickness of 152 μm. These devices has a P-buffer layer with doping concentration of 1.0×10^{17} cm^{-3} and thickness of 1 μm. A two zone JTE edge termination with total width of 400 microns was employed to achieve this breakdown voltage.

17.3.2 On-State Voltage Drop

The minority carrier (electron) distribution profiles for the p-channel asymmetric 4H-SiC IGBT structure can be expected to be governed by the same high-level injection physics previously described for the n-channel structure in section 17.1. In this section, the characteristics of the 20-kV p-channel asymmetric 4H-SiC IGBT structure will be described using the results of numerical simulations[20].

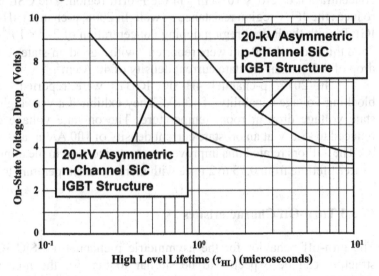

Fig. 17.24 On-state voltage drop for the 20-kV asymmetric p-channel IGBT structure: drift region lifetime dependence.

The on-state characteristics of the 20-kV asymmetrical p-channel 4H-SiC IGBT structure were obtained by using a gate bias voltage of 10 volts for the case of various values for the lifetime in the drift region. This device structure has a buffer layer doping concentration of 5×10^{16} cm^{-3} and thickness of 5 microns. The on-state voltage drop increases as expected with reduction of the lifetime

(τ_{p0}, τ_{n0}). The on-state voltage drop at a hole lifetime (τ_{p0}) value of 2 microseconds is found to be 4.65 volts at an on-state current density of 25 A/cm^2.

The variation of the on-state voltage drop as a function of the lifetime in the drift region obtained by using the numerical simulations is compared with that for the n-channel device structure in Fig. 17.24. It can be clearly seen that the n-channel 4H-SiC IGBT structure has a lower on-state voltage drop than that of the p-channel 4H-SiC device structure for all lifetime values. This is consistent with other analyses reported in the literature[29] and measured results on fabricated devices[27].

In early planar p-channel IGBTs reported in 2006, the on-state voltage drop was found[24] to be very high at room temperature. This was significantly improved[25] by 2007 by the addition of a 'current suppressing layer (CSL)' which has a doping concentration of 5-8 × 10^{15} cm^{-3} just under the N-base regions compared with the doping concentration of 2-6 × 10^{14} cm^{-3} in the P-drift region. The CSL layer reduces the JFET region resistance as well. In 2008, p-channel 4H-SiC IGBTs with the P-drift region doping concentration of 2-3 × 10^{14} cm^{-3} and thickness of 100 μm were reported[26] with a good on-state voltage drop of 4.5 V at an on-state current density of 40 A/cm^2.

In 2014, p-channel 4H-SiC IGBTs were reported[28] with blocking voltage capability of 15 kV. They exhibited a very high on-state voltage drop at room temperature. The on-state voltage drop reduced to -8.5 V at an on-state current density of 100 A/cm^2 at   250 °C. The authors related this improvement to an increase in the minority carrier lifetime from 0.75 to 2 μsec with increase in temperature to 250 °C.

17.3.3 Turn-Off Characteristics

The turn-off behavior for the asymmetric p-channel 4H-SiC IGBT structure can be expected to be similar to that for the n-channel structure as discussed in section 17.1.3. The same theoretical analysis should therefore be applicable for both devices. In this section, the turn-off characteristics of the 20-kV p-channel asymmetric silicon carbide IGBT structure will be described using only the results of numerical simulations[20].

17.3.4 Lifetime Dependence

From an applications perspective, the optimization of the power losses for the IGBT structure requires performing a trade-off between the on-state voltage drop and the switching losses. One approach to achieve this is by adjusting the lifetime in the drift (P-base) region. A reduction of the lifetime in the drift region also alters the lifetime in the P-buffer layer in the case of silicon devices. However, the relationship between the lifetime in the drift region and the buffer layer has not yet been established for silicon carbide devices. Consequently, it will be assumed that the lifetime in the P-buffer layer is the same as that in the P-base region for silicon carbide structures. The turn-off behavior for the asymmetric p-channel 4H-SiC IGBT structure can be expected to be similar to that for the n-channel structure as discussed in section 17.1.3. The same theoretical analysis should therefore be applicable for both devices. In this section, the turn-off characteristics of the 20-kV p-channel asymmetric silicon carbide IGBT structure will be described using only the results of numerical simulations[20].

The numerical simulations[20] showed a decrease in the time taken for the collector voltage to increase to the reach-through voltage when the lifetime in the P-base region is reduced. In the simulation results, the reach-through voltage remains independent of the lifetime in the N-base region as predicted by the analytical model. After reach-through of the space-charge-region occurs, the collector voltage increases linearly with time. The [dV/dt] values for the collector voltage transients increase with reduced lifetime in the drift region as predicted by the analytical model. The [dV/dt] values are 0.83, 1.2, 2.5, 3.3, and 5.0 × 10^9 V/s for high-level lifetime values of 10, 6, 4, 2, and 1 microseconds, respectively.

The numerical simulations of the 20-kV asymmetric p-channel 4H-SiC IGBT structure also show[20] a substantial increase in the collector current fall-time when the lifetime increases. For all the lifetime values, the collector current decays exponentially with time as predicted by the analytical model. The collector current fall-time values are 15, 10, 7, 3.5 and 1.8 microseconds, for high-level lifetime values of 10, 6, 4, 2, and 1 microseconds, respectively. The turn-off time for the p-channel 4H-SiC IGBT has been found[30] to be much (2-times) longer than that for the n-channel structure. This is due to the larger current gain of the internal NPN transistor that is supporting the bipolar current flow.

Good switching performance with an inductive load was demonstrated[31] for 4H-SiC p-channel IGBTs in 2008. The devices with 100 mm thick P-drift regions has a turn-off time of ~1 µsec. The voltage rise-time consisted of two phases as discussed above in the analytical models. The switching performance of p-channel 4H-SiC IGBTs with breakdown voltage of 14.8 kV was reported[28] in 2014 for the case of inductive load turn-off. However, the data was taken using a 16-kV 4H-SiC P-i-N rectifier with a large junction capacitance across the inductor. This results in soft-switching behavior with the turn-off [dV/dt] controlled by the capacitance[32]. The collector voltage then increases linearly with time unlike in the case of hard switching. The turn-off time was 5 µsec for a collector DC supply voltage of 6 kV.

17.3.5 Switching Energy Loss

The power loss incurred during the turn-off switching transient limits the maximum operating frequency for the IGBT structure. Power losses during the turn-on of the IGBT structure are also significant but strongly dependent on the reverse recovery behavior of the fly-back rectifiers in circuits. Consequently, it is common practice to use only the turn-off energy loss per cycle during characterization of IGBT devices. The turn-off losses are associated with the voltage rise-time interval and the current fall-time interval. The energy loss for each event can be computed by integration of the power loss, as given by the product of the instantaneous current and voltage. During the voltage rise-time interval, the anode current is constant while the voltage increases in a non-linear manner as a function of time until reach-through occurs. In order to simplify the analysis, the energy loss during this interval will be computed using Eq. [17.58]. During the second phase of the voltage rise-time, the collector voltage increases linearly with time while the collector current is constant. The energy loss during this interval can be computed using Eq. [17.59]. For the typical switching waveforms for the 20-kV asymmetric p-channel 4H-SiC IGBT structure with a negative collector supply voltage of 12,000 volts, the energy loss per unit area during the collector voltage rise-time is found to be 0.24 Joules/cm^2 if the on-state current density is 25 A/cm^2.

During the collector current fall-time interval, the collector voltage is constant while the current decreases exponentially with time. The energy loss during the collector current fall-time interval can be computed using Eq. [17.60]. For the typical switching waveforms for the 20-kV asymmetric p-channel 4H-SiC IGBT structure with a negative collector supply voltage of 12,000 volts, the energy loss per unit area during the collector current fall-time is found to be 0.30 Joules/cm^2 if the on-state current density is 25 A/cm^2. The total energy loss per unit area ($E_{OFF,V} + E_{OFF,I}$) during the turn-off process for the

20-kV asymmetric p-channel 4H-SiC IGBT structure is found to be 0.54 Joules/cm^2.

Using the results obtained from the numerical simulations, the on-state voltage drop and the total energy loss per cycle can be computed for the 20-kV p-channel asymmetric 4H-SiC IGBT structure. These values are plotted in Fig. 17.25 to create a trade-off curve to optimize the performance of the 20-kV asymmetric p-channel 4H-SiC IGBT structure by varying the lifetime in the N-base region. Devices used in lower frequency circuits would be chosen from the left-hand-side of the trade-off curve while devices used in higher frequency circuits would be chosen from the right-hand-side of the trade-off curve. For comparison purposes, the trade-off curve for the 20-kV n-channel asymmetric 4H-SiC IGBT structure is also shown in Fig. 17.25. It can be concluded that the performance of the 20-kV n-channel asymmetric 4H-SiC IGBT structure is far superior to that of the p-channel 4H-SiC device. Consequently, more attention has been given to develop the n-channel 4H-SiC IGBT structure than the p-channel device.

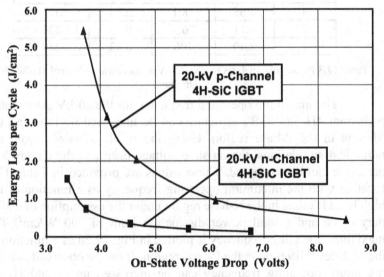

Fig. 17.25 Trade-off curve for the 20-kV asymmetric p-channel 4H-SiC IGBT structure: lifetime in P-base region.

17.3.6 Maximum Operating Frequency

The maximum operating frequency for operation of the 20-kV asymmetric p-channel 4H-SiC IGBT structure can be obtained by

GaN and SiC Power Devices

combining the on-state and switching power losses (see Eq. [17.61]). In the case of the baseline 20-kV asymmetric p-channel 4H-SiC IGBT device structure with a high-level lifetime of 4 microseconds in the P-base region, the on-state voltage drop is 4.65 volts at an on-state current density of 25 A/cm^2. For the case of a 50% duty cycle, the on-state power dissipation contributes 58 W/cm^2 to the total power loss. For this lifetime value, the energy loss per cycle during the voltage rise-time obtained from the numerical simulations[20] is 1.44 Joules/cm^2 and the energy loss per cycle during the current fall-time obtained from the numerical simulations[20] is 0.60 Joules/cm^2. Using a total turn-off energy loss per cycle of 2.04 Joules/cm^2 in   Eq. [17.85] yields a maximum operating frequency of only 70 Hz.

High-Level Lifetime (μs)	On-State Voltage Drop (Volts)	On-State Power Dissipation (W/cm^2)	Energy Loss per Cycle (J/cm^2)	Maximum Operating Frequency (Hz)
10	3.69	46.1	5.44	28
6	4.12	51.5	3.17	47
4	4.65	58.1	2.04	69
2	6.11	76.3	0.95	131
1	8.49	106	0.43	219

Table 17.3 Power loss analysis for the 20-kV asymmetric p-channel 4H-SiC IGBT structure.

The maximum operating frequency for the 20-kV asymmetric p-channel 4H-SiC IGBT structure can be increased by reducing the lifetime in the P-base region. Using the results obtained from the numerical simulations, the on-state voltage drop and the energy loss per cycle can be computed. These values are provided in Table 17.3 together with the maximum operating frequency as a function of the high level lifetime in the N-base region under the assumption of a 50% duty cycle and a total power dissipation limit of 200 W/cm^2. The maximum operating frequency is plotted in Fig. 17.26 as a function of the high-level lifetime in the P-base region. It can be observed that the maximum operating frequency can be increased up to 200 Hz by reducing the high-level lifetime to 1 microsecond. For comparison purposes, the maximum operating frequency for the 20-kV n-channel asymmetric 4H-SiC IGBT structure is also shown in Fig. 17.26. It can be concluded that the maximum operating frequency of the 20-kV n-channel asymmetric 4H-SiC IGBT structure is far greater to that of the p-channel 4H-SiC device.

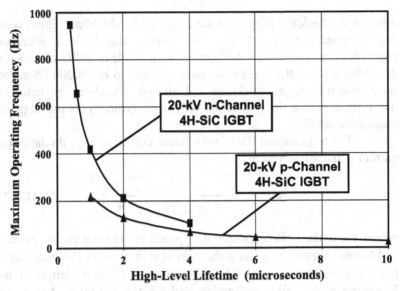

Fig. 17.26 Maximum operating frequency for the 20-kV asymmetric p-channel 4H-SiC IGBT structure.

17.4 Latch-Up and S-O-A

The issue to destructive failure from latch-up of the parasitic thyristor within the silicon IGBT structure needed to be addressed during the initial stages of its development[33]. With increasing maturity in its design and manufacturing process, the immunity from latch-up is now taken for granted. However, this cannot be assumed for the 4H-SiC IGBTs now under development.

For an n-channel IGBT with linear cell geometry, the latch-up current density is given by:

$$J_{L,n-Ch} = \frac{V_{bi}}{\alpha_{PNP,ON} \, \rho_{SP} \, L_{N+} \, p} \qquad [17.76]$$

where V_{bi} is the built-in potential of junction between the N^+ emitter and P-base regions, $\alpha_{PNP,ON}$ is the current gain of the P-N-P transistor, ρ_{SP} is the sheet resistance of the P-base region, L_{N+} is the length of the N^+ emitter in the cell cross-section, and p is the cell pitch. Fortunately, the built-in potential for the junction in 4H-SiC is three times bigger than for silicon which increases the latch-up current level. Unfortunately, the sheet resistance of the P-base region is high for 4H-SiC IGBTs due to incomplete ionization of P-type dopants despite the use of a P^+ shielding region. In the case of a P^+ shielding region with doping concentration of 1×10^{18} cm^{-3}, the ionized acceptor concentration is only 2.2×10^{16} cm^{-3}. Using a hole mobility of 54.8 for this doping level, the sheet resistance is found to be 50,000 Ohms/sq for a typical P^+ region thickness of 1 micron. The sheet resistance of the P-base region in n-channel IGBTs is typically two-orders of magnitude smaller.

For a p-channel IGBT with linear cell geometry, the latch-up current density is given by:

$$J_{L,p-Ch} = \frac{V_{bi}}{\alpha_{NPN,ON} \, \rho_{SN} \, L_{P+} \, p}$$

[17.77]

where V_{bi} is the built-in potential of junction between the P^+ emitter and N-base regions, $\alpha_{NPN,ON}$ is the current gain of the N-P-N transistor, ρ_{SN} is the sheet resistance of the N-base region, L_{P+} is the length of the P^+ emitter in the cell cross-section, and p is the cell pitch. Again, the built-in potential for the junction in 4H-SiC is three times bigger than for silicon which increases the latch-up current level. Fortunately, the sheet resistance of the N-base region is low for 4H-SiC IGBTs due to the use of an N^+ shielding region despite incomplete ionization of N-type dopants. In the case of a N^+ shielding region with doping concentration of 1×10^{18} cm^{-3}, the ionized donor concentration is 5.25×10^{17} cm^{-3}. Using an electron mobility of 324 for this doping level, the sheet resistance is found to be 370 Ohms/sq for a typical N^+ region thickness of 1 micron. The sheet resistance of the N-base region in p-channel IGBTs is similar in magnitude.

The latch-up performance of 4H-SiC IGBTs has not been reported in the literature. However, it has been shown that the p-channel 4H-SiC IGBT has excellent reverse-biased-safe-operating-area[31]. Devices designed with a breakdown voltage of 7.5 kV were switched off from an on-state current density of 200 A/cm^2 with a DC supply voltage of 4-kV without failure.

17.5 Trench-Gate Structure

As in the case of silicon IGBTs[16,34], it can be expected that trench gate 4H-SiC IGBT would also have improved on-state characteristics. However, as in the case of 4H-SiC trench-gate power MOSFETs discussed in chapter 12, a high electric field develops across the gate oxide during blocking high voltages leading to device failure. This problem can be solved by using a P^+ shielding region at the bottom on the trenches[3] for n-channel devices.

Trench-gate 4H-SiC p-channel IGBTs were reported[35] in 2003 without the use of shielding regions under the trench gate oxide. These devices exhibited very low breakdown voltage of -85 volts in spite of using a drift region with doping level of 5×10^{15} cm^{-3} and thickness of 12 μm. The threshold voltage for the devices was very high (~–30 V). The devices had a large knee voltage of – 7.1 V and an on-state voltage drop of –11.25 V at low on-state current density of 1 A/cm^2.

17.6 Summary

The physics of operation and design principles for the asymmetric 4H-SiC IGBT structure have been described in this chapter. The analysis, which includes both n-channel and p-channel structures, demonstrates that the 20-kV n-channel asymmetric 4H-SiC IGBT structure offers excellent characteristics for utility applications. However, the usual method used to design the drift region of the asymmetric IGBT leads to a high [dV/dt] during switching. A design methodology for the drift region doping and thickness is provided to avoid a high [dV/dt] during the turn-off event for the silicon carbide IGBT structure.

The characteristics of both n-channel and p-channel 4H-SiC IGBTs reported in the literature have been provided to describe the progress with this technology. The analytical models derived in this chapter are consistent with the experimental results. The highest blocking voltage achieved for 4H-SiC IGBTs is 27.5 kV making them a candidate for utility applications.

References

[1] B.J. Baliga, "Semiconductors for High Voltage Vertical Channel Field Effect Transistors", Journal of Applied Physics, Vol. 53,   pp. 1759-1764, 1982.

[2] Q. Zhang et al., "10 kV Trench Gate IGBTs on 4H-SiC", IEEE International Symposium on Power Semiconductor Devices and ICs, pp. 159-162, 2005.

[3] B.J. Baliga, "Silicon Carbide Switching Device with Rectifying Gate", U.S. Patent 5,396,085, Filed December 28, 1993, Issued March 7, 1995.

[4] B.J. Baliga, "Silicon Carbide Power Devices", World Scientific Press, 2006.

[5] Q. Zhang *et al.*, "9 kV 4H-SiC IGBTs with 88 mΩ-cm^2 of $R_{diff,on}$", Material Science Forum Vols. 556-557, pp. 771-774, 2007.

[6] Q. Zhang *et al.*, "12-kV p-Channel IGBTs with Low On-Resistance in 4H-SiC", IEEE Electron Device Letters, Vol. 29, pp. 1027-1029, 2008.

[7] Y. Sui, X. Wang and J.A. Cooper, "High-Voltage Self-Aligned, p-Channel DMOS-IGBTs in 4H-SiC", IEEE Electron Device Letters, Vol. 28, pp. 728-730, 2007.

[8] T. Tamaki *et al.*, "Optimization of On-State and Switching Performances for 15-20 kV 4H-SiC IGBTs", IEEE Transactions on Electron Devices, Vol. 55, pp. 1920-1927, 2008.

[9] W. Sung *et al.*, "Design and Investigation of Frequency Capability of 15-kV 4H-SiC IGBT", IEEE International Symposium on Power Semiconductor Devices and ICs, pp. 271-274, 2009.

[10] M.K. Das *et al.*, "A 13-kV 4H-SiC n-channel IGBT with Low $R_{diff,on}$ and Fast Switching", International Conference on Silicon Carbide and Related Materials, October 2007.

[11] Q. Zhang *et al.*, "SiC Power Devices for Microgrids", IEEE Transactions on Power Electronics, Vol. 25, pp. 2889-2896, 2010.

[12] S. Ryu *et al.*, "Ultra High Voltage (>12 kV), High Performance 4H-SiC IGBTs", IEEE International Symposium on Power Semiconductor Devices and ICs, pp. 257-260, 2012.

[13] S. Ryu *et al.*, "Ultra High Voltage IGBTs in 4H-SiC", IEEE Workshop on Wide Bandgap Power Devices and Applications", pp. 36-39, 2013.

[14] E.V. Brunt *et al.*, "22 kV, 1 cm2, 4H-SiC n-IGBTs with improved Conductivity Modulation", IEEE International Symposium on Power Semiconductor Devices and ICs, pp. 358-361, 2014.

[15] R.J. Callanan *et al.*, "Recent Progress in SiC DMOSFETs and JBS Diodes at CREE", IEEE Industrial Electronics Conference, pp. 2885-2890, 2008.

[16] B.J. Baliga, "Fundamentals of Power Semiconductor Devices", Chapter 9, Springer-Science, New York, 2008.

[17] B.J. Baliga and M.S. Adler, "Measurement of Carrier Lifetime Profiles in Diffused Layers of Semiconductors", IEEE Transactions on Electron Devices, Vol. ED-25, pp. 472-477, 1978.

[18] Y. Yonezawa *et al.*, "Low Vf and Highly Reliable 16 kV Ultrhigh Voltage SiC Flip-Type n-channel Implantation and Epitaxial IGBT", IEEE International Electron Devices Meeting, pp. 6.6.1-6.6.4, 2013.

[19] J.W. Palmour, "Silicon Carbide Power Device Development for Industrial Markets", IEEE International Electron Devices Meeting, pp. 1.1.1-1.1.8, 2014.

[20] B.J. Baliga, "Advanced High Voltage Power Device Concepts", Chapters 5 and 7, Springer-Science, New York, 2011.

[21] G. Wang *et al.*, "Static and Dynamic Performance Characterization and Comparison of 15 kV SiC MOSFET and 15 kV SiC n-IGBTs", IEEE International Symposium on Power Semiconductor Devices and ICs, pp. 229-232, 2015.

[22] A. Kadavelugu *et al.*, "Understanding dV/dt of 15 kV SiC n-IGBT and its Control using Active Gate Driver", IEEE Energy Conversion Congress and Exposition, pp. 2213-2220, 2014.

[23] A. Kadavelugu *et al.*, "Experimental Switching Frequency Limits of 15 kV SiC N-IGBT Module", IEEE Energy Conversion Congress and Exposition Asia, 2013.

[24] Q. Zhang *et al.*, "Design and Fabrication of High Voltage IGBTs on 4H-SiC", IEEE International Symposium on Power Semiconductor Devices and ICs, pp. 1-4, 2006.

[25] Q. Zhang *et al.*, "New Improvement Results on 7.5 kV 4H-SiC p-IGBTs with $R_{diff,on}$ of 26 mΩ-cm^2 at 25 °C", IEEE International Symposium on Power Semiconductor Devices and ICs, pp. 281-284, 2007.

[26] A. Agarwal *et al.*, "Prospects of Bipolar Power Devices in Silicon Carbide", IEEE Industrial Electronics Conference, pp. 2879-2884, 2008.

[27] S-H. Ryu *et al.*, "High Performance, Ultra High Voltage 4H-SiC IGBTs", IEEE Energy Conversion Congress and Exposition, pp. 3603-3608, 2012.

[28] T. Deguchi *et al.*, "Static and Dynamic Performance Evaluation of > 13 kV SiC P-Channel IGBTs at High Temperatures", IEEE International Symposium on Power Semiconductor Devices and ICs, pp. 261-264, 2014.

[29] J.A. Cooper *et al.*, "Power MOSFETs, IGBTs, and Thyristors in SiC", IEEE International Electron Devices Meeting, pp. 7.2.1-7.2.4, 2014.

[30] J.W. Palmour *et al.*, "SiC Power Devices for Smart Grid Systems", IEEE International Power Electronics Conference, pp. 1006-1013, 2010.

[31] Q. Zhang *et al.*, "Design and Characterization of High-Voltage 4H-SiC p-IGBTs", IEEE Transactions on Electron Devices, Vol. 55, pp. 1912-1919, 2008.

[32] A. Kadavelugu *et al.*, "Zero Voltage Switching Characterization of 12 kV SiC N-IGBTs", IEEE International Symposium on Power Semiconductor Devices and ICs, pp. 350-353, 2014.

[33] B.J. Baliga, "The IGBT Device: Physics, Design, and Applications of the Insulated Gate Bipolar Transistor", Elsevier Press, Amsterdam, 2015.

[34] H.R. Chang and B.J. Baliga, "500-V n-Channel Insulated Gate Bipolar Transistor with a Trench Gate Structure", IEEE Transactions on Electron Devices, Vol. ED-36, pp. 1824-1829, 1989.

[35] R. Singh *et al.*, "High Temperature SiC Trench Gate p-IGBT", IEEE Transactions on Electron Devices, Vol. 50, pp. 774-784, 2003.

Chapter 18

Synopsis

The motivation for the development of wide bandgap semiconductor based power devices arises from the improvements in efficiency for applications. The impact of improving the efficiency of power electronics on energy savings has been recognized for a long time[1]. This paper points out that adoption of adjustable speed motor drives and compact fluorescent lamps can lead to huge energy savings, referred to a 'negawatts'. The commercialization of the silicon Insulated Gate Bipolar Transistor (IGBT) has already led to enormous savings in electricity and gasoline[2,3]. The myriad applications for the IGBT in various sectors of the economy has been described[4]. During the past 25 years, this technology has produced an electricity savings of over 73,000 terra-watt-hours and a gasoline savings of 1.5 trillion pounds. A reduction in carbon dioxide emissions by over 100 trillion pounds has resulted from the improved efficiencies.

The introduction of wide bandgap semiconductor based power devices that can replace the silicon IGBT offers the promise to make even further improvements in the efficiency if their cost becomes competitive. These benefits have been promoted in many papers over the years[5,6,7]. The revenue for SiC power devices reached \$ 95 million in 2013. The main applications for the SiC power devices at present are in power factor correction and photovoltaics. In the case of GaN devices, the main applications at present are in point-of-load DC-DC converters. The use of wide bandgap power devices in electric and hybrid electric vehicles is promising in the future because of simplification of the cooling system.

The potential applications for wide bandgap semiconductor based power devices will be reviewed in this concluding chapter. The applications for silicon carbide power devices are more mature because of earlier commercialization of Schottky (JBS) power rectifiers followed by power MOSFETs. The applications for GaN power devices has been growing rapidly due to their improved high frequency capability. The prospective applications for SiC and GaN power devices are: (a) motor control for the industrial sector; (b) uninterruptible power supplies for data centers; (c) renewable energy (solar and wind) power systems; (d) electric and hybrid electric vehicles; and (e) power factor correction.

18.1 SiC Device Applications

The first SiC power devices to be commercialized were JBS rectifiers described in chapter 6. The very first high voltage SiC Schottky rectifiers were reported in 1992 with low on-state voltage drop and absence of large reverse recovery currents[8]. The Junction Barrier controlled Schottky (JBS) rectifier concept, first proposed[9] for silicon devices in the 1980s, was applied to SiC devices in the 1990s to reduce the leakage current[10,11]. Commercial 4H-SiC JBS rectifiers became available in the early 2000s making their applications feasible. The SiC JBS rectifier has very favorable characteristics for applications due to its smaller on-state voltage drop and significantly smaller reverse recovery losses when compared with silicon Pi-i-N rectifiers. They have found various applications leading to a market exceeding $150 million in 2015. Examples of applications for these devices will be first provided in this chapter.

   The development of SiC power MOSFETs took a much longer time frame due to problems with the MOS interface. Although the first high performance planar SiC power MOSFET was reported[12] in 1997, the commercialization of these devices did not occur until 2010. The SiC power MOSFET has smaller on-state voltage drop and switching losses than silicon IGBTs for blocking voltage up to at least 5-kV. SiC power MOSFETs are commercially available with voltage ratings of 600-V, 900-V, 1200-V, and 1700-V with products under development for higher voltages in 2016. This has encouraged their use in various applications to improve the efficiency even though their cost is still much greater than that of silicon IGBTs. Examples of the applications for these devices will be provided next in this chapter.

18.1.1 Boost PFC Converters using SiC JBS Rectifiers

One major application space for SiC JBS rectifiers is in power factor correction (PFC) circuits[13]. The performance of the boost PFC circuit is limited by reverse-recovery related power losses in the silicon

P-i-N power rectifier. This power loss can be mitigated by performing soft-switching but this introduces snubber inductors and capacitors. Utilizing the SiC JBS rectifier eliminates the reverse-recovery problem allowing operation at significantly higher frequencies. This allows reduction of the size of the boost inductor while maintaining a high efficiency. A bridgeless boost PFC converter with SiC power rectifiers is becoming the next generation of high performance front-end converter in applications. These PFC converters can be used in computers and uninterruptible power supplies. The size of the EMI filters, boost inductor, and heat sink has been shown[14,15] to be reduced by using SiC diodes operating at 1 MHz. The SiC JBS rectifier has also been used with a silicon COOLMOS transistor to achieve high performance in PFC circuits[16,17]. Enhanced performance of continuous conduction mode active PFC circuits has been demonstrated by using SiC diodes[18].

18.1.2 Photovoltaic Inverters using SiC JBS Rectifiers

A typical grid connected photovoltaic system contains the solar cells for generation of electrical energy, a buck-boost DC-DC converter followed by a DC-to-AC inverter, and a LCL filter[19]. SiC JBS rectifiers have been used as boost diodes in the converter stage with silicon IGBTs since 2005 in PV applications and as flyback diodes in the inverter stage[20]. The replacement of silicon P-i-N rectifiers with SiC Schottky rectifiers has been found to increase the efficiency by 2%. An increase in efficiency by 0.8% at the same power level has been reported[21] by replacing silicon P-i-N rectifiers with SiC Schottky rectifiers. However, the output power can be increased by 85% by using the SiC diodes and the switching frequency can be increased from 16 kHz to 48 kHz. This reduces the size and weight of passive components.

18.1.3 Automotive Traction Inverters using SiC JBS Rectifiers

Another potential high volume application space for SiC diodes is electric and hybrid electric vehicles. Here, the SiC JBS rectifiers are used as flyback diodes across silicon IGBTs as first proposed in 1994[22]. The power losses were demonstrated to be reduced by 33.6% by replacing silicon P-i-N rectifiers with SiC Schottky rectifiers in a 55-kW three-phase automotive traction inverter[23].

Many hybrid electric vehicles, such as the Prius from Toyota, make use of a DC-to-DC boost converter between the battery and the electric motors. It has been demonstrated that the power losses can be reduced by using the Si COOLMOS transistor and the SiC JBS diode to achieve an efficiency of 99 percent[24].

18.1.4 Photovoltaic Inverters using SiC MOSFETs

An all SiC device based PV inverter solution can be achieved due to the availability of 1.2-kV SiC power MOSFETs. The most prevalent silicon IGBT based PV inverter topology is the three-level, Dual-Neutral-Point-Clamped (3L-DNPC) topology[19]. The approach requires three IGBT modules each with 4 IGBTs and 2 diodes. The operating frequency is limited to 16-kHz due to the high switching losses in the IGBTs. The low operating frequency increases the size of the DC link capacitor and the LCL output filter increasing their cost.

With a PV inverter made using SiC power MOSFETs, the topology can be simplified to a two-level, full bridge with split capacitor (2L-FB)[19]. The switching frequency can be increased to 50-kHz by using the SiC devices due to their smaller switching losses. This reduces the size, weight and cost of the inverter because the number of device is reduced to 6. Cost of the SiC-MOSFET based 2L-FB PV inverter has been calculated to be 8.8% less than that for a Si-IGBT based 3L-DNPC inverter due to the reduced cost for the PCB, heat-sink, DC-link capacitors, and the LCL-filter despite the higher total semiconductor device cost.

The advantages of replacing silicon IGBTs with SiC power MOSFETs have been identified encouraging implementation into PV systems. A 2% improvement in efficiency has been demonstrated by replacing silicon IGBTs with SiC power MOSFETs in Japan[25].

The reduction of cost of PV inverters by replacing silicon IGBTs with SiC power MOSFETs has been analyzed[26]. The authors conclude that SiC-based PV inverters can be operated only at switching frequencies up to 50-kHz due to electromagnetic interference. The inverter efficiency increase by using SiC power MOSFETs and SiC diodes is 0.8% for the single phase case and 2% for the three-phase case. These improvements provide a modest annual cost savings of 20-30 Euros per year. The power density for the PV inverter can be improved from 0.38 kW/kg with silicon IGBTs to 1 kW/kg by using an all-SiC implementation[27]. This is attractive to commercial roof-top solar panel installers.

The inverter cost for PV installations in Europe has been analyzed for the case of silicon and SiC components[28]. The authors concluded that the cost of SiC-based PV inverters is higher than that of Si-based inverters by 226-293% due to higher prices for SiC devices. They project equal cost for both inverters can be achieved if the cost of SiC devices is reduced in half. The SiC-based inverters will achieve lower levelized cost of generated electricity by 4-7% and higher energy production by 1.6-4%.

18.1.5 Automotive Traction Inverters using SiC MOSFETs

Hybrid-Electric and Electric vehicles require power management from the battery. Many stages of DC-to-DC and DC-to-AC are inherent in the automobile. The silicon IGBT has been extensively used for the development of all the hybrid-electric and electric cars on the road today[4]. The cost of power semiconductor devices is a significant fraction of the power electronics in the vehicles[29]. In addition, a significant (33 to 50 %) size reduction of the power conversion unit is projected[30] by replacing silicon devices with SiC power MOSFETs. Furthermore, a similar reduction in the size of capacitors and inductors is achieved by the higher frequency of operation enabled by SiC power MOSFETs. An air-cooled inverter with power density of 60-kW/L can be designed by using the SiC power MOSFETs.

Although the low switching power losses of SiC power MOSFETs when compared with Si IGBTs is attractive for improving the efficiency, large voltage oscillations of the DC bus voltage have been observed[31]. Actual power losses in the SiC power MOSFETs have been measured to be ten-times smaller than for Si IGBT as projected in 1994[22]. The impact of these reduced power losses has been estimated to extend the range of the electric vehicle by 5 percent[32]. The high voltages generated at the DC bus due to oscillations can also damage the winding insulation in motors[33]. A gain in efficiency by about 2 percent has been reported by replacing Si IGBTs and PiN diodes with SiC power MOSFETs and JBS diodes in an electric vehicle drive-train[34].

The Toyota Motor Company has studied the benefits of replacing Si IGBTs and PiN rectifiers with SiC power MOSFETs and JBS diodes[35]. The Si devices are replaced with the SiC devices in the power control unit (PCU) consisting of the boost converter and inverter stages. The SiC devices could be operated at a higher frequency due to reduced switching losses. Driving tests using the Japanese JC08 test cycle showed an increase in fuel efficiency by

5 percent. The larger switching frequency is expected to produce a reduction in PCU size by 80 percent.

18.1.6 Locomotive Traction Inverters using SiC MOSFETs

The benefits of using SiC power MOSFETs and JBS diodes in place of the Si IGBT and PiN rectifier have been shown to apply to railway traction applications as well[36]. In this application, small mass and volume are a premium. Reducing converter losses is also important to minimize the size of the cooling system. Increasing operating frequency allows reduction of the size of the passive filtering components.

It was demonstrated that the SiC module could be operated at 15 kHz with the same power losses as in the Si module at 1 kHz. At a fixed frequency of 10-kHz, the power loss with the SiC module was reduced by 40 percent compared with the Si module.

18.1.7 Industrial Motor Drives using SiC MOSFETs

Silicon IGBTs have become the workhorse for industrial motor drives[4]. Silicon carbide devices offer significant reduction in power losses for this application[22]. A motor drive has been implemented using SiC devices for an Allen Bradley 50 hp 3-phase induction motor[37]. An improvement in efficiency by 2.5% was observed when the inverter was operated at 8-kHz and 5% at 16-kHz. The main problem with using SiC devices was found to be the high dV/dt which can damage the motor winding insulation. This can be mitigated by increasing the gate drive resistance at the expense of larger power losses.

The economics of replacing Si devices with SiC devices has also been examined for various countries with differing electricity cost[37]. Based upon an efficiency gain of 0.8%, the annual energy cost savings for China, USA, Japan, Germany and Denmark are $ 80, $ 120, $ 270, $ 360, and $ 420, respectively. Based upon a cost differential of $ 200 for the SiC drive, a payback period of 29, 19, 9, 7, and 6 years was computed for these countries.

The viability and advantages of SiC power MOSFETs for industrial drives has been analyzed taking into account not only reduction in the switching power losses but other considerations such as short circuit withstand capability and vulnerability to cosmic ray induced failures[38]. It was found that short-circuit withstand time was inversely proportional to the drain bias voltage as expected from analytical models[39]. Derating of the SiC power MOSFETs was found to be necessary to avoid cosmic ray induced failures as observed for Si IGBTs.

18.1.8 Data Centers using SiC MOSFETs

Large data centers consume 120 terra-watt-hours of electricity annually and operate at a low efficiency of 67% due to many power conversion stages[40]. The power distribution efficiency can be improved to 73% by using a DC bus architecture that makes use of SiC and GaN power devices. The 400-V DC architecture consists of a front end rectifier (FER) stage to create the 400-V DC bus, an intermediate bus converter (IBC) on the server motherboard to reduce the voltage to 10-V DC, and a point-of-load (POL) converter to reduce the voltage to 1-V for the chips.

SiC power devices are used for the FER stage to obtain a power loss reduction by 50%, a volume reduction by 20% and a weight reduction by 10%. An efficiency of 98% was observed using the SiC devices.

18.2 GaN Device Applications

The development of GaN power devices was initially hampered by lack of availability of high quality GaN substrates with large diameters. This barrier was overcome by successful growth of GaN and AlGaN layers on silicon substrates in the 1990s. High voltage GaN lateral transistors that utilized a 2D-electron gas were reported[41] by 2010 with low specific on-resistance. These devices became commercially available only a few years later enabling their use in applications. Examples of applications for GaN devices are provided next in the chapter.

18.2.1 Point-Of-Load Converters

Delivery of power to microprocessors and other chips in servers requires point-of-load converters that reduce the 12-V bus voltage to 1-V DC required for the chips[40]. These converters are operated at high frequency to shorten the response time and shrink the passive components on the board. This requires power switches with a better ($R_{ON} * Q_{GD}$) figure-of-merit (FOM). The GaN lateral HEMT devices discussed in chapter have been reported to have a FOM 2-3 times better than silicon power MOSFETs[42,43]. With a GaN device operating at 5-MHz a POL has been demonstrated with a power density of 1100 W/in^3. A 4% improvement in the peak efficiency was also obtained by replacing the Si devices with the GaN devices.

18.2.2 Monolithic Motor Drive

The lateral configuration of the GaN HFET structure allows integration of multiple devices on a single chip to create a monolithic motor control chip[44]. This was implemented using 600-V enhancement-mode GIT devices with iron ion implantation to create isolation regions between them. An efficiency of 93% (1% greater than for a discrete element IGBT/PiN diode case) was obtained at 20-W power level. This offers a very compact motor drive option.

18.2.3 Discrete Motor Drive

The cascode GaN HFET device has been compared with a Si IGBT device for the case of a 400-V DC bus[45]. The body diode of the Si MOSFET in the cascode was utilized as the fly-back diode. This approach allowed the operation of the GaN devices at 100-kHz which is not possible for the Si IGBT module with Si PiN fly-back diodes. No difference in the drive efficiency was observed at full-load of 2 kW. An efficiency gain of 1 % was observed at light load conditions.

18.2.4 Automotive Applications

Companies developing hybrid electric and electric vehicles have shown interest in GaN power devices. It was concluded that GaN HEMT devices could offer low specific on-resistance and high temperature (300 °C) operating capability[46]. However, the performance of the devices in 2006 was far below system requirements. It has been projected based on simulations that a GaN inverter would reduce power losses by a factor of 10-times when compared with a Si IGBT inverter[47].

18.2.5 Uninterruptible Power Supply

The 600-V cascode GaN switch and GaN diodes have been evaluated for use in UPS applications[48]. A DC-DC boost converter and a half-bridge inverter were built for comparison with Si IGBT designs. The authors concluded that the converter switching frequency could be increased from 25 kHz to 200 kHz. This reduced the size of the magnetics and capacitors by 70% which led to a 20% reduction in size of the UPS. The GaN devices improved the efficiency by 6% in the on-line mode and 3% in the on-battery mode. However, the GaN devices did not have the short-circuit capability exhibited by the Si IGBTs.

18.3 Summary

A great deal of interest has been generated in the power electronic community by the availability of SiC and GaN power devices. The GaN devices have been found to be attractive for blocking voltages in the 200-V to 600-V range. For larger blocking voltages, SiC power devices offers significant reduction in power losses and the ability to increase the circuit operating frequency when compared with existing bipolar silicon devices. This produces many system advantages, namely smaller size for the passive components and output filters as well as higher efficiency. The adoption of this wide bandgap semiconductor technology in applications is constrained by concerns regarding reliability and the higher cost of the devices. Recent studies indicate that the devices have been engineered to provide good reliability. Major programs are underway around the world to reduce the manufacturing cost[49]. This bodes a promising future for a technology that was envisioned more than 35 years ago.

References

[1] A.B. Lovins, "The Negawatt Revolution", Across the Board – The Conference Board Magazine, Vol. XXVII, Issue 9, pp. 18-23, 1990.

[2] B.J. Baliga, "The Role of Power Semiconductor Devices on Creating a Sustainable Society", IEEE Applied Power Electronics Conference, Invited Plenary Paper, March 18, 2013.

[3] B.J. Baliga, "Social Impact of Power Semiconductor Devices", IEEE International Electron Devices Meeting, Abstract 2.1.1, pp. 20-23, December 2014.

[4] B.J. Baliga, "The IGBT Device: Physics, Design, and Applications of the Insulated Gate Bipolar Transistor", Elsevier Press, Amsterdam, 2015.

[5] I. Omura, "Future Role of Power Electronics", International Conference on Integrated Power Electronics Systems, Paper 14.2, March 2010.

[6] N. Kaminski, "SiC and GaN Devices – Competition or Coexistence?", IEEE Integrated Power Electronic Systems Conference, pp. 1-11, 2012.

[7] A. Bindra, "Wide-Bandgap-Based Power Devices", IEEE Power Electronics Magazine, pp. 42-47, March 2015.

[8] M. Bhatnagar, P.M. McLarty and B.J. Baliga, "Silicon Carbide High Voltage (400 V) Schottky Barrier Diode", IEEE Electron Device Letters, Vol. EDL-13, pp. 501-503, 1992.

[9] B.J. Baliga, "The Pinch Rectifier: A Low Forward Drop High Speed Power Diode", IEEE Electron Device Letters, Vol. EDL-5, pp. 194-196, 1984.

[10] R. Held, N. Kaminsi and E. Niemann, "SiC Merged p-n/Schottky Rectifiers for High Voltage Applications", Silicon Carbide and Related Materials – 1997, Material Science Forum, Vol. 264-268, pp. 1057-1060, 1998.

[11] F. Dahlquist et al., "Junction barrier Schottky Diodes in 4H-SiC and H-SiC", Silicon Carbide and Related Materials – 1997, Material Science Forum, Vol. 264-268, pp. 1061-1064, 1998.

[12] P.M. Shenoy and B.J. Baliga, "The Planar 6H-SiC ACCUFET", IEEE Electron Device Letters, Vol. EDL-18, pp. 589-591, 1997.

[13] M.M. Javanovic and Y. Jang, "State-of-the-Art, Single-Phase, Active Power-Factor-Correction Techniques for High-Power Applications – An Overview", IEEE Transactions on Industrial Electronics, Vol. 52, pp. 701-708, 2005.

[14] M. Hernando et al., "Comparing Si and SiC Diodes Performance in a Commercial AC-to-DC Rectifier with Power Factor Correction", IEEE Power Electronics Specialists Conference, Vol. 4, pp. 1979-1982, 2003.

[15] P-O. Jeannin et al., "1 MHz Power Factor Correction Boost Converter with SiC Schottky Diode", IEEE Industrial Applications Society Meeting, pp. 1267-1272, 2004.

[16] B. Lu et al., "Performance Evaluation of CoolMOS and SiC Diode for Single-Phase Power factor Correction Applications", IEEE Applied Power Electronics Conference, Vol. 2, pp. 651-657, 2003.

[17] L. Lorenz, G. Deboy and I. Zverev, "Matched Pair of CoolMOS Transistor with SiC-Schottky Diode – Advantages in Applications", IEEE Transactions on Industrial Applications, Vol. 40, pp. 1265-1272, 2004.

[18] W-S. Choi and S-M. Young, "Effectiveness of a SiC Schottky Diode for Super-Junction MOSFETs on Continuous Conduction Mode PFC", IEEE International Symposium on Power Electronics, Electrical Drives, Automation and Motion, pp. 562-567, 2010.

[19] C. Sintamarean *et al.*, "Wide-band Gap Devices in PV Systems – Opportunities and Challenges", IEEE International Power Electronics Conference, pp. 1912-1919, 2014.

[20] A. Mumtaz, "Photovoltaic Systems: Technology Trends and Challenges through to 2020", IEEE International Symposium on Power Semiconductor Devices and ICs Short Course, 2014.

[21] E. Theodossu, "Efficiency Improvement in Booster Power Modules with siC Components", Bodo Power Systems, pp. 32-35, March 2014.

[22] B.J. Baliga, "Power Semiconductor Devices for Variable Frequency Drives", Proceedings of the IEEE, Vol. 82, pp. 1112-1122, 1994.

[23] B. Ozpineci *et al.*, "A 55-kW Three-Phase Inverter with Si IGBT and SiC Schottky Diodes", IEEE Transactions on Industrial Applications, Vol. 45, pp. 278-285, 2009.

[24] W. Martinez *et al.*, "Efficiency Optimization of a Single-Phase Boost DC-DC Converter for Electric Vehicle Applications", IEEE Industrial Electronics Society Annual Conference, pp. 4279-4285, 2014.

[25] R. Ammo and H. Fujita, "Analysis and Reduction of Power Losses in PV Converters for Grid Connection to Low-Voltage Three-Phase Three-Wire Systems", IEEE International Power Electronics Conference, pp. 2027-2034, 2014.

[26] B. Burger, D. Kranzer and O. Statler, "Cost Reduction of PV-Inverter with SiC-DMOSFETs", IEEE International Conference on Integrated Power Systems, pp. 1-5, 2008.

[27] J. Mookken, B. Agrawal and J. Liu, "Efficient and Compact 50kW Gen2 SiC Device based PV String Inverter", IEEE Power Electronics, Intelligent Motion, Renewable Energy and Energy Management Conference, pp. 780-786, 2014.

[28] S. Saridakis, E. Koutroulis and F. Blaabjerg, "Optimization of SiC-Based H5 and Conergy-NPC Transformerless PV Inverters", IEEE

Journal of Emerging and Selected Topics on Power Electronics, Vol. 3, pp. 555-567, 2015.

[29] M. Marz et al., "Power Electronics Systems Integration for Electric and Hybrid Vehicles", IEEE Conference on Integrated Power Electronic Systems, Paper 6.1, 2010.

[30] K. Hamada, "Power Semiconductor Device and Module Technologies for Hybrid Vehicles", IEEE International Symposium on Power Semiconductor Devices and ICs Short Course, 2014.

[31] P. Shamsi et al., "Performance Evaluation of Various Semiconductor Technologies for Automotive Applications", IEEE Applied Power Electronics Conference, pp. 3061-2066, 2013.

[32] K. Kumar et al., "Impact of Sic MOSFET Traction Inverters on Compact-Class Electric Car Range", IEEE International Conference on Power Electronics, pp. 1-6, 2014.

[33] M.J. Scott et al., "Design Considerations for Wide Bandgap based Motor Drive Systems", IEEE Electric Vehicles Conference, pp. 1-6, 2014.

[34] S. Jahdi et al., "IEEE Journal of Emerging and Selected Topics in Power Electronics, Vol. 2, pp. 517-528, 2014.

[35] K. Hamada et al., "SiC – Emerging Power Device Technology for Next-Generation Electrically Powered Environmentally Friendly Vehicles", IEEE Transactions on Electron Devices, Vol. 62, pp. 278-285, 2015.

[36] J. Fabre, P. Ladoux and M. Piton, "Characterization and Implementation of Dual-SiC MOSFET Modules for Future Use in Traction Converters", IEEE Transactions on Power Electronics, Vol. 30, pp. 4079-4090, 2015.

[37] J. Rice and J. Mookken, "Economics of High Efficiency SiC MOSFET based 3-ph Motor Drive", Power Conversion and Intelligent Motion, pp. 1003-1010, 2014.

[38] A. Bolotnikov et al., "Overview of 1.2kV – 2.2kV SiC MOSFETs targeted for Industrial Power Conversion Applications", IEEE Applied Power Electronics Conference, pp. 2445-2452, 2015.

[39] B.J. Baliga, "Fundamentals of Power Semiconductor Devices", Springer-Science, pp. 960-964, New York, 2008.

[40] Y. Cui et al., "High Efficiency Data Center Power Supply using Wide Bandgap Power Devices", IEEE Applied Power Electronics Conference, pp. 3437-3443, 2014.

[41] N. Ikeda et al., "GaN Power Transistors on Si Substrates for Switching Applications", Proceedings of the IEEE, Vol. 98, pp. 1151-1161, 2010.

[42] F.C. Lee and Q. Li, "High-Frequency Integrated Point-of-Load Converters: Overview", IEEE Transactions on Power Electronics, Vol. 28, pp. 4127-4136, 2013.

[43] S. Li, D. Reusch and F.C. Lee, "High-Frequency High Power Density 3-D Integrated Gallium-Nitride-Based Point of Load Module Design", IEEE Transactions on Power Electronics, Vol. 28, pp. 4216-4225, 2013

[44] Y. Uemoto *et al.*, "GaN Monolithic Inverter IC using Normally-Off Gate Injection Transistors with Planar Isolation on Si Substrates", IEEE International Electron Devices Meeting, pp. 7.6.1-7.6.4, 2009.

[45] K. Shirabe *et al.*, "Efficiency Comparison between Si-IGBT-based Drive and GaN-based Drive", IEEE Transactions on Industry Applications, Vol. 50, pp. 566, 572, 2014.

[46] H. Ueda *et al.*, "Wide-Bandgap Semiconductor Devices for Automotive Applications", CS MANTECH Conference, pp. 37-40, 2006.

[47] C. Assad and H. Mureau, :GaN-over-Si: The Promising Technology for Power Electronics in Automotive", Conference on Integrated Power Electronic Systems, Paper P20, 2012.

[48] R. Mitova *et al.*, "Investigations of 600-V GaN HEMT and GaN Diode for Power Converter Applications", IEEE Transactions on Power Electronics, Vol. 29, pp. 2441-2452, 2014.

[49] A. Agarwal *et al.*, "Wide Band gap Semiconductor Technology for Energy Efficiency", Material Science Forum, Vol. 858, pp. 797-802, 2016.

Homework Problems

The contents of this book were used to teach a graduate course at North Carolina State University in the Fall 2015 semester. The homework assignments given to the students are provided here for the assistance of other instructors who wish to use this book in their class.

Home Work 1: Material Properties and Breakdown

Problem 1: Determine the intrinsic carrier concentration for silicon, 4H-SiC, and GaN at 300, 400 and 500 °K. Provide your answers in a Tabular form for comparing the values at each temperature.

Problem 2: Calculate the built-in potential for silicon, 4H-SiC, and GaN at 300, 400 and 500 °K using a doping concentration of 1×10^{19} cm^{-3} on the P-side and 1×10^{16} cm^{-3} on the N-side of the junction. Provide your answers in a Tabular form for comparing the values at each temperature.

Problem 3: Determine the electric field at which the impact ionization coefficient for silicon, 4H-SiC, and GaN using Baliga's Power Laws becomes 10^4 cm^{-1}. Provide your answers in a Tabular form for comparing the values.

Problem 4: Determine the mobility for electrons in Silicon, 4H-SiC, and GaN at doping concentrations of: 1×10^{14} cm^{-3}, 1×10^{15} cm^{-3}, 1×10^{16} cm^{-3}, 1×10^{17} cm^{-3}, 1×10^{18} cm^{-3}, 1×10^{19} cm^{-3}. Provide your answers in a Tabular form with values for each mobility.

Problem 5: Name the deep levels that determines lifetime in 4H-SiC. What are the positions of these deep levels? What is the threshold energy required to create these deep levels by electron irradiation? Determine the 160 keV Electron Fluence required to obtain a lifetime of 0.2 microseconds.

Problem 6: Calculate the breakdown voltage for a junction termination using the single optimally located floating field ring with a depth of 1 microns for a 4H-SiC drift region with doping concentration of 1 $\times$ 10^{16} cm^{-3}. Determine the spacing for the single optimally located floating field ring. What is the mask dimension required for this design?

Problem 7: What is the optimum dose for a single zone JTE region for 4H-SiC? What is its minimum width for a breakdown voltage of 3000 volts?

Problem 8: What is the Positive Bevel Angle required to reduce the Surface Electric Field to 40 percent of the Bulk value? You should use the analytical Model B for your calculations.

Problem 9: Calculate the width of the drift region for a GaN punch-through diode to achieve a breakdown voltage of 5000 volts if the drift region doping concentration is 1×10^{14} cm^{-3}. Compare this to the width of the drift region for a GaN non-punch-through diode to achieve the same breakdown voltage of 5000 volts.

Home Work 2: Schottky and JBS Rectifiers

Problem 1: Calculate the on-state voltage drop for a 4H-SiC Schottky barrier rectifier designed to block 1000V. You can assume the following: (1) parallel-plane breakdown voltage; (2) On-state current density of 200 A/cm^2; (c) Barrier height of 1.6 eV; (d) Operation at room temperature (300 °K); (e) Substrate thickness of 350 microns and resistivity of 0.020 Ω-cm (f). Provide the voltage drop across the Schottky barrier and the series resistance as well as the total voltage drop.

Problem 2: Calculate the on-state voltage drop for a GaN Schottky barrier rectifier designed to block 1000V. You can assume the following: (1) parallel-plane breakdown voltage; (2) On-state current density of 200 A/cm^2; (c) Barrier height of 0.70 eV; (d) Operation at room temperature (300 °K); (e) Substrate thickness of 350 microns and resistivity of 0.010 Ω-cm (f). Provide the voltage drop across the Schottky barrier and the series resistance as well as the total voltage drop.

Problem 3: A 4H-SiC Schottky barrier rectifier is designed to block 1000-V. (a) Calculate the leakage current density without Schottky barrier lowering and tunneling. (b) Calculate the leakage current density with Schottky barrier lowering but without tunneling. (c) Calculate the leakage current density with Schottky barrier lowering and tunneling. (d) What is the barrier reduction in eV due to the image force? Use the following assumptions: (1) parallel-plane breakdown voltage; (2) Reverse bias voltage of 800V; (3) Barrier height of 1.6 eV; (4) No impact ionization; (5) No generation or diffusion current.

Problem 4: A GaN Schottky barrier rectifier is designed to block 1000-V. (a) Calculate the leakage current density without Schottky barrier lowering and tunneling. (b) Calculate the leakage current density with Schottky barrier lowering but without tunneling. (c) Calculate the leakage current density with Schottky barrier lowering and tunneling. (d) What is the barrier reduction in eV due to the image force? Use the following assumptions: (1) parallel-plane breakdown voltage; (2) Reverse bias voltage of 800V; (3) Barrier height of 0.7 eV; (4) No impact ionization; (5) No generation or diffusion current.

Problem 5: A 4H-SiC JBS rectifier is designed to block 1000V. The P-N junction depth is 1 micron. The cell pitch (p) is 2 micron. The P-region width (s) is 1 micron. (a) Calculate the On-State Voltage Drop. What is the contribution from the Schottky contact and from the series resistance? (b) Calculate the difference in on-state voltage drop of the JBS rectifier and the on-state voltage drop of the Schottky rectifier in Problem 1. You can assume the following: (1) parallel-plane breakdown voltage; (2) On-state current density of 200 A/cm^2; (3) Barrier height of 1.6 eV; (4) Operation at room temperature (300 °K); (5) Substrate thickness of 350 microns and resistivity of 0.020 Ω-cm; (6) Depletion width at on-state calculated using an on-state voltage drop of 1.5 volts; (7) Doping of the P$^+$ region is 1×10^{19} cm^{-3}.

Problem 6: A 4H-SiC JBS rectifier is designed to block 1000V. The P-N junction depth is 1 micron. The cell pitch (p) is 2 micron. The P-region width (s) is 1 micron. (a) Calculate the leakage current at 300 °K for a reverse bias of 800V. (b) Calculate the ratio of the leakage current for the Schottky rectifier in Problem 2 to that for the JBS rectifier. You can assume the following: (1) parallel-plane break-down voltage; (2) On-state current density of 200 A/cm^2; (3) Barrier height of 1.6 eV; (4) Operation at room temperature (300 °K); (5) Substrate thickness of 350 microns and resistivity of 0.020 Ω-cm; (6) Depletion width at on-state calculated using an on-state voltage drop of 1.5 volts; (7) Doping of the P^+ region is 1×10^{19} cm^{-3}.

Home Work 3: P-i-N and MPS Rectifiers

Problem 1: Design a 4H-SiC P-i-N rectifier with reverse blocking voltage of 15-kV. The drift region has a doping concentration of 1×10^{14} cm^{-3}. The lifetime in the drift region is 1 microsecond. (a) What is the thickness of the drift region? (b) What is the on-state voltage drop at a current density of 100 A/cm^2? (c) What is the stored charge in the drift region? (d) The diode is switched off with a ramp rate of 5 $\times 10^7$ A/cm^2-s. The reverse recovery occurs with the space charge layer extending through the entire drift region before the voltage reaches the supply voltage of 10 kV. What is the reverse recovery time (t_{RR})? (e) What is the peak reverse recovery current density under the conditions in part (d)? (f) What is the voltage at which the space charge layer penetrates the entire drift region? You can assume the following: (1) parallel-plane breakdown voltage; (d) Operation at room temperature (300 °K).

Problem 2: Design a 4H-SiC MPS rectifier with reverse blocking voltage of 15-kV. The drift region has a doping concentration of 1×10^{14} cm^{-3}. The lifetime in the drift region is 10 microseconds. (a) What is the thickness of the drift region? (b) What is the on-state voltage drop at a current density of 100 A/cm^2? Provide the components: Voltage drop across the Schottky contact; Voltage drop across the middle region; Voltage drop across the drift/N^+ substrate. (c) What is the stored charge in the drift region? (d) The diode is switched off with a ramp rate of 5×10^7 A/cm^2-s. The reverse recovery occurs with the space charge layer extending through the entire drift region before the voltage reaches the supply voltage of 10 kV. What is the reverse recovery time (t_{RR})? (e) What is the peak reverse recovery current density under the conditions in part (d)? (f) What is the voltage at which the space charge layer penetrates the entire drift region? You can assume the following: (1) parallel-plane breakdown voltage; (2) Operation at room temperature (300 °K); (3) Schottky barrier height of 2.0 eV; (4) Cell pitch (p) of 3 microns; (5) Schottky contact width of 1 micron; (6) P-N junction depth of 1 micron.

Home Work 4: Silicon GD-MOSFET and SJ-MOSFET

Problem 1: Design a silicon GD-MOSFET structure with blocking voltage of 150-V. (a) What is the length of the Source Electrode in the trench? (b) What is the optimum thickness of the trench oxide around the source electrode? (c) What is the optimum doping gradient? (d) Calculate the specific on-resistance for the device. Provide all the components: channel resistance, accumulation resistance, drift resistance in the trench portion. You can neglect the second component of the drift region resistance and the substrate resistance. (e) Compare the GD-MOSFET specific on-resistance to that for the ideal specific on-resistance for the one-dimensional case. You can assume the following: (1) breakdown voltage is not limited by edge termination; (2) Operation at room temperature (300 °K); (3) Inversion layer mobility of 200 cm^2/V-s; (4) Accumulation layer mobility of 1000 cm^2/V-s; (5) Gate Oxide thickness of 500 angstroms; (6) mesa width of 1 micron; (7) trench width of 2 microns; (8) P-base junction depth of 1 micron; (9) N^+ source depth of 0.2 microns; (10) gate electrode depth of 1.2 microns; (11) threshold voltage of 2 volts; (12) gate bias of 10 volts; (13) initial doping in the mesa region of 1×10^{16} cm^{-3}.

Problem 2: Design a silicon SJ-MOSFET structure with blocking voltage of 500-V. (a) What is the length of the drift region (P & N columns)? (b) What is the optimum dose for the P and N drift regions? (c) What is the optimum doping concentration for the P and N drift regions if their width is 10 microns? (d) Calculate the specific on-resistance for the device. Provide all the components: channel resistance, accumulation resistance, drift resistance in the portion with P-regions. You can neglect the second component of the drift region resistance and the substrate resistance. (e) Compare the SJ-MOSFET specific on-resistance to that for the ideal specific on-resistance for the one-dimensional case. You can assume the following: (1) breakdown voltage is not limited by edge termination; (2) Operation at room temperature (300 °K); (3) Inversion layer mobility of 200 cm^2/V-s; (4) Accumulation layer mobility of 1000 cm^2/V-s; (5) Gate Oxide thickness of 500 angstroms; (6) P-base junction depth of 1 micron; (7) N^+ source depth of 0.2 microns; (8) threshold voltage of 2 volts; (9) gate bias of 10 volts.

Home Work 5: SiC Power MOSFETs

Problem 1: Design an optimized 4H-SiC Shielded Planar-Gate Power MOSFET linear cell structure to obtain a blocking voltage of 1000 volts. The edge termination limits the breakdown voltage to 80% of the parallel-plane value. (a) Determine the drift region doping concentration assuming all the blocking voltage is supported by the drift region. (b) Determine the drift region thickness assuming all the blocking voltage is supported by the drift region. (c) Calculate the P-base doping concentration (assuming it is uniformly doped) to obtain a threshold voltage of 5 volts. The gate oxide thickness is 500 angstroms. The fixed charge in the gate oxide is 2×10^{11} cm^{-2}. Ignore work function difference for gate electrode. (d) Calculate the depletion width within the P-base region when the structure is supporting 1000 volts if the P-base region is not shielded. Calculate the depletion width within the P+ shielding region (doping concentration of 1×10^{18} cm^{-3}) when the structure is supporting 1000 volts. (e) Determine the optimum gate length to obtain the minimum specific on-resistance for your linear cell design using the parameters provided at the end of the problem. (f) What is the minimum specific on-resistance for your optimum linear cell design using the parameters provided at the end of the problem? Provide the components (R_{ch}, R_A, R_J, R_D, R_{SUB}) in absolute values of mΩ-cm^2 and as a percentage of the total. (g) Provide a graph of the specific on-resistance versus the gate length (ranging from 4 to 10 microns) showing all the components, including the total. (h) Calculate the specific on-resistance for the ideal drift region for blocking 1000 volts in mΩ-cm^2. Compare your design to this value by taking the ratio of your design value to the ideal value. Parameters and Assumptions: (1) Inversion Mobility = 20 cm^2/Vs. (2) Accumulation Mobility = 100 cm^2/Vs. (3) Cell Polysilicon window width = 5 microns (4) N$^+$ Substrate resistivity of 0.02 Ohm-cm and thickness of 200 microns. (5) Gate Bias = 15 volts. (6) K factor for accumulation spreading = 0.6. (7) P-Base Extension beyond Polysilicon Gate Edge = 1.2 microns. (8) N$^+$ Source Extension beyond Polysilicon Gate Edge = 0.2 microns. (8) P$^+$ Junction Depth = 1 micron

Problem 2: Design a 4H-SiC Shielded Trench-Gate Power MOSFET linear cell structure to obtain a blocking voltage of 1000 volts. The edge termination limits the breakdown voltage to 80% of the parallel-plane value. Use the drift region doping concentration from Problem

1. Use the drift region thickness from Problem 1. (a) What is the specific on-resistance for your device using the parameters provided at the end of the problem? Provide the components (R_{ch}, R_{J1}, R_{J2}, R_D, R_{SB}) in absolute values of mΩ-cm^2 and as a percentage of the total. (b) Calculate the specific on-resistance for the ideal drift region for blocking 1000 volts in mΩ-cm^2. Compare your design to this value by taking the ratio of your design value to the ideal value. Parameters and Assumptions: (1) P-base junction depth = 0.7 microns. (2) N+ source junction depth = 0.2 microns. (3) Trench depth = 1.5 microns. (4) P$^+$ region thickness/junction depth = 0.2 microns. (5) Gate oxide thickness = 500 angstroms. (6) Inversion Mobility = 20 cm^2/Vs. (7) Accumulation Mobility = 100 cm^2/Vs. (8) N$^+$ Substrate resistivity of 0.02 Ohm-cm and thickness of 200 microns. (9) Gate Bias = 15 volts; Threshold voltage = 5 volts. (10) Mesa width = 2.5 microns. (11) Trench width = 1 micron. (12) JFET region doping = 5×10^{16} cm^{-3}.

Home Work 6: GaN HEMT

Problem 1: Calculate the Ideal Specific On-resistance for the GaN HEMT structure for breakdown voltages of 100, 200, 500, 1000, and 2000 volts. State the assumptions made for calculating the resistances.

Problem 2: Design a GaN HEMT device with a breakdown voltage of 500 volts. (a) Determine the gate-to-drain drift region length assuming an average electric field obtained by the Furukawa Electric research. (b) What is the total specific on-resistance for your design using the parameters provided at the end of the problem? Provide the components (R_{SC}, R_{GS}, R_G, R_{GD}, R_{DC}) in absolute values of mΩ-cm^2 and as a percentage of the total. (c) Compare your design to ideal specific on-resistance from Problem 1. Parameters and Assumptions: (1) Source Width (L_S) = 5 microns. (2) Drain Width (L_D) = 10 microns. (3) Gate Width (L_G) = 1 micron. (4) Gate-Source distance (L_{GS}) = 1 micron. (5) Specific Contact Resistance = 1×10^{-5} Ω-cm^2.

Home Work 7: SiC BJTs and GTOs

Problem 1: Consider a 4H-SiC N$^+$PN-N$^+$ bipolar power transistor structure with uniformly doped emitter, base, and collector drift regions. The emitter region has a doping concentration of 1×10^{19} cm^{-3}

and thickness of 1 micron. The base region has a doping concentration of 2×10^{17} cm^{-3} and thickness of 1 micron. The collector drift region has a doping concentration of 1×10^{15} cm^{-3} and thickness of 40 microns. The Shockley-Read-Hall (Low-Level, High-Level, and Space-Charge Generation) Lifetime is 10 ns in the Emitter region, 0.1 μsec in the Base region, 1 μsec in the Collector Drift region. You can ignore band-gap narrowing and auger recombination. Use ionized impurity concentrations whenever appropriate. (a) Determine the emitter injection efficiency for the transistor (excluding the Webster Effect). (b) Calculate the Base Transport Factor. (c) Determine the common-base and common-emitter current gains at low current levels (excluding Surface Recombination). (d) Determine the common-emitter current gain at a collector current of 100 A/cm^2 (assume low-level injection in the base) including Surface Recombination with a velocity of 4000 cm/s. To account for surface recombination, use:

$$J_{pT} = \frac{J_C}{\beta_{SR}} = (J_{p,E} + J_{p,B}) + J_{p,SR}$$ where J_{pT} is the total base current; $J_{p,E}$

is the hole current due to injection into the emitter; $J_{p,B}$ is the hole current due to recombination in the base; and $J_{p,SR}$ is the hole current due to surface recombination. The hole current due to surface recombination can be obtained using: $J_{p,SR} = K_{SR} \cdot v_{SR}$ where K_{SR} is the surface recombination coefficient (value of 5×10^{-4} Coulombs/cm^3) and v_{SR} is the surface recombination velocity. (e) Calculate the open-emitter breakdown voltage (BV$_{CBO}$). Make sure to take punch-through into account for the drift region. You can use the critical electric field for the doping concentration in the Collector Drift region. (f) Calculate the open-base breakdown voltage (BV$_{CEO}$). Use $n = 6$ in the equation for the Multiplication factor. Use the Current Gain from part (c). (g) Determine the depletion region penetration in the base region. Check if it is less than the P-base thickness. (h) Calculate the on-state specific resistance for the transistor. (i) Determine the Webster current density for the transistor. Calculate the common emitter current gain at a current density of 200 A/cm^2.

Problem 2: Consider a 4H-SiC P$^+$NP-P(BL)N$^+$ power GTO structure with uniformly doped P$^+$ Anode, N-Base, P- Drift and N$^+$ Cathode regions. The P$^+$ anode region has a doping concentration of 1×10^{19} cm^{-3} and thickness of 1 micron. The N-base region has a doping concentration of 2×10^{17} cm^{-3} and thickness of 2 microns. The P-drift region has a doping concentration of 1×10^{14} cm^{-3} and thickness of 80 microns. The P Buffer Layer has a doping concentration of 5×10^{16} cm^{-3} and thickness of 4 microns. The N$^+$ cathode region has a doping concentration of 5×10^{19} cm^{-3} and thickness of 300 microns. The Shockley-Read-Hall (Low-Level, High-Level, and Space-Charge Generation) Lifetime is 10 ns in the N$^+$ Cathode and P$^+$ Anode regions, 0.1 μsec in the N-Base region and 1 μsec in the P- Drift region. You can ignore band-gap narrowing and auger recombination. The structure has a linear cell geometry with an emitter width (W$_{KS}$) of 200 microns and length (Z) of 1 cm. Use ionized impurity concentrations whenever appropriate. (a) Determine the forward blocking voltage for the GTO ignoring open-base transistor action, i.e. use the diode breakdown assumption. (b) Determine the on-state voltage drop at a cathode current of 2 Amperes. (c) Determine the storage time if the device is turned-off at a current gain of 2 with a constant gate drive current. Use the 2D Model in the textbook. (d) Determine the voltage rise-time for the GTO if the device is operated using a DC supply voltage of 5000 V. Provide the electron concentration in the space-charge region used for your analysis. (e) Determine the Maximum Controllable Current assuming a one-dimensional parallel-plane breakdown voltage for the cathode-base junction.

Home Work 8: SiC IGBTs

Problem 1: (a) Determine the width of the N-drift region for an n-channel 4H-SiC IGBT to obtain a blocking voltage of 15 kV if it's doping concentration is 1.5×10^{14} cm^{-3}. The N-Buffer layer has a doping concentration of 5×10^{16} cm^{-3} and its thickness is 5 microns. The lifetime (Low-Level, High-Level, Space-Charge-Generation) in the N-drift layer is 2 microsecond. Scale the lifetime in the N-Buffer layer using a reference doping of 5×10^{16} cm^{-3}. The P$^+$ Collector region has a doping concentration of 1×10^{19} cm^{-3} and its thickness is 10 microns. (b) Calculate the injected hole concentration at the P$^+$ collector/N-Buffer Layer junction for the planar-gate asymmetric n-Channel 4H-SiC IGBT structure under the operating conditions defined in part (e). Assume a diffusion length for electrons of 1 micron in the P$^+$ collector region. (c) What is the hole concentration in the N-buffer layer at the interface between the N-drift and N-buffer regions? (d) What is the hole concentration in the N-drift region at the interface between the N-drift and N-buffer regions? (e) Determine the on-state voltage drop at an on-state current density of 50 A/cm^2 for the planar-gate asymmetric n-channel 4H-SiC IGBT structure using the two-dimensional model. Provide the values for the voltage drop across the P$^+$/N junction, the N-base region, and the MOSFET (using only the channel resistance). (f) The planar-gate asymmetric n-channel 4H-SiC IGBT structure is switched off under inductive load conditions from the on-state operating conditions defined in part (e). (g) Calculate the voltage rise-time to reach a collector DC supply voltage of 10 kV. (h) Calculate the reach-through voltage. (i) Calculate the reach-through time. (j) Calculate the [dV/dt] during the second phase of the voltage waveform. (k) What is the current fall-time? (l) Obtain the total energy loss per cycle. Provide the energy loss during the voltage rise-time and the current fall-time. Use the following parameters: (1) Cell pitch of 10 microns. (2) Channel length of 1.0 microns. (3) Inversion mobility of 15 cm^2/V-s. (4) Gate oxide thickness of 500 angstroms. (5) Gate bias of 15 volts. (6) Threshold voltage of 5 volts. (7) Lifetime (Low-Level, High-Level, Space-Charge-Generation) in the N-drift layer of 2 microsecond. (8) All the dopants are fully ionized.

Problem 2: (a) Determine the width of the P-drift region for a p-channel 4H-SiC IGBT to obtain a blocking voltage of 15 kV if it's doping concentration is 1.5×10^{14} cm^{-3}. The P-Buffer layer has a doping concentration of 5×10^{16} cm^{-3} and its thickness is 5 microns. The lifetime (Low-Level, High-Level, Space-Charge-Generation) in the P-drift layer is 2 microsecond. Scale the lifetime in the P-Buffer layer using a reference doping of 5×10^{16} cm^{-3}. The N$^+$ Collector region has a doping concentration of 1×10^{19} cm^{-3} and its thickness is 10 microns. (b) Calculate the injected electron concentration at the N$^+$ collector/P-Buffer Layer junction for the planar-gate asymmetric p-Channel 4H-SiC IGBT structure under the operating conditions defined in part (e). Assume a diffusion length for holes of 1 micron in the N$^+$ collector region. (c) What is the electron concentration in the P-buffer layer at the interface between the P-drift and P-buffer regions? (d) What is the electron concentration in the P-drift region at the interface between the P-drift and P-buffer regions? (e) Determine the on-state voltage drop at an on-state current density of 50 A/cm^2 for the planar-gate asymmetric p-channel 4H-SiC IGBT structure using the two-dimensional model. Provide the values for the voltage drop across the P$^+$/N junction, the N-base region, and the MOSFET (using only the channel resistance). (f) The planar-gate asymmetric p-channel 4H-SiC IGBT structure is switched off under inductive load conditions from the on-state operating conditions defined in part (e). (g) Calculate the voltage rise-time to reach a collector DC supply voltage of 10 kV. (h) Calculate the reach-through voltage. (i) Calculate the reach-through time. (j) Calculate the [dV/dt] during the second phase of the voltage waveform. (k) What is the current fall-time? (l) Obtain the total energy loss per cycle. Provide the energy loss during the voltage rise-time and the current fall-time. Use the following parameters: (1) Cell pitch of 10 microns. (2) Channel length of 1.0 microns. (3) Inversion mobility of 15 cm^2/V-s. (4) Gate oxide thickness of 500 angstroms. (5) Gate bias of 15 volts. (6) Threshold voltage of 5 volts. (7) Lifetime (Low-Level, High-Level, Space-Charge-Generation) in the N-drift layer of 2 microsecond.

Index

Printed in the United States
By Bookmasters